P9-DUC-517

MATHEMATICS
ITS POWER AND UTILITY

MATHEMATICS
ITS POWER AND UTILITY

Karl J. Smith   Santa Rosa Junior College

BROOKS/COLE PUBLISHING COMPANY
Monterey, California

Contemporary Undergraduate Mathematics Series,
Robert J. Wisner, Editor

Brooks/Cole Publishing Company
A Division of Wadsworth, Inc.

Printed in the United States of America

10 9 8 7 6 5 4 3 2

Library of Congress Cataloging in Publication Data:

Smith, Karl J.
  Mathematics: its power and utility.

  Includes index.
  1. Mathematics—1961–   .   I. Title.
QA39.2.S598   1983   510   82-12932

ISBN 0-534-01190-X

Subject Editor: Craig Barth
Production Service: Phyllis Niklas
Production Coordinator: Joan Marsh
Interior Design: Janet Bollow
Illustrations: Carl Brown
Cartoons: Ryan Cooper
Typesetting: Jonathan Peck, Typographer

Cover: Judith Karelitz is a pioneer in the resurging field of kaleidoscope making. The image here is in her device called the Karascope. Based on polarized-light principles, the Karascope does not have the symmetry of traditional kaleidoscopes, so the image appears to spiral toward the eye of the viewer. © 1982 Wayne Sorce.

To Hal Andersen,
a mentor and a friend

Preface

This book was written to display some of the power and some of the utility of mathematics. Over the years, first as a student, and later as a teacher, I have gradually come to appreciate not only the usefulness of mathematics, but also the intrinsic power of the subject. However, as my involvement with mathematics has grown, so also has my awareness of others' avoidance of the subject. After several years of teaching, I went back to school and found that my fellow students' greatest fears were of courses with any mathematical content. It was not really their fear that surprised me, but rather the intensity of their fear.

In the last few years there has been much attention given to *math anxiety* and *math avoidance,* particularly with reference to reentry women. Sheila Tobias, a leading authority on the subject, states in the Preface of her book *Overcoming Math Anxiety* (New York: W. W. Norton & Company, 1978):

> Although I realized by then that many of the choices women make are really made for them, I did not see for a very long time that women are predestined to study certain subjects and pursue certain occupations not only because these areas are "feminine" but because girls are socialized not to study math.

However, she continues:

> Four years ago, when I began, I hypothesized that mathematics anxiety and mathematics avoidance were feminist issues. Now I am not so sure. Observing men has shown me

that some men as well as the majority of women have been denied the pleasures and the power that competence in math and science can provide.

Now, as a teacher, I am faced with transferring some knowledge of the power, the beauty, and the utility of mathematics to my students. But I can't instruct until my students have overcome their initial fears and are ready to listen. Mathematics cannot be presented in the abstract to these students. Mathematics cannot be presented simply because it is there. Mathematics cannot be presented because it might be useful—someday. Mathematics cannot be presented with the usual trite and stereotyped problems found in most elementary mathematics textbooks.

It is with these issues in mind that I've written this book. It is divided into two parts. The first, called "The Power of Mathematics," develops some ideas in arithmetic, algebra, and geometry. The second part, called "The Utility of Mathematics," develops each section around some situation with which I hope the student will be able to identify.

The main theme throughout the book is *problem solving*. In "The Power of Mathematics" we begin by discussing math anxiety and how to formulate the problem. The most difficult first step for many students is to determine exactly what the problem is. All too often we try to solve a problem before we are even sure about what we are trying to solve. Techniques from arithmetic, algebra, and geometry are all applied to problem solving.

These techniques of problem solving are then used in the second part of the book, "The Utility of Mathematics." Each topic in this part of the book was selected because of its usefulness in everyday concerns. The topics include personal money management; working with interest; installment buying; using a credit card; inflation; buying a car, home, or insurance; probability; contests; games; statistics; surveys; advertising; computers; and the influence of computers on our lives.

The material in this book can be adapted to almost any arrangement. Chapter 1 on problem solving is prerequisite for all the subsequent chapters. Chapter 2 on calculators and arithmetic is also required for the rest of the book, but may be treated lightly or skipped by those familiar with its contents. Chapter 4 on algebra is a prerequisite only for the material in Chapters 7 and 13. Chapter 14 on computers can be treated any time after Chapter 2. Other relationships among chapters are shown in the following chart. Notice that the chapters are, for the most part, independent of one another, so you have a great deal of flexibility in the order or manner in which you treat the topics in this book.

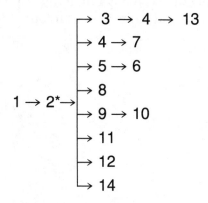

*The material in this chapter may be assumed depending on the background of the students.

I am most grateful to those who have assisted me in the development of this material: to my students who made many valuable suggestions about the material, and especially to those students who were bold enough to share their fears and anxieties with me; to my colleagues who shared ideas and teaching suggestions; to the reviewers who offered many valuable suggestions. In particular, I would like to thank Elton Beougher, Fort Hays State University; James W. Brown, Northern Essex Community College; Ben V. Flora, Jr., Moorehead State University; Elton Fors, Northern State College, South Dakota; Roy E. Garland, Millersville State College; Armando Gingras, Metropolitan State College, Denver; Glenn E. Johnston, Moorehead State University; Keith S. Joseph, Metropolitan State College, Denver; Genevieve M. Knight, Hampton Institute; Glen E. Mills, Pensacola Junior College; and George C. Sethares, Bridgewater State College.

The production staff at Brooks/Cole also deserves special credit. In particular, Craig Barth was most helpful not only in doing extensive market research for this book but also in suggesting material and for taking an active part in the development of this book. He went far beyond the job requirements of a good editor. I am grateful to Phyllis Niklas for her meticulous editing and checking of the manuscript. Her suggestions were invaluable. The production coordinator, Joan Marsh, also took a personal interest in the success of this book and deserves a special thank you.

Most of all, my thanks to my family. To my wife, Linda, for her unending love and support, and to our children, Melissa and Shannon.

Karl J. Smith
Sebastopol, California

CREDITS

Page 357. Finance charge statement courtesy of Sears, Roebuck and Co.

Page 357. Photo courtesy of Pontiac Motor Division.

Page 375. Photo © Eric A. Roth/The Picture Cube.

Page 384. Random digit table reprinted with permission from *CRC Standard Mathematical Tables,* Student Edition, page 565 of 15th Edition. Copyright The Chemical Rubber Company, CRC Press, Inc.

Page 386. The dice in Problem 11 were designed by Bradley Efron of Stanford University. The game of WIN was suggested by Joseph Smyth of Santa Rosa Junior College.

Page 386. "Le Trente-et-un, ou la maison de prêt sur nantissement" by Darcis. Reprinted by permission of Bibliothèque Nationale, Paris.

Page 393. Photos by Richard Hagberg.

Page 402. Taken from the Dear Abby column. Copyright © 1974, Universal Press Syndicate. Reprinted with permission. All rights reserved.

Pages 403–404. Monte Carlo example from *How to Take a Chance* by Darrell Huff. Illustrated by Irving Geis. By permission of W. W. Norton & Company, Inc.

Page 404. From *Mathematical Circles Revisited* by H. Eves. Copyright 1971 by Prindle, Weber & Schmidt, Inc. Reprinted by permission.

Page 412. Mortality tables courtesy of the Society of Actuaries, Chicago.

Page 417. Lottery tickets courtesy of Hospital's Trust, Dublin, Ireland; Ohio State Lottery Commission; Pennsylvania State Lottery Bureau; New Hampshire Sweepstakes Commission; Massachusetts State Lottery Commission; Michigan Bureau of State Lottery.

Page 419. 1972 Reader's Digest Sweepstakes © 1972 The Reader's Digest Association, Inc. Reprinted by permission.

Page 423. Photo © Arthur Tress/Photo Researchers, Inc.

Page 435. Advertisement © 1971 by SAAB-Scania of America, Inc. Reprinted by permission.

Page 436. *Peanuts* cartoon © 1961 United Feature Syndicate, Inc. Reprinted by permission.

Page 447. Cartoon by David Pascal. Reprinted by permission.

Page 453. Photo used by permission of United Press International.

Page 459. Photo © Barbara Alper/Stock, Boston, Inc.

Page 467. Avis trademark, Dollar Rent-A-Car System trademark, and Hertz Registered Service Mark reprinted by permission.

Page 475. *B.C.* cartoon reprinted by permission of Johnny Hart and Field Enterprises.

Page 483. Photo © Glenn R. Steiner, Photography.

Page 484. From the film *A Computer Glossary* (made for IBM by Charles and Ray Eames with Glen Fleck, 1968). Reprinted by permission.

Page 485. Photo courtesy of Radio Shack, A Division of Tandy Corp.

Page 486. Cartoon by J. McGrath Scheepsma. Reprinted by permission.

Page 487. Photos of digital computer courtesy of Digital Equipment Corporation.

Page 487. Photo of CRT terminal courtesy of Radio Shack, A Division of Tandy Corp.

Page 494. News article © 1972 United Press International. Advertisement courtesy of General Motors Research Laboratories. News article © 1973 Newhouse News Service. Reprinted by permission.

Page 495. Poem from *Page in a Modern Bestiary* by David H. H. Diamond. Reprinted with permission of DATAMATION magazine. © Copyright by Technical Publishing Company, A Dun & Bradstreet Company, 1976—all rights reserved.

Page 497. Flowchart reprinted by permission from The General Library Information Leaflet No. 1, University of California, Davis.

Page 499. Cartoon by Dav Holle. Reprinted by permission.

Contents

Part I
Foundations: The Power of Mathematics

CHAPTER 1
INTRODUCTION TO PROBLEM SOLVING   3

1.1 Math Anxiety   4
1.2 Formulating the Problem   10
1.3 Mathematical Patterns   21
1.4 Inductive and Deductive Reasoning   30
1.5 Review Problems   36

CHAPTER 2
CALCULATORS AND ARITHMETIC   39

2.1 Whole Numbers   40
2.2 Fractions and Decimals   52
2.3 Common Fractions   64
2.4 Adding and Subtracting Fractions   74
2.5 Review Problems   87

CHAPTER 3
THE WORLD OF NUMBERS   89

3.1 Signed Numbers   90
3.2 Exponents and Roots   102

3.3 Large Numbers   115
3.4 Earthquakes, Rock Concerts, and Altitudes   124
3.5 Review Problems   132

CHAPTER 4
ALGEBRA AND PROBLEM SOLVING   133

4.1 Symbol Shock   134
4.2 Equations   142
4.3 Solving Equations   150
4.4 Problem Solving with Algebra   154
4.5 Review Problems   166

CHAPTER 5
THE METRIC WORLD   167

5.1 Measuring Length   168
5.2 Measuring Capacity   177
5.3 Measuring Weight and Temperature   182
5.4 Converting Units   189
5.5 Review Problems   200

CHAPTER 6
GEOMETRY AND PROBLEM SOLVING   201

6.1 Polygons and Angles   202
6.2 Perimeter   207
6.3 Area   216
6.4 Volume   228
6.5 Review Problems   240

CHAPTER 7
PERCENTS AND PROBLEM SOLVING   243

7.1 Ratio and Proportion   244
7.2 Problem Solving with Proportions   250
7.3 Percent   257
7.4 Problem Solving with Percents   266
7.5 Review Problems   272

Part II
Applications: The Utility of Mathematics

CHAPTER 8
PERSONAL MONEY MANAGEMENT   275

8.1 Checking Accounts   276
8.2 Discount, Sale Price, and Sales Tax   289
8.3 Household Budgeting   296
8.4 Income Taxes   304
8.5 Review Problems   313

CHAPTER 9
INTEREST   315

9.1 Simple Interest   316
9.2 Working with Simple Interest   323
9.3 Compound Interest   329
9.4 Inflation   340
9.5 Review Problems   344

CHAPTER 10
CONSUMER APPLICATIONS   345

10.1 Installment Buying   346
10.2 Buying with a Credit Card   351
10.3 Buying a New Automobile   357
10.4 Buying a Home   363
10.5 Review Problems   374

CHAPTER 11
PROBABILITY   375

11.1 Probability Experiments   376
11.2 Introduction to Probability   386
11.3 Probability Models   399
11.4 Mathematical Expectation   408
11.5 Review Problems   421

CHAPTER 12
STATISTICS   423

12.1   Frequency Distributions and Graphs   424
12.2   Descriptive Statistics   436
12.3   The Normal Curve   447
12.4   Sampling   453
12.5   Review Problems   457

CHAPTER 13
GRAPHS   459

13.1   Ordered Pairs and the Cartesian Coordinate System   460
13.2   Lines   467
13.3   Parabolas   471
13.4   Exponential and Logarithmic Graphs   475
13.5   Review Problems   481

CHAPTER 14
COMPUTERS   483

14.1   Introduction to Computers   484
14.2   BASIC Programming   495
14.3   Communicating with a Computer   506
14.4   Programming Repetitive Processes   519
14.5   Review Problems   526

ANSWERS TO SELECTED PROBLEMS   529

INDEX   553

Foundations: The Power of Mathematics

PART 1

Most people view mathematics as a series of techniques of use only to the scientist, the engineer, or the specialist. In fact, the majority of our population could be classified as *math-avoiders*, who consider the assertion that mathematics can be creative, beautiful, and significant not only an "impossible dream," but something they don't even want to discuss.

Nevertheless, the first part of this book attempts to develop an appreciation for mathematics by displaying the intrinsic power of the subject. We begin by looking at some of the causes and effects of *math anxiety*. We will take natural steps, small at first, then a little larger as you gain confidence, to review and learn about calculators, fractions, large numbers, earthquakes, symbol shock, equations, metrics, geometry, and percents. The discussions of all these topics will involve the idea of problem solving. Mathematics cannot be presented simply because it is there; it must be concrete and down-to-earth, and the best way to learn an appreciation of mathematics is to successfully apply your own ingenuity to solve a new problem.

At each turn of the page I hope you will find something new and *interesting* to you. I want you to participate and become involved with the material. I want you to experience what I mean when I speak of the beauty of mathematics. Bertrand Russell has described what I hope unfolds for you as you read this book:

> Mathematics, rightly viewed, possesses . . . supreme beauty—a beauty cold and austere, like that of sculpture, without appeal to any part of our weaker nature, without the gorgeous trappings of painting or music, yet sublimely pure, and capable of a stern perfection such as only the greatest art can show. The pure spirit of delight, the exaltation, the sense of being more than man, which is the touchstone of the highest excellence, is to be found in mathematics as surely as in poetry.

Chapter 1
Introduction to Problem Solving

Chapter 2
Calculators and Arithmetic

Chapter 3
The World of Numbers

Chapter 4
Algebra and Problem Solving

Chapter 5
The Metric World

Chapter 6
Geometry and Problem Solving

Chapter 7
Percents and Problem Solving

Introduction to Problem Solving

1.1 Math Anxiety

There are many reasons for reading a book, but the best reason is because you want to read it. Although you are probably reading this first page because you were requested to do so by your instructor, it is my hope that in a short while you will be reading this book because you want to read it.

This book was written for people who are math-avoiders, or people who think they can't work math problems, or people who think they are never going to use math. Do you see yourself making any of these statements?

Do you feel that you can be reasonably successful in other subjects, but have an inability to do math? Do you make career choices based on avoidance of mathematics courses? If so, you have *math anxiety*. If you reexamine your negative feelings toward mathematics, you *can* overcome them. In this book, I'll constantly try to help you overcome these feelings.

Sheila Tobias, an educator, feminist, and founder of an organization called Overcoming Math Anxiety, has become one of our nation's leading spokespersons on math anxiety.

She is not a mathematician, and in fact describes herself as a math-avoider. She has written a book titled *Overcoming Math Anxiety* (New York: W. W. Norton & Company, 1978; available in paperback). I recommend this book to anyone who has ever said "I'm no good at numbers." In this book, she describes a situation characterizing anxiety (p. 45):

> Paranoia comes quickly on the heels of the anxiety attack. "Everyone knows," the victim believes, "that I don't understand this. The teacher knows. Friends know. I'd better not make it worse by asking questions. Then everyone will find out how dumb I really am." This paranoid reaction is particularly disabling because fear of exposure keeps us from constructive action. We feel guilty and ashamed, not only because our minds seem to have deserted us but because we believe that our failure to comprehend this one new idea is proof that we have been "faking math" for years.

The reaction described in this paragraph sets up a vicious cycle. The more we avoid math, the less able we feel, and the less able we feel, the more we avoid it. The cycle can also work in the other direction. What do you like to do? Chances are, if you like it, you do it. The more you do something, the better you become at it. In fact, you've probably thought "I like to do it, but I don't get to do it as often as I'd like to." This is the normal reaction toward something you like to do. In this book, I have attempted to break the negative cycle concerning math and replace it with a positive cycle. However, I will need your help and willingness to try.

The central theme in this book is problem solving. Through problem solving I'll try to dispel your feelings of panic. Once you find that you are capable of doing mathematics, we'll look at some of its foundations and uses. There are no prerequisites for this book, and as we progress through the book, I'll include a review of the math you never quite learned in school—from fractions, decimals, percentages, and metrics, to algebra and geometry. I hope to answer the questions that, perhaps, you were embarrassed to ask.

I hope you will enjoy reading this book, but if you feel an anxiety attack coming—STOP and put it aside for a while. Talk to your instructor, or call me. My telephone number is

(707) 829-0606

You can generally reach me between 7 and 8 A.M. (Pacific Time Zone). I care about your progress with the course, and I'd like to hear your reactions to this book. You can write to me at the following address:

Karl Smith
Mathematics Department
Santa Rosa Junior College
1501 Mendocino Avenue
Santa Rosa, CA 95401

At the end of each section in this book there is a problem set. This first problem set is built around twelve math myths. These myths come from another book on math anxiety, *Mind Over Math*, by Stanley Kogelman and Joseph Warren (New York: Dial Press, 1978), which I highly recommend. These commonly believed myths have resulted in false impressions about how math is done, and they need to be dispelled.

Math Anxiety Bill of Rights*
by Sandra L. Davis

1. I have the right to learn at my own pace and not feel put down or stupid if I'm slower than someone else.
2. I have the right to ask whatever questions I have.
3. I have the right to need extra help.
4. I have the right to ask a teacher or TA for help.
5. I have the right to say I don't understand.
6. I have the right not to understand.
7. I have the right to feel good about myself regardless of my abilities in math.
8. I have the right not to base my self-worth on my math skills.
9. I have the right to view myself as capable of learning math.
10. I have the right to evaluate my math instructors and how they teach.
11. I have the right to relax.
12. I have the right to be treated as a competent adult.
13. I have the right to dislike math.
14. I have the right to define success in my own terms.

*From *Overcoming Math Anxiety*, by Sheila Tobias, pp. 236–237.

Problem Set 1.1

In Problems 1–12 comment on each math myth. There are no right or wrong answers to these questions, but they will help you gain insight into your own attitudes, as well as begin to dispel some false notions you might have about the subject. The answers to selected questions are found in the back of the book.

1. Myth 1: Men are better in math than women.
2. Myth 2: Math requires logic, not intuition.
3. Myth 3: You must always know how you got the answer.
4. Myth 4: Math is not creative.
5. Myth 5: There is a best way to do a math problem.
6. Myth 6: It's always important to get the answer exactly right.
7. Myth 7: It's bad to count on your fingers.
8. Myth 8: Mathematicians do problems quickly, in their heads.
9. Myth 9: Math requires a good memory.
10. Myth 10: Math is done by working intensely until the problem is solved.
11. Myth 11: Some people have a "math mind" and some don't.
12. Myth 12: There is a magic key to doing math.
13. Summarize your math experiences in elementary school.
14. Summarize your math experiences in high school.

Mind Bogglers

These are optional problems that are loosely related to the material in the text, and they may require research outside the text. This set of Mind Bogglers is designed to show you the variety of ways that numbers are used. You will probably be familiar with different parts of these problems, but nobody will be familiar with them all. (My thanks to James Smart of San Jose University for the use of these problems.)

15. Answer the following questions about numbers:
    a. A metronome is often used to help piano students. What does a metronome setting of 120 mean?
    b. On most sophisticated cameras the lens opening is indicated by a set of numbers called *f-numbers*. If your camera is set at *f*/8, what should be the new setting if you want to let in twice as much light?
    c. What do the numbers 20–30 mean as a result of a vision test?

    d. What do the numbers 25–10–5 on a sack of fertilizer mean?

16. Answer the following questions about numbers:
    a. What does a bicycle size of 26 inches mean?
    b. What does a tire size of H70-15 mean?
    c. What does a gasoline octane rating of 93 mean?
    d. How long is an eightpenny nail in inches?

17. Answer the following questions about numbers:
    a. Golf clubs are named by using numbers and material, such as number 2 wood or number 5 iron. As the numbers increase, does the loft increase or decrease?
    b. A number $2\frac{1}{2}$ pencil contains more of what common substance mixed with the graphite in the lead than does a number 2 pencil?
    c. What does 80 proof whiskey mean?
    d. Several numerals appear on the face of paper money, such as the serial number and the denomination. In addition, a numeral, such as 2, 7, or 12, appears four times, once in each quarter of the face. What does this number indicate?

18. Answer the following questions about numbers:
    a. In printing, the height of type is ordinarily given in points. How high is 24-point type in inches?
    b. In textbooks, ordinarily on the page with the copyright information, there often appears a sequence of numerals beginning with 10. What would the sequence 10 9 8 7 6 5 4 indicate?
    c. What does 1140 on an AM radio dial mean?
    d. The size of a juice or vegetable can is given by numbers such as 1, 2, or 3. A number 2 can contains how many cups?

Bibliography on Math Anxiety*

Aiken, L. R. "Two scales of attitude toward mathematics." *Journal for Research in Mathematics Education*, 1974, **5**, 67–71.

Aiken, L. "Review of the literature on attitudes toward mathematics." *Review of Educational Research*, 1976, **46**, 293–311.

Betz, N. E. "Prevalence, distribution, and correlates of math anxiety in college students." *Journal of Counseling Psychology*, 1978, **25**, 441–448.

*My thanks to Jim Daniel, University of Texas at Austin, for this bibliography.

Daniel, J. W. "Math anxiety." *Discovery, Research and Scholarship at University of Texas at Austin*, September 1978.

Degnan, J. A. "General anxiety and attitudes toward mathematics in achievers and underachievers in mathematics." *Graduate Research in Education and Related Disciplines*, 1967, **3**, 49–62.

Donady, B., and S. Tobias. "Math anxiety." *Teacher Magazine*, November 1977.

Donady, B., and S. Tobias. "Counseling the math anxious." *Journal of the National Association for Women Deans, Administrators and Counselors*, 1977, **41**.

Dreger, R. M., and L. R. Aiken. "The identification of number anxiety in a college population." *Journal of Educational Psychology*, 1975, **48**, 344–351.

Dutton, W. H. "Attitude change of prospective elementary school teachers toward arithmetic." *Arithmetic Teacher*, 1962, **9**, 418–424.

D'Zurilla, T., and M. Goldfried. "Problem solving and behavior modification." *Journal of Abnormal Psychology*, 1971, **78**, 107–126.

Ernest, J. "Mathematics and sex." *American Mathematical Monthly*, October 1976, **83**.

Fennema, E., and J. A. Sherman. "Fennema–Sherman mathematics attitudes scales: Instruments designed to measure attitudes toward the learning of mathematics by females and males." *JSAS Catalogue of Selected Documents in Psychology*, 1976, **6**, 31.

Fox, L., E. Fennema, and J. Sherman (Eds.). *Women and Mathematics: Research Perspectives for Change*. NIE Papers in Education and Work: No. 8 (Washington, D.C.: U.S. Department of Health, Education and Welfare, 1977).

Hendel, D. D., and S. O. Davis. "Effectiveness of an intervention strategy for reducing mathematics anxiety." *Journal of Counseling Psychology*, 1978, **25**, 429–434.

Hilton, P. "Some thoughts in math anxiety." *Ontario Mathematics Gazette*, 1978.

Kogelman, S., and J. Warren. *Mind Over Math* (New York: Dial Press, 1978).

Lazarus, A. (Ed.). *Multimodal Behavior Therapy* (New York: Springer, 1976).

Lazarus, M. "Mathophobia: Some personal speculations." *The Principal*, 1974, **53**(2). (See also whole issue, "Reckoning with math," February 1974.)

Lazarus, M. "Rx for mathophobia." *Saturday Review*, June 28, 1975, 46.

Meichenbaum, D., and D. Turic. *Cognitive-Behavioral Management of Anxiety, Anger, and Pain* (New York: Brunner/Mazel, 1976).

Mitzman, B. "Seeking a cure for 'mathephobia.'" *American Education*, March 1976.

Naiman, A. "Memoirs of a female mathophobe." *The Principal*, 1974, **53**(2). (See also whole issue, "Reckoning with math," February 1974.)

Richardson, F. C. "Anxiety management training: A multimodal approach." In A. A. Lazarus (Ed.), *Multimodal Behavior Therapy* (New York: Springer, 1976).

Richardson, F. C., and R. M. Suinn. "The mathematics anxiety rating scale: Psychometric data." *Journal of Counseling Psychology*, 1972, **19**, 551–554.

Richardson, F. C., and R. M. Suinn. "A comparison of traditional systematic desensitization, accelerated massed desensitization, and anxiety management training in the treatment of mathematics anxiety." *Behavior Therapy*, 1973, **4**, 212–218.

Richardson, F. C., and R. L. Woolfolk. "Mathematics anxiety." In I. Sarason (Ed.), *Test Anxiety: Theory, Research, and Applications* (Hillsdale, N.J.: A. Erlbaum Assoc., 1980).

Sells, L. "Mathematics—A critical filter." *The Science Teacher*, February 1978, **45**(2).

Smith, W. H. "Treatments of mathematics anxiety: Anxiety management training; systematic desensitization; and self-help bibliotherapy." Unpublished doctoral dissertation, The University of Texas at Austin, 1979.

Suinn, R. M. A program of systematic desensitization for mathematics anxiety, cassette tapes (Colorado: Colorado State University, 1972).

Suinn, R. M., and F. C. Richardson. "Anxiety management training: Non-specific behavior therapy control." *Behavior Therapy*, 1971, **2**, 498–519.

Suinn, R. M., C. A. Edie, J. Nicoleti, and P. R. Spinelli. "The MARS, a measure of mathematics anxiety: Psychometric data." *Journal of Clinical Psychology*, 1972, **28**, 373–375.

Tobias, S. "Math anxiety: What is it and what can be done about it?" *Ms.*, September 1976, 56–69.

Tobias, S. *Overcoming Math Anxiety* (New York: W. W. Norton, 1978).

Tobias, S. "Who's afraid of math and why?" *Atlantic Monthly*, September 1978.

Woolfolk, R. L., and F. C. Richardson. *Stress, Sanity and Survival* (New York: Monarch, Simon and Schuster, 1978).

1.2 Formulating the Problem

Read the story given below. No questions are asked, but try to imagine yourself sitting in a living room with several others who have the same feelings you do about math. Your job is to read the story and make up a problem you know how to solve from any part of the story. You should have a pencil and paper and you can have as much time as you want, and nobody will look at what you are doing, but I want you to keep track of your feelings as you read the story and follow the directions.

> On the way to the market, which is 12 miles from home, I stopped at the drugstore to pick up a get-well card. I selected a series of cards with puzzles on them. The first one said "A bottle and a cork cost $1.10 and the bottle is a dollar more than the cork. How much is the bottle and how much is the cork?" I thought that would be a good card for Joe, so I purchased it for $1.25, along with a six-pack of cola for $1.79. The total bill was $3.22, which included 6% sales tax. My next stop was the market, which was exactly 3.4 miles from the drugstore. I bought $15.65 worth of groceries and paid for it with a $20 bill. I deposited the change in a charity bank on the counter, and left the store. On the way home I bought 8.5 gallons of gas for $13.60. Since I had gone 238 miles since my last fill-up, I was happy with the mileage on my new car. I returned home and made myself a ham and cheese sandwich.

Have you spent enough time on the story? Take time to reread it (spend at least 10 minutes with this exercise). Can you summarize your feelings? If my experiences in doing this exercise with my students apply to your setting, I would guess that you encountered some difficulty, some discomfort, perhaps despair or anger, or even indifference. Most students tend to focus on the more difficult questions (perhaps a miles per gallon problem) instead of following the directions to formulate a problem that will give you no difficulty. How about the question "What is the round-trip distance from home to market and back?" *Answer:*

$$2 \times 12 \text{ miles} = 24 \text{ miles}$$

Math anxiety builds on focusing on what you can't do rather than what you can do. This leads to anxiety and frustration. Do you know what is the most feared thing in our society? It is the fear of speaking in public. And the fear of letting others know you are having trouble with this problem is related to that fear of speaking in public.

If you focus on a problem that is too difficult, you will be facing a blank wall. This applies to all hobbies or subjects. If you play tennis or golf, has your game improved since you started? If you don't play these games, how do you think you would feel trying to learn in front of all your friends? Do you think you would feel foolish?

Mathematicians don't start with complicated problems. If a mathematician runs into a problem that she can't solve, she will probably rephrase the problem into a simpler related problem that she can't solve. This problem is, in turn, rephrased into yet a simpler problem, and the process continues until the problem is manageable.

For example, calculate

$$\frac{999,999,999 \times 999,999,999}{1+2+3+4+5+6+7+8+9+8+7+6+5+4+3+2+1}$$

This problem is terrible! But how about the following example?

$$\frac{1 \times 1}{1}$$

This is equal to 1. Next, try

$$\frac{22 \times 22}{1 + 2 + 1}$$

This is a little more difficult, but $22 \times 22 = 484$, so

$$\frac{22 \times 22}{1 + 2 + 1} = \frac{484}{4} = 121$$

Continue the pattern:

$$\frac{333 \times 333}{1 + 2 + 3 + 2 + 1} = \frac{110,889}{9} = 12,321$$

Do you see a pattern? What is the next problem?

$$\frac{444 \times 444}{1 + 2 + 3 + 4 + 3 + 2 + 1}$$

Four fours times four fours; what is the next number in this pattern? 55555×55555 (five fives)

Count from one up to four and then back down; what is the next number down here? $1 + 2 + 3 + 4 + 5 + 4 + 3 + 2 + 1$

Does the problem with nines fit into this pattern?

Now, look at the pattern of the answers:

1

1 2 1 ⟵———— *Answer with twos; compare with the denominator by eliminating the plus signs.*

1 2 3 2 1 ⟵———— *Answer with threes; compare with the denominator.*
Next answer? 1 2 3 4 3 2 1

Can you predict the answer to the original problem with the nines?

12,345,678,987,654,321

Let's consider some steps in problem solving, and then relate these steps to the story given at the beginning of this section.

1. Understand the problem; restate it in your own words.
2. Devise a plan.
3. Carry out the plan.
4. Look back; does the solution make sense?

Steps in Problem Solving

In the story at the beginning of this section there is a question on a get-well card. Do you remember it? "A bottle and a cork cost $1.10 and the bottle is a dollar more than the cork. How much is the bottle and how much is the cork?" You might have said the bottle cost $1.00 and the cork 10¢; the total is $1.10 alright, but did you look back to see if the solution makes sense? Does the bottle cost a dollar more than the cork?

$$\begin{pmatrix} \text{COST OF} \\ \text{BOTTLE} \end{pmatrix} - \begin{pmatrix} \text{COST OF} \\ \text{CORK} \end{pmatrix}$$
$$\$1.00 \quad - \quad \$0.10 \quad = \$0.90$$

The correct answer is

$$\begin{pmatrix} \text{COST OF} \\ \text{BOTTLE} \end{pmatrix} - \begin{pmatrix} \text{COST OF} \\ \text{CORK} \end{pmatrix}$$
$$\$1.05 \quad - \quad \$0.05 \quad = \$1.00$$

14   CHAPTER 1   INTRODUCTION TO PROBLEM SOLVING

Let's apply this procedure for solving a problem to the map shown in Figure 1.1.

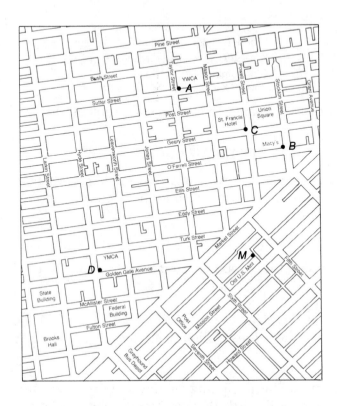

FIGURE 1.1   Portion of a map of San Francisco

Joan lives at the YWCA (point *A*) and works at Macy's (point *B*). She usually walks to work. How many different routes can Joan take if she doesn't backtrack; that is, if she always travels toward her destination? Where would you begin with this problem? Don't panic, but consider the steps in problem solving one at a time.

Step 1. Understand the problem. Can you restate it in your own words? Can you trace out one or two possible paths?

Step 2. Devise a plan. Simplify the question asked. Consider the simplified drawing shown in Figure 1.2.

Step 3. Carry out the plan. Count the number of ways it is possible to arrive at each point:

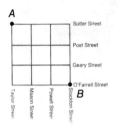

FIGURE 1.2   Detail of Figure 1.1

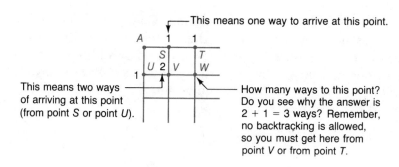

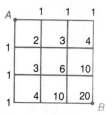

FIGURE 1.3   Map with solution

Now, you can fill in all the possibilities on Figure 1.2 as shown in Figure 1.3.

Step 4. Look back. Does the answer 20 different routes make sense? Do you think you could draw all 20 routes on the map?

EXAMPLE 1

How many different ways could Joan get from the YWCA (point *A*) to the St. Francis Hotel (point *C*)?

Solution

Draw a simplified version of Figure 1.1, as shown. You can see that there are 6 different paths.

EXAMPLE 2

How many different ways could Joan get from the YWCA (point *A*) to the U.S. Mint (point *M*)?

Solution

If the streets are irregular or if there are obstructions, you can still count blocks in the same fashion, as shown in Figure 1.4.

There are 52 paths from point *A* to point *M* (no backtracking). ∎

FIGURE 1.4

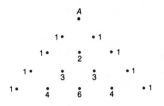

Let's formulate a general solution. Consider a map with starting point *A*. Do you see the pattern for building the figure shown here? Each new row is found by adding the two previous numbers. This pattern is known as *Pascal's triangle*. Notice in Figure 1.5 that the rows and diagonals are numbered for easy reference.

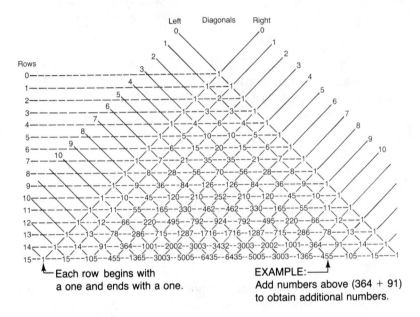

FIGURE 1.5   Pascal's triangle

Each row begins with a one and ends with a one.

EXAMPLE:
Add numbers above (364 + 91) to obtain additional numbers.

How does this apply to Joan's trip from the YWCA to Macy's? It is 3 blocks down and 3 blocks over. Look at Figure 1.5 and count out these blocks as shown in Figure 1.6:

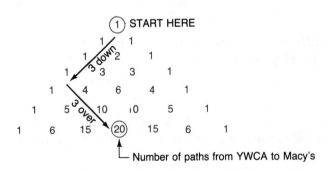

FIGURE 1.6

Number of paths from YWCA to Macy's

EXAMPLE 3

How many different ways could Joan get from the YWCA (point *A* in Figure 1.1) to the YMCA (point *D*)?

Solution

Look at Figure 1.1; from point *A* to point *D* it is 7 blocks down and 3 blocks left. Use Figure 1.5 to see that there are 120 paths:

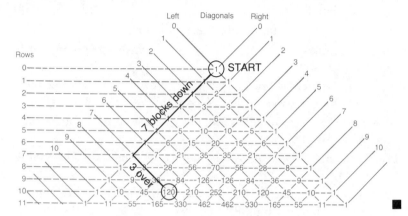

If there are alleys or irregular paths, you will need to use the map as shown in Example 2 rather than Pascal's triangle as shown in Example 3.

Problem Set 1.2

1. Read the following story and make up a problem you can solve from some part of the story.

   Yesterday, I purchased five calves at the auction for $95 each, so my herd now consists of nineteen cows, one bull, twenty-six steers, and thirteen calves. The auction yard charged me $35 to deliver the calves to my ranch, but I figured it was a pretty good deal since I live 42 miles from the auction yard. Today, twelve tons of hay were delivered and I paid $780 for it plus $10 a ton delivery charge. Yes sir, if my crops do well, this will be a very good year.

2. Describe some of your feelings as you worked Problem 1.

3. Consider Pascal's triangle shown in Figure 1.5. Name several patterns you notice about the triangle.

4. Describe the location of the numbers 1, 2, 3, 4, 5, . . . in Pascal's triangle.

5. Describe the location of the numbers 1, 5, 15, 35, 70, . . . in Pascal's triangle.

6. **a.** What is the sum of the numbers in row 1 of Pascal's triangle?

    b. What is the sum of the numbers in row 2 of Pascal's triangle?

    c. What is the sum of the numbers in row 3 of Pascal's triangle?

    d. What is the sum of the numbers in row 4 of Pascal's triangle?

7. How many 4¢ stamps are there in a dozen?

8. If you take 7 cards from a deck of 52 cards, how many cards do you have?

9. Oak Park cemetery in Oak Park, New Jersey, will not bury anyone living west of the Mississippi. Why?

10. Two U.S. coins total 30¢, yet one of these coins is not a nickel. What are the coins?

11. How many outs in a baseball game that lasts the full 9 innings?

12. If posts are spaced 10 feet apart, how many posts are needed for 100 feet of straight line fence?

Use the following map to determine the number of different paths from point A to the point indicated in Problems 13–19.

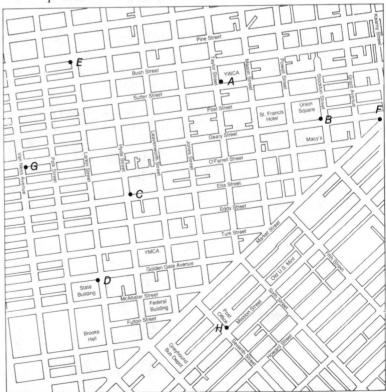

13. *B*   **14.** *C*   **15.** *D*   **16.** *E*

17. *F*   **18.** *G*   **19.** *H*

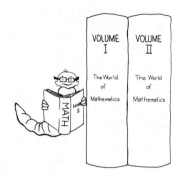

20. Two volumes of Newman's *The World of Mathematics* stand side-by-side in order on a shelf. A bookworm starts at page i of Volume I and bores its way in a straight line to the last page of Volume II. Each cover is 2 mm thick, and the first volume is $\frac{17}{19}$ as thick as the second volume. The first volume is 38 mm thick without its cover. How far does the bookworm travel?

21. A boy cyclist and a girl cyclist are 10 miles apart and pedaling toward each other. The boy's rate is 6 miles per hour, and the girl's rate is 4 miles per hour. There is also a friendly fly zooming continuously back and forth from one bike to the other. If the fly's rate is 20 miles per hour, how far does the fly fly by the time the cyclists reach each other?

A friendly fly

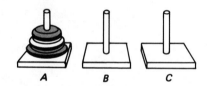

10 miles

22. A farmer has to get a fox, a goose, and a bag of corn across a river in a boat that is large enough only for him and one of these three items. If he leaves the fox alone with the goose, the fox will eat the goose. If he leaves the goose alone with the corn, the goose will eat the corn. How does he get all the items across the river?

23. When my daughter was 2, she had a toy that consisted of colored rings of different sizes, as shown in the figure. Suppose we wish to move the "tower" from stand *A* to stand *C*. To make this interesting, let's agree to the following rules: (1) move only one ring at a time; (2) at no time may a larger ring be placed on a smaller ring. According to these rules, we can move the tower from stand *A* to *C* in seven moves. Explain how this can be done.

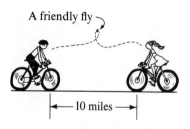

Tower problem

24. If you add another ring to the tower described in Problem 23, the tower can be moved in fifteen moves. Explain how.

Mind Bogglers

25. If you add a fifth ring to the tower described in Problems 23 and 24, how many moves will be required? [*Hint:* Look for a pattern.]

3 rings: 7 moves   and $7 = 2^3 - 1$
4 rings: 15 moves   and $15 = 2^4 - 1$
5 rings: ? moves*

*2^3 means $2 \times 2 \times 2$ and 2^4 means $2 \times 2 \times 2 \times 2$.

26. Problems 23–25 are examples of a famous problem called the *Tower of Hanoi*. The ancient Brahman priests were to move a pile of 64 such disks of decreasing size, after which the world would end. This task would require $2^{64} - 1$ moves. Try to estimate how long this would take at the rate of one move per second.

27. Problem 26 on the Tower of Hanoi raises some interesting mathematical questions. Do some research on this problem and answer the following questions:

    a. How soon after you begin to solve the puzzle will you move each of the disks for the first time?

    b. With what frequency will you change a given disk's position after you have moved it the first time?

    c. How many times will you move each of the disks in the course of rebuilding the tower consisting of n disks?

    d. What is the most efficient means of solving this puzzle?

References: Schuh, Frederick, *The Masterbook of Mathematical Recreations* (New York: Dover, 1968).

             Schwager, Michael, "Another look at the Tower of Hanoi," *The Mathematics Teacher*, September 1977, 528–533.

28. Pascal's triangle. Do some more research on Pascal's triangle, and see how many properties you can discover. You might begin by answering these questions:

    a. What is a binomial expansion?

    b. How is the binomial expansion related to Pascal's triangle?

    c. How are Fibonacci numbers related to Pascal's triangle?

    d. What relationship do the patterns in the figures have to Pascal's triangle?

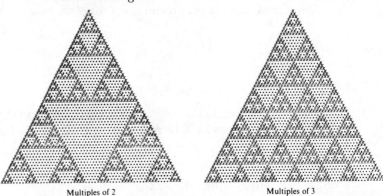

Multiples of 2   Multiples of 3

Some references you might check are "Mathematical games department," *Scientific American*, December 1966, **215**(6) and January 1967, **216**(1), and "Probability," by Mark Kac, *Scien-*

tific American, September 1964, **211**(3). An article with bibliography is "Pascal's triangle," by Karl Smith, *The Two-Year College Mathematics Journal*, Winter 1973, **4**.

1.3 Mathematical Patterns

Patterns are sometimes used as part of an IQ test. An organization called *Mensa* restricts its membership to people who have scored at or above the 98th percentile on a standardized IQ test. The test on this page is one of Mensa's "quickie" tests. You might want to see how well you can do.

ARE YOU A GENIUS?

EACH problem is a series of some sort—that is, a succession of either letters, numbers or drawings—with the last item in the series missing. Each series is arranged according to a different rule and, in order to identify the missing item, you must figure out what that rule is.

Now, it's your turn to play. Give yourself a maximum of 20 minutes to answer the 15 questions. If you haven't finished in that time, stop anyway. In the test problems done with drawings, it is always the top row or group that needs to be completed by choosing one drawing from the bottom row.

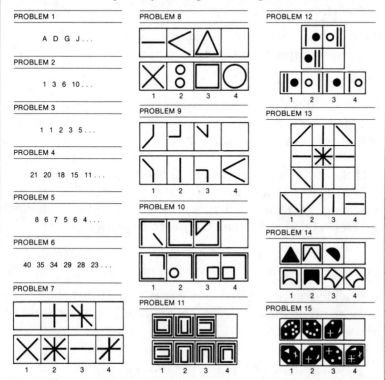

PROBLEM 1

A D G J . . .

PROBLEM 2

1 3 6 10 . . .

PROBLEM 3

1 1 2 3 5 . . .

PROBLEM 4

21 20 18 15 11 . . .

PROBLEM 5

8 6 7 5 6 4 . . .

PROBLEM 6

40 35 34 29 28 23 . . .

PROBLEM 7

PROBLEM 8

PROBLEM 9

PROBLEM 10

PROBLEM 11

PROBLEM 12

PROBLEM 13

PROBLEM 14

PROBLEM 15

SCORING
Give yourself 1 point for each correct answer. If you completed the test in 15 minutes or less, give yourself an additional 4 points.

15–3
11–4; 12–4; 13–2; 14–3;
6–22; 7–2; 8–3; 9–2; 10–4;
1–M; 2–15; 3–8; 4–6; 5–5;

IF YOU SCORED:
15 to 19 points: You are exceptionally intelligent—a perfect Mensa candidate!

13 or 14 points: This should put you in the upper 2% of the population.

10, 11, or 12 points: An honorable score.

Less than 10 points: Forget about joining Mensa. But you're in good company—many world-famous figures don't have exceptional IQs either! And whatever you scored, don't take this test too seriously.

For more information about Mensa, write: Mensa, Dept. IA, 1701 West 3rd St., Brooklyn, N.Y. 11223.

It used to be thought that IQ tests measured "innate intelligence" and that a person's IQ score was fairly constant. Today, it is known that this is not the case. IQ test scores can be significantly changed by studying the types of questions asked. For example, let's focus on one type of question on these tests.

Number Sequence

> A **number sequence** or **progression** is a collection of numbers arranged in order so that there is a first term, a second term, a third term, and so on.

A sequence may stop at a particular place, or it may continue indefinitely. It should be understood that sequences are arranged from left to right with each number, or term, separated by a comma. In this section we'll look at three categories of progressions.

EXAMPLE 1

What is the pattern for the sequence 1, 4, 7, 10, 13, . . . ? Fill in the next three terms.

Solution

There is a difference of 3 between each two terms. That is, 3 is added to each term in order to find the next term. The next three terms are 16, 19, and 22.   ■

If the difference between each two terms of a sequence is the same, it is called the **common difference**. The common difference in Example 1 is 3. If a sequence has a common difference, it is called an **arithmetic sequence**.

EXAMPLE 2

Tell which of the given sequences are arithmetic. If the sequence is arithmetic, give the common difference and the next term.

a. 6, 9, 12, 15, . . .   **b.** 66, 72, 78, 84, . . .
c. 4, 11, 18, 24, . . .   **d.** 113, 322, 531, 740, . . .

Solution

a. Check the differences:

$$9 - 6 = 3 \quad 12 - 9 = 3 \quad 15 - 12 = 3$$
$$6, \qquad 9, \qquad 12, \qquad 15, \ldots$$

The common difference is 3; the next term is found by adding 3:

$$15 + 3 = 18$$

b. Check the differences:

$$72 - 66 = 6 \quad 78 - 72 = 6 \quad 84 - 78 = 6$$
$$66, \qquad 72, \qquad 78, \qquad 84, \ldots$$

The common difference is 6; the next term is $84 + 6 = 90$.

c. Check the differences:

$$11 - 4 = 7 \quad 18 - 11 = 7 \quad 24 - 18 = 6$$
$$4, \qquad 11, \qquad 18, \qquad 24, \ldots$$

There is no common difference, so this sequence is not arithmetic. This example illustrates a common mistake among students, which is not to carefully check *all* the differences to see that they are the same.

d. Check the differences:

$$322 - 113 = 209 \quad 531 - 322 = 209 \quad 740 - 531 = 209$$
$$113, \qquad 322, \qquad 531, \qquad 740, \ldots$$

The common difference is 209; the next term is $740 + 209 = 949$. ∎

Example 3 illustrates the second category of sequences that we will consider in this book.

EXAMPLE 3

What is the pattern for the sequence 2, 4, 8, 16, 32, . . . ? Fill in the next three terms.

Solution

This is *not* an arithmetic sequence because there is no common difference. However, if each term is *divided* by the preceding term, the results are the same. This number found by division is called a **common ratio**. If a sequence has a common ratio, then the sequence is called a **geometric sequence**. Additional terms of a geometric sequence can be found by multiplying successive terms by the common ratio. For this example the common ratio is 2, so the next three terms are

$$32 \times 2 = 64$$
$$64 \times 2 = 128$$
$$128 \times 2 = 256$$ ■

EXAMPLE 4

Tell which of the given sequences are geometric. If the sequence is geometric, give the common ratio and the next term.

a. 1, 3, 9, 27, . . . **b.** 3, 12, 48, 182, . . .
c. 4, 20, 100, 500, . . . **d.** 5, 50, 500, 5000, . . .

Solution

a. Check the ratios:

$$\textbf{1,}\,^{3 \div 1 = 3}\,\textbf{3,}\,^{9 \div 3 = 3}\,\textbf{9,}\,^{27 \div 9 = 3}\,\textbf{27,}\ldots$$

The common ratio is 3; the next term is found by multiplying by 3:

$$27 \times 3 = \textbf{81}$$

b. Check the ratios:

$$\textbf{3,}\,^{12 \div 3 = 4}\,\textbf{12,}\,^{48 \div 12 = 4}\,\textbf{48,}\,^{182 \div 48 \neq 4}\,\textbf{182,}\ldots$$

There is no common ratio, so this sequence is not geometric.

c. Check the ratios:

$$\textbf{4,}\,^{20 \div 4 = 5}\,\textbf{20,}\,^{100 \div 20 = 5}\,\textbf{100,}\,^{500 \div 100 = 5}\,\textbf{500,}\ldots$$

The common ratio is 5; the next term is $500 \times 5 = \textbf{2500}$.

d. Check the ratios:

5, $^{50 \div 5 = 10}$ **50,** $^{500 \div 50 = 10}$ **500,** $^{5000 \div 500 = 10}$ **5000,...**

The common ratio is 10; the next term is **50,000.** ∎

To find the pattern in a sequence you might look for a pattern with the differences or ratios, but in general, there is no formula or procedure for finding these patterns.

EXAMPLE 5

Classify the following sequences as arithmetic, geometric, both, or neither of these types. If the sequence is arithmetic, give the common difference; if it is geometric, give the common ratio; and if it is neither, describe the pattern. Give the next term for each sequence.

a. 15, 30, 60, 120, ...   **b.** 15, 30, 45, 60, ...
c. 15, 30, 45, 75, ...   **d.** 15, 20, 26, 33, ...
e. 3, 3, 3, 3, ...   **f.** 15, 30, 90, 360, ...
g. 15, 30, 120, 960, ...   **h.** A, C, F, J, O, ...

Solution

a. Differences: 30 − 15 = 15   60 − 30 = 30   Not arithmetic
     15,   30,   60,   120,...
Ratios: 30 ÷ 15 = 2   60 ÷ 30 = 2   120 ÷ 60 = 2   Common
              ratio is 2

This is a geometric sequence; the next term is
120 × 2 = **240**.

b. Differences: 30 − 15 = 15   45 − 30 = 15   60 − 45 = 15   Common difference is 15
     15,   30,   45,   60,...

This is an arithmetic sequence; the next term is
60 + 15 = **75**.

c. Differences: 15   15   30   Not arithmetic
     15,   30,   45,   75,...
Ratios: 2   Not 2   Not geometric

The pattern is to add successive terms. This is a fairly common way of formulating a sequence:

$$15 + 30 = 45; \quad 30 + 45 = 75;$$
$$45 + 75 = \textbf{120} \text{ is the next term}$$

181164

d. Differences: 5   6   7   Not arithmetic
15,   20,   26,   33,...
Ratios: 20 ÷ 15 ≠ 26 ÷ 20   Not geometric

There is a pattern in the differences; that is, the differences form an arithmetic sequence, so the next difference is 8. Thus, the next term is $33 + 8 = \mathbf{41}$.

e. Differences: 0   0   0   Arithmetic
3,   3,   3,   3,...
Ratios: 1   1   1   Geometric

This pattern is both arithmetic and geometric.

f. Differences: 15   60   270   Not arithmetic
15,   30,   90,   360,...
Ratios: 2   3   4   Not geometric

There is a pattern of the ratios; the ratios form an arithmetic sequence, so the next ratio is 5. Thus, the next term is $360 \times 5 = \mathbf{1800}$.

g. Differences: 15   90   840   Not arithmetic
15,   30,   120,   960,...
Ratios: 2   4   8   Not geometric

There is a geometric pattern in the ratios; the next ratio is 16, so the next term is $960 \times 16 = \mathbf{15,360}$.

h. Not all patterns need be patterns of numbers, as this example illustrates. It is, of course, not arithmetic or geometric, but notice that after A there is one letter (B) omitted; then C; then two letters omitted; then F; then three letters omitted; J; four letters omitted; O; so the next gap should have five letters omitted; thus, the next letter in the pattern is **U**.   ■

The pattern of adding successive terms illustrated by Example 5c suggests a particular famous sequence that has a name. The sequence

$$1, 1, 2, 3, 5, 8, 13, 21, 34, 55, 89, 144, \ldots$$

is called the **Fibonacci sequence**. The numbers of this sequence are called **Fibonacci numbers**, and they have interested mathematicians for centuries—at least since the 13th century, when Leonardo Fibonacci wrote *Liber Abaci*, which discussed the advantages of Hindu–Arabic numerals over

Roman numerals. In this book, he had a problem that related Fibonacci numbers to the birth patterns of rabbits. Let's
see what he did. The problem was to find the number of
rabbits alive after a given number of generations. Suppose a
pair of rabbits will produce a new pair of rabbits in their
second month and thereafter will produce a new pair every
month. The new rabbits will do exactly the same. Starting
with one pair, how many pairs will there be in 10 months if
no rabbits die?

To solve this problem, we could begin by counting:

Number of months	Number of pairs	Pairs of rabbits (the shaded pairs are ready to reproduce in the next month)
START	1	
1	1	
2	2	
3	3	
4	5	
5	8	

Same pair
(rabbits never
die)

Instead of continuing in this fashion, Fibonacci looked for a
pattern:

$$1, 1, 2, 3, 5, 8, ?$$

Do you see a pattern for this sequence? (For each term, consider the sum of the two preceding terms.)

Using this pattern, Fibonacci was able to compute the number of rabbits alive after 10 months (it is the tenth term after the first 1), which is 89. (Continuing with the above sequence: . . . , 5, 8, 13, 21, 34, 55, 89, ...) He could also compute the number of rabbits after the first year or after any other interval. Without a pattern, the problem would indeed be a difficult one.

Many mathematicians, such as Verner Hoggatt and Brother Brosseau, have devoted large portions of their careers to the study of Fibonacci numbers. There is even a mathematical society called the Fibonacci Association, which publishes *The Fibonacci Quarterly* and operates a Fibonacci Bibliographical and Research Center.

You may wonder why there is so much interest in the Fibonacci sequence. Some of its applications are far-reaching. It has been used in botany, zoology, physical science, business, economics, statistics, operations research, archeology, fine arts, architecture, education, psychology, sociology, and poetry. An example of the occurrence of Fibonacci numbers in nature is illustrated by a sunflower. The seeds are arranged in spiral curves as shown in Figure 1.7. If we count the number of clockwise and counterclockwise spirals (13 and 21 in this example), they are successive terms in the Fibonacci sequence. This is true of all sunflowers and, indeed, of the seed head of any composite flower such as the daisy or aster.

Closely related to the Fibonacci numbers are the **Lucas numbers** (named after the 19th century French mathematician Lucas):

$$1, 3, 4, 7, 11, 18, 29, \underline{\quad ? \quad}$$

Can you fill in the blank? [*Hint:* This sequence begins differently but is generated in the same way as the Fibonacci sequence.]

What properties of Lucas or Fibonacci numbers can you discover? These numbers can also be found in Pascal's triangle. Can you find the Fibonacci numbers embedded in Pascal's triangle?

FIGURE 1.7 The arrangement of the pods (phyllotaxy) of a sunflower

Problem Set 1.3

In Problems 1–9 tell which of the given sequences are arithmetic. If the sequence is arithmetic, give the common difference and the next term.

1. 41, 45, 49, 53, . . .
2. 8, 18, 28, 38, . . .
3. 19, 25, 31, 37, . . .
4. 17, 24, 31, 37, . . .
5. 34, 51, 68, 84, . . .
6. 83, 102, 121, 142, . . .
7. 119, 122, 125, 128, . . .
8. 119, 222, 325, 428, . . .
9. 48, 300, 552, 804, . . .

In Problems 10–18 tell which of the given sequences are geometric. If the sequence is geometric, give the common ratio and the next term.

10. 2, 6, 18, 54, . . .
11. 2, 12, 72, 432, . . .
12. 1, 4, 16, 64, . . .
13. 3, 18, 108, 654, . . .
14. 6, 12, 24, 96, . . .
15. 3, 15, 45, 215, . . .
16. 5, 55, 555, 5555, . . .
17. 1, 10, 100, 1000, . . .
18. 8, 8, 8, 8, . . .

Classify the sequences in Problems 19–30 as arithmetic, geometric, both, or neither. If the sequence is arithmetic, give the common difference; if it is geometric, give the common ratio; and if it is neither, describe the pattern. Give the next term.

19. 2, 5, 8, 11, 14, . . .
20. 1, 2, 1, 1, 2, 1, 1, 1, 2, 1, 1, 1, . . .
21. 1, 2, 4, 7, 11, 16, . . .
22. 1, 3, 4, 7, 11, 18, 29, . . .
23. 3, 6, 12, 24, 48, . . .
24. 5, 15, 45, 135, 405, . . .
25. 100, 99, 97, 94, 90, . . .
26. 1, 4, 9, 16, 25, . . .
27. 2, 5, 2, 5, 5, 2, 5, 5, 5, . . .
28. 5, 15, 25, 35, . . .
29. A, A, B, Ь, C, Ɔ, D, . . .
30. A, C, E, G, I, K, . . .
31. What are Fibonacci numbers?
32. Write out the first ten terms of the Lucas sequence.

Mind Bogglers

Fill in the missing terms in Problems 33–37.

33. 225, 625, 1225, 2025, _____
34. 1, 8, 27, 64, 125, _____
35. 8, 5, 4, 9, 1, _____

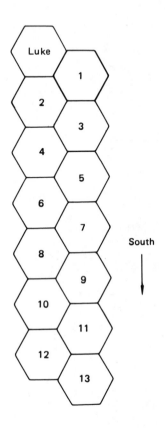

36. A, E, F, H, I, K, L, M, N, _____

37. dog, three, lion, four, tiger, five, hippopotamus, _____
[*Hint:* Answer is not six.]

38. On a planet far, far away, Luke finds himself in a strange building with hexagon-shaped rooms. In his search for the princess, he always moves to an adjacent room and always moves south.
  a. How many paths are there to room 1? Room 2? Room 3? Room 4?
  b. How many paths are there to room 10?
  c. How many paths are there to room 20?

39. **Fibonacci numbers.** Write a short paper about Fibonacci numbers. You might like to check *The Fibonacci Quarterly*, particularly "A Primer on the Fibonacci Sequence," Parts I and II, in the February and April 1963 issues. The articles were written by Verner Hoggatt and S. L. Basin. See also a booklet by Verner Hoggatt, Jr., *Fibonacci and Lucas Numbers*, published in the Houghton Mifflin Mathematics Enrichment Series, 1969. An interesting article for teachers, "Fibonacci Numbers and the Slow Learner," by James Curl, appeared in the October 1968 issue of *The Fibonacci Quarterly*.

1.4 Inductive and Deductive Reasoning

The type of reasoning we've used in this chapter so far—first observing patterns and then predicting answers for complicated problems—is called **inductive reasoning**. It is a very important method of thought and is sometimes called the **scientific method**. It involves reasoning from particular facts or individual cases to a general **conjecture**—a statement you think may be true. That is, a generalization is made on the basis of some observed occurrences.

Inductive Reasoning

Inductive reasoning is reaching a conclusion by making particular observations.

The more individual occurrences you observe, the better able you are to make a correct generalization. The cartoon shows an incorrect generalization based on only two questions. The child reached the conclusion that the pattern was a sequence of garbage trucks. You must keep in mind that an inductive conclusion is always tentative and may have to be revised on the basis of new evidence. For example, consider this pattern:

$$
\begin{aligned}
1 \times 9 &= 9 \\
2 \times 9 &= 18 \\
3 \times 9 &= 27 \\
&\ \vdots \\
10 \times 9 &= 90
\end{aligned}
$$

You might have made the conjecture that the sum of the digits of the answer to any product involving a 9 is always 9. Certainly the first ten examples substantiate this speculation. But the conjecture is suspect because it is based on so few cases. It is shattered by the very next case, where we get $11 \times 9 = 99$.

On the other hand, we can make a prediction about the exact time of sunrise and sunset tomorrow. This is also an example of inductive reasoning, since our prediction is based on a large number of observed cases. Thus, there is a very high probability that we will be successful in our prediction.

Although mathematicians often proceed by inductive reasoning to formulate new ideas, they are not content to stop at the "probable" stage. So they often formalize their predictions into theorems and then try to prove these theorems *deductively*.

Deductive reasoning produces results that are *certain* within the logical system being developed. For example, consider the following argument:

1. If you take the *Times*, then you are well-informed.
2. You take the *Times*.
3. Therefore, you are well-informed.

If you accept statements 1 and 2 as true, then you **must** accept statement 3 as true. Statements 1 and 2 are called the **premises** of the argument, and statement 3 is called the **conclusion**. Such reasoning is called **deductive reasoning**, and if the conclusion follows from the premises, the reasoning is said to be **valid**.

Deductive Reasoning

> **Deductive reasoning** is reaching a conclusion by using a formal structure based on a set of **undefined** terms and on a set of accepted unproved **axioms** or **premises**. The conclusions are said to be **proved** and are called **theorems**.

Logic began to flourish during the classical Greek period. Aristotle (384–322 B.C.) was the first person to study the subject systematically, and he and many other Greeks searched for universal truths that were irrefutable. The logic of this period, referred to as **Aristotelian logic**, is still used today. The example of deductive reasoning given above is an example of Aristotelian logic.

The second great period for logic came with the use of symbols to simplify complicated logical arguments. This began when the great German mathematician Gottfried von Leibniz (1646–1716), at the age of 14, attempted to reform Aristotelian logic. He called his logic the **universal characteristic** and wrote, in 1666, that he wanted to create a general method in which truths of reason would be reduced to a calculation so that errors of thought would appear as computational errors.

However, the world took little notice of Leibniz' logic, and it wasn't until George Boole (1815–1864) completed his book, *An Investigation of the Laws of Thought*, that deductive logic entered its third and most important period. Boole considered various mathematical operations by separating them from the other commonly used symbols. This idea was popularized by Bertrand Russell (1872–1970) and Alfred North Whitehead (1861–1947) in their monumental *Principia Mathematica*. In this work, they began with a few assumptions and three undefined terms and built a system of symbolic logic. From this, they then formally developed the theorems of arithmetic and mathematics.

Problem Set 1.4

1. In your own words explain inductive reasoning. Give an original example of when you have used inductive reasoning or heard it being used.

2. In your own words explain deductive reasoning. Give an original example of when you have used deductive reasoning or heard it being used.

3. What is the universal characteristic?

4. Does the following *B.C.* cartoon illustrate inductive or deductive reasoning? Explain your answer.

THE YOUNG LADY in charge of the checkroom in a fancy restaurant was known for her memory. She never used the usual markers to identify the hats and coats.

One day a guest decided to put her to a test, and when he received his hat said, "Sally, how do you know this is my hat?"

"I don't sir," was the response.

"But why did you return it to me?" asked the guest.

"Because," said Sally, "it's the one you gave to me."

5. Does the news story in the margin illustrate inductive or deductive reasoning? Explain your answer.

Problems 6–9 refer to the lyrics of "By the Time I Get to Phoenix." Tell whether each answer you give is arrived at inductively or deductively.

By the Time I Get to Phoenix

By the time I get to Phoenix she'll be risin'.
She'll find the note I left hangin' on her door.
She'll laugh when she reads the part that says I'm leavin',
'Cause I've left that girl so many times before.

By the time I make Albuquerque she'll be workin'.
She'll probably stop at lunch and give me a call.
But she'll just hear that phone keep on ringin'
Off the wall, that's all.

By the time I make Oklahoma she'll be sleepin'.
She'll turn softly and call my name out low.
And she'll cry just to think I'd really leave her,
'tho' time and time I've tried to tell her so,
She just didn't know,
I would really go.

Words and Music by Jim Webb

6. What is the basic direction (north, south, east, or west) the person is traveling?

7. What method of transportation or travel is the person using?

8. What is the probable starting point of this journey?

9. List five facts you know about each person involved.

Problems 10–14 refer to the lyrics of "Ode to Billy Joe." Tell whether each answer you give is arrived at inductively or deductively.

Ode to Billy Joe*

It was the third of June, another sleepy, dusty, delta day.
I was out choppin' cotton and my brother was balin' hay;
And at dinnertime we stopped and walked back to the house to eat,
And mama hollered at the back door, "Y'all remember to wipe your feet."
Then she said, "I got some news this mornin' from Choctaw Ridge,
Today Billy Joe McAllister jumped off the Tallahatchee Bridge."

* ©1967 Northridge Music Company. Administered by Artista Music, Inc. USED BY PERMISSION.

Papa said to Mama, as he passed around the black-eyed peas,
"Well, Billy Joe never had a lick o' sense, pass the biscuits please,
There's five more acres in the lower forty I've got to plow,"
And Mama said it was a shame about Billy Joe anyhow.
Seems like nothin' ever comes to no good up on Choctaw Ridge,
And now Billy Joe McAllister's jumped off the Tallahatchee Bridge.

Brother said he recollected when he and Tom and Billy Joe,
Put a frog down my back at the Carroll County picture show,
And wasn't I talkin' to him after church last Sunday night,
I'll have another piece of apple pie, you know, it don't seem right
I saw him at the sawmill yesterday on Choctaw Ridge,
And now you tell me Billy Joe's jumped off the Tallahatchee Bridge.

Mama said to me, "Child, what's happened to your appetite?
I been cookin' all mornin' and you haven't touched a single bite,
That nice young preacher Brother Taylor dropped by today,
Said he'd be pleased to have dinner on Sunday, Oh, by the way,
He said he saw a girl that looked a lot like you up on Choctaw
    Ridge
And she an' Billy Joe was throwin' somethin' off the Tallahatchee
    Bridge."

A year has come and gone since we heard the news 'bout Billy Joe,
Brother married Becky Thompson, they bought a store in Tupelo,
There was a virus goin' 'round, Papa caught it and he died last
    spring,
And now Mama doesn't seem to want to do much of anything.
And me I spend a lot of time pickin' flowers up on Choctaw Ridge,
And drop them into the muddy water off the Tallahatchee Bridge.

Words and Music by Bobbie Gentry

10. How many people are involved in this story? List them by name and/or description.
11. Who "saw him at the sawmill yesterday"?
12. In which state is the Tallahatchee Bridge located?
13. On what day or days of the week could the death not have taken place? On what day of the week was the death most probable?
14. What was it that she was throwing off the bridge?

Mind Bogglers

Problems 15–20 do not take any special knowledge, but they do take clear thinking. Some will seem very easy while others will seem very difficult. You are not expected to be able to answer them all, but you are expected to try.

15. If you had only one match and entered a room in which there was a kerosene lamp, an oil burner, and a wood-burning stove, which would you light first?

16. Do they have a 4th of July in England?

17. Some months have 30 days and some have 31 days. How many have 28 days?

18. A man built a house with four sides and it is rectangular in shape. Each side has a southern exposure. A big bear came wandering by. What color is the bear?

19. If a doctor gave you 3 pills and told you to take one every half hour, how long would they last you?

20. A woman gives a beggar 50¢. The woman is the beggar's sister, but the beggar is not the woman's brother. How come?

1.5 Review Problems

The answers to all these review questions are given in the back of the book in order to help you review the material. In case you have difficulty with a certain problem, the section from which it was taken is shown in the margin.

Section 1.1   **1.** In Problem Set 1.1 we listed 12 myths concerning mathematics. Turn to these myths on page 7 and select the one that is the most difficult for you to accept as a myth. Explain why you think you have trouble accepting this statement as a myth.

Section 1.2   **2.** In how many ways can a person in San Francisco walk from the corner of Webster and Turk Streets to the corner of Geary and Steiner Streets by traveling six blocks?

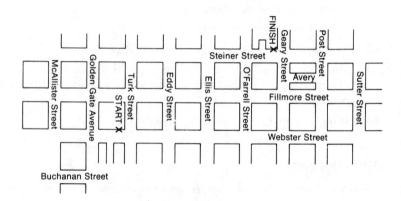

3. Compute $111,111,111 \times 111,111,111$. Do not use direct multiplication. Find

Section 1.3

$$1 \times 1$$
$$11 \times 11$$
$$111 \times 111$$
$$\vdots$$

and look for a pattern.

In Problems 4–7:

Section 1.3

a. Classify each sequence as arithmetic, geometric, both, or neither.
b. Give the common difference, common ratio, or describe the pattern.
c. Find the next term in the sequence.

4. 10,000, 1000, 100, 10, . . .   **5.** 12, 22, 32, 42, . . .

6. 10, 20, 30, 50, . . .   **7.** 65, 76, 87, 98, . . .

8. In your own words, explain the differences or similarities between inductive and deductive reasoning.

Section 1.4

9. An electric locomotive is traveling north at the rate of 40 miles per hour. It is being chased from the south by a wind blowing at 80 miles per hour. Will the smoke from the locomotive be blown ahead of the train at the rate of 40 miles per hour?

Sections 1.2, 1.4

10. There is a 10-foot steel ladder attached to the side of a boat. The rungs on the ladder are exactly one foot apart with the bottom rung resting exactly on the top of the water. If the tide rises one foot every half hour, how long will it take to cover the first three rungs of the ladder?

Sections 1.2, 1.4

Calculators and Arithmetic

CHAPTER 2

LET'S SEE... ONE...
TWO... THREE...

**Natural Numbers
and Whole Numbers**

2.1 Whole Numbers

In mathematics, we generally focus our attention on some particular sets of numbers. The simplest of these sets is the set we use to count objects and is the first set that a child learns. It is called the set of **counting numbers** or **natural numbers**.

The set of numbers {1, 2, 3, 4, . . .} is called the set of **counting numbers** or the set of **natural numbers**. If the number zero is included, then the set {0, 1, 2, 3, 4, . . .} is referred to as the set of **whole numbers**.

Addition, subtraction, multiplication, and division are called the **elementary operations** for the whole numbers, and it is assumed that you understand these operations.

However, certain agreements in dealing with these operations are necessary. Consider this arithmetic example:

Find: $2 + 3 \times 4$

There are two possible approaches.

Left to right: $\quad$ **$2 + 3$** $\times 4 =$ **5** $\times 4$
$\qquad\qquad\qquad\qquad\qquad = \mathbf{20}$

Multiplication first: $\quad 2 +$ **3×4** $= 2 +$ **12**
$\qquad\qquad\qquad\qquad\qquad = 14$

Since there are different answers to this arithmetic example, it is necessary for us all to agree on one method or the other. At this point, you might be thinking "Why on earth would I start at the right and do the multiplication first?" Consider the following example.

EXAMPLE 1

Suppose you sold a $2 soccer benefit ticket on Monday and three $4 tickets on Tuesday. What is the total amount collected?

Solution

$$\begin{pmatrix} \text{SALES ON} \\ \text{MONDAY} \end{pmatrix} + \begin{pmatrix} \text{SALES ON} \\ \text{TUESDAY} \end{pmatrix} = \begin{pmatrix} \text{TOTAL} \\ \text{SALES} \end{pmatrix}$$

$$\$2 \quad + \quad 3 \times \$4 \quad = \begin{pmatrix} \text{TOTAL} \\ \text{SALES} \end{pmatrix}$$

CORRECT:	NOT CORRECT:
Multiplication first	*Left to right*
$2 + 3 \times 4 = 2 + 12$	$2 + 3 \times 4 = 5 \times 4$
$= 14$	$= 20$

Do you see why the correct solution to Example 1 requires multiplication before addition? Consider another example.

EXAMPLE 2

Suppose you purchase a chair for $100. If the sales tax is 6% (write as .06), what is the total, including tax?

Solution

$$\begin{pmatrix} \text{PRICE OF} \\ \text{CHAIR} \end{pmatrix} + \quad (\text{TAX}) \quad = (\text{TOTAL PRICE})$$

$$\$100 \quad + .06 \times \$100 = (\text{TOTAL PRICE})$$

CORRECT:
Multiplication first
$$100 + .06 \times 100 = 100 + 6$$
$$= 106$$

NOT CORRECT:
Left to right
$$100 + .06 \times 100 = 100.06 \times 100$$
$$= 10,006$$

If the operations are mixed, we agree to do multiplication and division *first* (from left to right), and *then* addition and subtraction from left to right.

EXAMPLE 3

Perform the indicated operations.

a. $7 + 2 \times 6 = 7 + \mathbf{12}$ Multiplication first

$= \mathbf{19}$ Addition next

b. $3 \times 5 + 2 \times 5 = \mathbf{15} + \mathbf{10}$ Multiplication first

$= \mathbf{25}$ Addition next

c. $1 + 3 \times 2 + 4 - 3 + 6 \times 3 = 1 + \mathbf{6} + 4 - 3 + \mathbf{18}$

$= \mathbf{7} + 4 - 3 + 18$

$= \mathbf{11} - 3 + 18$

$= \mathbf{8} + 18$

$= \mathbf{26}$ ∎

If the order of operations is to be changed from this agreement, then parentheses are used to indicate this change.

EXAMPLE 4

Perform the indicated operations.

a. $10 + 6 + 2 = (10 + 6) + 2$ Work from left to right;

$= \mathbf{16} + 2$ parentheses are *not necessary*

$= \mathbf{18}$ for this example.

b. $10 + (6 + 2) = 10 + \mathbf{8}$ Parentheses change the

$= \mathbf{18}$ order of operations.

Notice that parts a and b illustrate different problems giving the same answer.

c. $10 - 6 - 2 = (10 - 6) - 2$

$= \mathbf{4} - 2$

$= \mathbf{2}$

d. $10 - (6 - 2) = 10 - \mathbf{4}$

$= \mathbf{6}$ ∎

We can now summarize the correct order of operations including the use of parentheses.

> 1. Parentheses first
> 2. Multiplication and division, reading from left to right
> 3. Addition and subtraction, reading from left to right

Order of Operations

You can remember this by using the mnemonic "**P**lease **M**ind **D**ear **A**unt **S**ally." The first letters will remind you of: "**P**arentheses, **M**ultiplication and **D**ivision, **A**ddition and **S**ubtraction."

EXAMPLE 5

Perform the indicated operations.

a. $12 + 9 \div 3 = 12 + 3$
$= 15$

b. $(12 + 9) \div 3 = 21 \div 3$
$= 7$

c. $4 \times (3 + 2) = 4 \times 5$
$= 20$

d. $4 \times 3 + 4 \times 2 = 12 + 8$
$= 20$

e. $2 \times 15 + 9 \div 3 - 7 \times 2 = 30 + 3 - 14$
$= 33 - 14$
$= 19$

f. $2 \times (15 + 9) \div 3 - 7 \times 2 = 2 \times (24) \div 3 - 7 \times 2$
$= 48 \div 3 - 7 \times 2$
$= 16 - 7 \times 2$
$= 16 - 14$
$= 2$ ∎

Look at Examples 5c and 5d. They illustrate a combined property of addition and multiplication called the **distributive law for multiplication over addition:**

$$4 \times (3 + 2) = 4 \times 3 + 4 \times 2$$

Number
outside
parentheses

Number outside
parentheses is **distributed**
to *each* number inside parentheses.

This property holds for all whole numbers.

EXAMPLE 6

Write out each of the given expressions without parentheses by using the distributive property.

a. $8 \times (7 + 10) = 8 \times 7 + 8 \times 10$
b. $3 \times (400 + 20 + 5) = 3 \times 400 + 3 \times 20 + 3 \times 5$ ■

CALCULATOR COMMENT

The rest of this section is optional for those who have, or expect to have, a calculator. Calculators are classified according to their ability to do different types of problems, as well as by the type of logic they use to do the calculations. The problem of selecting a calculator is further compounded by the multiplicity of brands from which to choose. Therefore, to choose a calculator and to learn to use it require some sort of instruction.

The different levels of calculators are distinguished primarily by their price.

1. **Four-function calculators (under $10).** These calculators have a keyboard consisting of the numerals (see Figure 2.1) and the four arithmetic operations, or functions: addition $\boxed{+}$, subtraction $\boxed{-}$, multiplication $\boxed{\times}$, and division $\boxed{\div}$.
2. **Four-function calculators with memory ($10–$20).** Usually no more expensive than four-function calculators, these offer a memory register, $\boxed{\text{M}}$, $\boxed{\text{STO}}$, or $\boxed{\text{M}^+}$. The more expensive models may have more than one memory register. Memory registers allow you to store partial calculations for later recall. Some checkbook models will even remember the total when they are turned off.
3. **Scientific calculators ($20–$50).** These calculators add additional mathematical functions, such as square root $\boxed{\sqrt{}}$, trigonometric $\boxed{\text{SIN}}$, $\boxed{\text{COS}}$, and $\boxed{\text{TAN}}$, logarithmic $\boxed{\text{LOG}}$, and exponential $\boxed{\text{EXP}}$. Depending on the particular brand, a scientific model may have other keys as well.
4. **Special-purpose calculators ($40–$400).** Special-use calculators for business, statistics, surveying, medicine, or even gambling and chess are available.

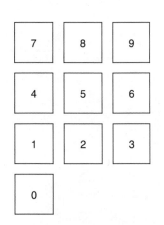

FIGURE 2.1 Standard calculator keyboard, additional keys depend on brand and model

5. Programmable calculators ($50–$600). These calculators allow the insertion of different cards that "remember" the sequence of steps for complex calculations.

For most nonscientific purposes, a four-function calculator with memory will be sufficient for everyday usage. If you anticipate taking several mathematics and/or science courses, then you'll find a scientific calculator to be a worthwhile investment.

There are essentially three types of logic used by these calculators: **arithmetic**, **algebraic**, and **RPN**. In this section, we discussed the correct order of operations, which means that the correct value for

$$2 + 3 \times 4$$

is 14 (multiply first). An algebraic calculator will "know" this fact and will give the correct answer, whereas an arithmetic calculator will simply work from left to right to obtain the incorrect answer, 20. Therefore, if you have an arithmetic-logic calculator, you will need to be careful about the order of operations. Some arithmetic-logic calculators provide parentheses ⎡(⎤⎡)⎤ so that operations can be grouped, as in

⎡2⎤⎡+⎤⎡(⎤⎡3⎤⎡×⎤⎡4⎤⎡)⎤⎡=⎤

but then you must remember to insert the parentheses.

The last type of logic is RPN. A calculator using this logic is characterized by ⎡ENTER⎤ or ⎡SAVE⎤ keys and does not have an equal key ⎡=⎤. With an RPN calculator, the operation symbol is entered after the numbers have been entered. These three types of logic can be illustrated by the problem $2 + 3 \times 4$.

Arithmetic logic	Algebraic logic	RPN logic
3	2	2
×	+	ENTER
4	3	3
=	×	ENTER
+	4	4
2	=	×
=		+

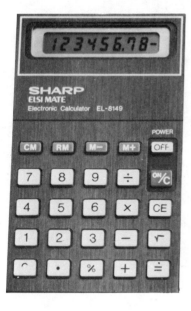

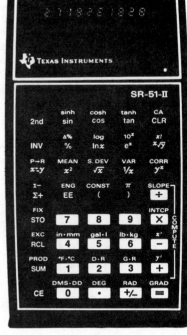

The Sharp calculator is an example of a calculator with arithmetic logic.

The Hewlett-Packard calculator uses RPN logic.

This Texas Instruments SR-51-II is an example of a calculator with algebraic logic.

FIGURE 2.2 Calculators with different types of logic. A calculator uses RPN logic if it has a SAVE or ENTER key. The test problem [2] [+] [3] [×] [4] [=] will distinguish arithmetic and algebraic logic. If the display is 14, it is algebraic; if the display is 20, it is arithmetic.

In this book, we will illustrate the examples using algebraic logic, and, if you have a calculator with RPN logic, you can use your owner's manual to change the examples to RPN. We will also indicate the keys to be pushed by drawing boxes around the numbers and operational signs as shown above, and we'll use only the keys on a four-function calculator with memory. Numerals for calculator display will be designated as shown below:

$$0 \quad 1 \quad 2 \quad 3 \quad 4 \quad 5 \quad 6 \quad 7 \quad 8 \quad 9$$

EXAMPLE 7

Show the calculator steps for $14 + 38$.

Solution

Be sure to turn your calculator on, or clear the machine if it is already on. A clear button is designated by [c], and the display will show 0 after the clear button is pushed. You will need to check these steps every time you use your calculator, but after a while it becomes automatic. We will not remind you of this on each example.

Action	Display	
[1]	1.	Here, we show each numeral in a
[4]	14.	single box, which means you key in
[+]	14.	one numeral at a time, as shown.
[38]	38.	But from now on, this will be shown
[=]	52.	as [14] 14 ∎

Notice that every time you push [+], a subtotal is shown in the display. After completing Example 7 you can either continue with the same problem or start a new problem. If the next button pressed is an operation button, the result 52 will be carried over to the new problem. If the next button pressed is a numeral, the 52 will be lost and a new problem started. For this reason, it is not necessary to press [c] to clear between problems. The button [CE] is called the *clear entry* key and is used if you make a mistake keying in a number and don't want to start over with the problem. For example, if you want 2 + 3 and push

$$[2]\ [+]\ [4]$$

you can then push

$$[CE]\ [3]\ [=]$$

to obtain the correct answer. This is especially helpful if you are in the middle of a long problem.

EXAMPLE 8

Show the calculator steps and display for 4 + 3 × 5 − 7.

Solution

Action:	[4]	[+]	[3]	[×]	[5]	[−]	[7]	[=]
Display:	4.	4.	3.	3.	5.	19.	7.	12.

If you have an algebraic-logic calculator, your machine will perform the correct order of operations. If it is an arithmetic-logic calculator, it will give the incorrect answer 28.

∎

Problem Set 2.1

Perform the indicated operations in Problems 1–11.

1. a. $5 + 6 \times 2$ **b.** $20 - 4 \times 2$

2. a. $12 \div 6 + 3$ **b.** $12 + 6 \div 3$

3. a. $15 + 6 \div 3$ **b.** $(15 + 6) \div 3$

4. a. $4 \times 3 + 4 \times 5$ **b.** $4 \times (3 + 5)$

5. a. $2 + 15 \div 3 \times 5$ **b.** $20 - 8 \div 4 \times 2 + 3$

6. $2 + 3 \times 4 - 12 \div 2$ **7.** $15 \div 5 \times 2 + 6 \div 3$

8. $2 \times 18 + 9 \div 3 - 5 \times 2$ **9.** $2 \times (18 + 9) \div 3 - 5 \times 2$

10. $4 \times (12 - 8) \div 2$ **11.** $4 \times 12 - 8 \div 2$

Write out the expressions in Problems 12–21 without parentheses by using the distributive property.

12. $3 \times (4 + 8)$ **13.** $7 \times (9 + 4)$

14. $12 \times (4 + 6)$ **15.** $7 \times (70 + 3)$

16. $8 \times (50 + 5)$ **17.** $6 \times (90 + 7)$

18. $4 \times (300 + 20 + 7)$ **19.** $6 \times (500 + 30 + 3)$

20. $5 \times (800 + 60 + 4)$ **21.** $4 \times (700 + 10 + 5)$

22. How many hours are there in 360 days?

23. How many pages are necessary to make 1,850 copies of a manuscript that is 487 pages long?

24. In 1981, the Internal Revenue Service allowed a $1,000 deduction for each dependent. If a family has 4 dependents, what is the allowed deduction?

25. If your monthly salary is $1,543 per month, what is your annual salary?

26. If your car gets 23 miles per gallon, how far can you go on 15 gallons of gas?

CALCULATOR PROBLEMS

In Problems 27–40 perform the indicated operations on your calculator.

27. $716 - 5 \times 91$ **28.** $143 + 12 \times 14$

29. $8 \times 14 + 8 \times 86$ **30.** $15 \times 27 + 15 \times 73$

31. $12 \times 63 + 12 \times 27$ **32.** $19 \times 250 + 19 \times 750$

33. $5 + 3 \times 7 + 6 - 5 + 8 \times 4$

34. $12 + 6 \times 9 - 5 \times 2 + 5 \times 14$

35. $27 \times 450 + 27 \times 550$

36. $23 \times 237 + 23 \times 763$

37. $862 + 328 \times 142 - 168$

38. $1214 - 18 \times 14 + 35 \times 8121$

39. $62 \times (48 - 12) + 13 \times (12 - 5)$

40. $12 \times (125 - 72) - 3 \times (18 - 3 \times 5)$

41. Compute:
    a. $1 \times 142,857$ b. $2 \times 142,857$
    c. $3 \times 142,857$ d. $4 \times 142,857$
    e. $5 \times 142,857$ f. $6 \times 142,857$
    g. Can you describe a pattern?

42. Compute:
    a. 123456789×9 b. 123456789×18
    c. 123456789×27 d. Can you describe a pattern?

43. Using the pattern in Problem 42, predict the following *without* direct calculation:
    a. 123456789×36 b. 123456789×45
    c. 123456789×72 d. 123456789×81

44. Enter your favorite number into a calculator (a counting number from 1 to 9). Multiply by 259; then multiply this result by 429. What is your answer? Try it for three different choices.

Mind Bogglers

45. Read pages 50–51. Use finger multiplication to do the following calculations:
    a. 3×9 b. 7×9 c. 27×9
    d. 48×9 e. 56×9

46. Read pages 50–51. Use finger multiplication to do the following operations:
    a. 7×7 b. 7×8 c. 6×9
    d. 8×8 e. 7×6

47. A method of finger calculation called *Chisanbop* has created a great deal of interest since it was demonstrated on the Johnny Carson show. It is described by one of its developers, Edwin Lieberthal, and the Director of the Psychological Research Laboratory, William Lamon, in "Chisanbop Finger Calculation," *California MathematiCs Journal*, Vol. 3, No. 2, October 1978, pp. 2–10. Write a report on Chisanbop or make a class demonstration on this method of finger calculation.

48. Read the article "Finger Multiplication" by Fred Balin in *The Arithmetic Teacher*, March 1979, pp. 34–37, and make a report on the article.

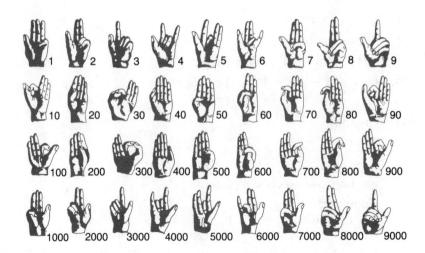

FIGURE 2.3 Aristophanes devised a complicated finger-calculating system in about 500 B.C., but it was very difficult to learn. The finger symbols illustrated here appeared in a manual published about two thousand years later, in 1520.

Finger Multiplication

A recent news article reported that a psychologist concluded that children hate math because teachers discourage them from counting on their fingers. However, finger counting can be very useful. Finger calculating is recorded as early as 500 B.C. (See Figure 2.3.)

Consider a rather simple method of using your fingers to multiply by 9. Place both hands as shown:

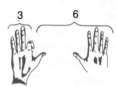

To multiply 4 × 9, simply bend the fourth finger from the left:

The answer is read as 36; the bent finger serves to distinguish between the first and second digits in the answer.

What about a two-digit number times 9? There is a procedure for this also, provided the tens digit is smaller than the ones digit. For example, to multiply 36 × 9, separate the third and fourth fingers from the left (as shown), since 36 has 3 tens:

Next, bend the sixth finger from the left, since 36 has 6 units. Now the answer can be read directly from the fingers. The answer is 324.

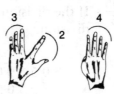

Another finger multiplication for numbers between 6 and 10 can be developed. Number the fingers on both hands as shown:

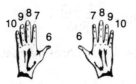

To multiply 8 × 6, place finger 8 on one hand against finger 6 on the other hand. Bend down all the fingers below the touching fingers as shown:

TENS DIGIT of product is found by adding the open fingers: 3 + 1 = 4

UNITS DIGIT of product is found by multiplying the bent fingers: 2 × 4 = 8

The answer is 48.

2.2 Fractions and Decimals

In the last section, the symbol ÷ was used for division (for example, $10 \div 5$). However, more often a division bar is used, as in $\frac{10}{5}$.

EXAMPLE 1

Perform the indicated operations.

a. $\frac{30}{6}$; $\frac{30}{6} = 5$ *Check by multiplication:* $6 \times 5 = 30$

b. $\frac{3,965}{305}$ If the division is lengthy, you may need to do long division:

$$
\begin{array}{r}
1\,3 \\
3\,0\,5\,\overline{)3\,9\,6\,5} \\
3\,0\,5 \\
\hline
9\,1\,5 \\
9\,1\,5 \\
\hline
0
\end{array}
$$
    Check: $305 \times 13 = 3{,}965$

Thus, $\frac{3,965}{305} = 13$.

c. $\frac{0}{5} = 0$ *Check by multiplication:* $5 \times 0 = 0$

d. $\frac{5}{0}$ In order to do this division, you would need to find a number so that when it is multiplied by 0 the result is 5. There is no such number. For this reason, we say **division by zero is impossible**.

e. $\frac{10}{3}$ There is no answer to this division in the set of whole numbers.   ■

The reason there is no answer in the set of whole numbers to $\frac{10}{3}$ in Example 1e is because there is no whole number that can be multiplied by 3 to give 10. If you do long division for this problem, there will be a **remainder** that is not 0:

$$\begin{array}{r} 3 \\ 3\overline{)1\,0} \\ \underline{9} \\ 1 \end{array} \longleftarrow \textit{Remainder}$$

What does this remainder mean? In the above example, the remainder 1 is still to be divided by 3, so it can be written as $\frac{1}{3}$. Such an expression is called a **fraction**. The word *fraction* comes from a Latin word meaning "to break." A fraction involves two numbers, one "upstairs," called the **numerator**, and one "downstairs," called the **denominator**. The denominator tells us into how many parts the whole has been divided, and the numerator tells us how many of those parts we have.

A **fraction** is a number that is the quotient of a whole number divided by a counting number. A fraction is usually written as a

NUMERATOR   ←— A whole number
————————— ←— Divided by
DENOMINATOR ←— A counting number (so that division by zero is excluded)

A fraction is called:

1. **A proper fraction** if the numerator is less than the denominator
2. **An improper fraction** if the numerator is greater than the denominator
3. **A whole number** if the denominator divides into the numerator with no remainder

Fraction

EXAMPLE 2

Classify each example as a proper fraction, an improper fraction, or a whole number.

a. $\frac{6}{7}$ Proper   **b.** $\frac{6}{8}$ Proper   **c.** $\frac{7}{6}$ Improper

d. $\dfrac{0}{4}$ Whole number   **e.** $\dfrac{6}{3}$ Whole number

f. $\dfrac{4}{0}$ None of these
(Don't forget, you can't divide by zero.)   ■

Improper fractions can also be written in a form called **mixed numbers** by first carrying out the division and leaving the remainder as a fraction. So, a mixed number has two parts: a counting number part and a proper fraction part.

EXAMPLE 3

Write $\dfrac{23}{5}$ as a mixed number.

Solution

$$
\begin{array}{r}
4 \\
5\overline{)2\ 3} \\
2\ 0 \\
\hline
3
\end{array}
$$

←— Counting number part

←— Remainder means 3 to be divided by 5, or $\frac{3}{5}$

$$\frac{23}{5} = 4 + \frac{3}{5} \ \ \text{or} \ \ 4\tfrac{3}{5}$$ ■

EXAMPLE 4

Write $4\tfrac{3}{5}$ as an improper fraction.

Solution

Reverse the procedure in Example 3. The remainder is 3, the quotient is 4, and the divisor is 5. Thus,

$$4 \times 5 + 3 = 23$$

Step 2. Add 3 $\ulcorner\!-23 = 4 \times 5 + 3$

$$4\tfrac{3}{5} = \frac{23}{5}$$

Step 1. 4×5 $\llcorner\!\!-\!\!-$ This is the divisor (or the denominator of the fraction).   ■

We find certain fractions to be of special interest in our study of arithmetic. These are fractions with denominators that are powers of 10 (10, 100, 1000, 10,000, . . .).

Tenths: $\dfrac{1}{10}, \dfrac{2}{10}, \dfrac{3}{10}, \dfrac{4}{10}, \dfrac{5}{10}, \dfrac{6}{10}, \dfrac{7}{10}, \dfrac{8}{10}, \dfrac{9}{10}, \dfrac{10}{10}, \dfrac{11}{10}, \cdots$

Hundredths: $\dfrac{1}{100}, \dfrac{2}{100}, \dfrac{3}{100}, \dfrac{4}{100}, \dfrac{5}{100}, \dfrac{6}{100}, \cdots$

Thousandths: $\dfrac{1}{1000}, \dfrac{2}{1000}, \dfrac{3}{1000}, \dfrac{4}{1000}, \dfrac{5}{1000}, \cdots$

Each of these fractions can be written in **decimal form**:

$$\frac{1}{10} = .1 \qquad \frac{1}{100} = .01 \qquad \frac{1}{1000} = .001$$

Sometimes zeros are inserted in front of the decimal points:

$$.1 = 0.1 \qquad .01 = 0.01 \qquad .001 = 0.001$$

You will often see this notation in books. The reason for inserting the first zero is so the decimal point won't be overlooked.

We will use division to show how this works. But first let's look again at the positional notation of our number system:

Millions	Hundred-thousands	Ten-thousands	Thousands	Hundreds	Tens	Units	Decimal point	Tenths, $\frac{1}{10}$	Hundredths, $\frac{1}{100}$	Thousandths, $\frac{1}{1000}$	Ten-thousandths, $\frac{1}{10,000}$	Hundred-thousandths, $\frac{1}{100,000}$	Millionths, $\frac{1}{1,000,000}$
1,	2	3	4,	5	6	7	.	8	9	0	1	2	3

Every whole number can be written in decimal form:

$0 = 0. = 0.0 = 0.00 = 0.000 = 0.0000 = 0.00000 = \cdots$
$1 = 1. = 1.0 = 1.00 = 1.000 = 1.0000 = \cdots$
$2 = 2. = 2.0 = 2.00 = 2.000 = \cdots$
$3 = 3. = 3.0 = 3.00 = \cdots$
$4 = 4. = 4.0 = \cdots$
$5 = 5. = \cdots$

$\vdots$

Sometimes zeros are placed after the decimal point or after the last digit to the right of the decimal point, as in

$$3.21 = 3.210 = 3.2100 = \cdots$$

These are called **trailing zeros**.

Now let's carry out division of $\frac{1}{10}$. Write 1 as 1.0 and use the ordinary process of division, except bring the decimal point straight up from the dividend to the quotient:

$$
\begin{array}{r}
.1 \quad \longleftarrow \text{Decimal point is carried straight up} \\
 \uparrow \quad \text{ from dividend to quotient} \\
10)\overline{1.0} \\
\underline{1\ 0} \\
0
\end{array}
$$

EXAMPLE 5

Write the given fractions as decimals by performing long division.

a. $\dfrac{3}{10}$

$$
\begin{array}{r}
.3 \\
10)\overline{3.0} \\
\underline{3\ 0} \\
0
\end{array}
$$

b. $\dfrac{7}{100}$

$$
\begin{array}{r}
.0\ 7 \\
100)\overline{7.0\ 0} \\
\underline{7\ 0\ 0} \\
0
\end{array}
$$

c. $\dfrac{137}{1000}$

$$
\begin{array}{r}
.1\ 3\ 7 \\
1000)\overline{1\ 3\ 7.0\ 0\ 0} \quad \longleftarrow \text{Place as many zeros here as are} \\
\underline{1\ 0\ 0\ 0} \quad\quad\quad\quad \text{necessary to complete the division.} \\
3\ 7\ 0\ 0 \\
\underline{3\ 0\ 0\ 0} \\
7\ 0\ 0\ 0 \\
\underline{7\ 0\ 0\ 0} \\
0
\end{array}
$$

Do you see a pattern? Look at the next example.

EXAMPLE 6

Write the given fractions in decimal form without doing any calculations.

a. $\dfrac{6}{10} = .6$ **b.** $\dfrac{6}{100} = .06$ **c.** $\dfrac{243}{1000} = .243$

d. $\dfrac{47}{10} = 4.7$ **e.** $\dfrac{47}{100} = .47$ **f.** $\dfrac{47}{1000} = .047$ ■

Fractions with denominators that are 10, 100, 1000, . . . are called **decimal fractions**, or simply **decimals**.

On the other hand, *any fraction*, even those with denominators that are not powers of 10, can be written in decimal form by dividing.

EXAMPLE 7

Change the given fractions to decimal form.

a. $\dfrac{3}{8}$;

$$
\begin{array}{r}
.3\,7\,5 \\
8\overline{)3.0\,0\,0} \\
\underline{2\,4} \\
6\,0 \\
\underline{5\,6} \\
4\,0 \\
\underline{4\,0} \\
0
\end{array}
$$

Remember, the decimal point is moved straight up from the dividend to the quotient. Otherwise, carry out the division in the usual fashion.

You may keep adding trailing zeros here as long as you wish.

Thus, $\dfrac{3}{8} = .375$.

b. $2\frac{3}{4}$ Since $2\frac{3}{4} = \dfrac{11}{4}$,   we divide 4 into 11.

$$
\begin{array}{r}
2.7\,5 \\
4\overline{)1\,1.0\,0} \\
\underline{8} \\
3\,0 \\
\underline{2\,8} \\
2\,0 \\
\underline{2\,0} \\
0
\end{array}
$$

An alternate method is to notice that

$2\frac{3}{4} = 2 + \frac{3}{4}$

Divide 4 into 3:

$$
\begin{array}{r}
.7\,5 \\
4\overline{)3.0\,0}
\end{array}
$$

and then add this to 2:

$2\frac{3}{4} = 2 + .75 = 2.75$

Thus, $2\frac{3}{4} = 2.75$.   ■

We noted in Example 7a that you could add as many zeros after the 3.0 as you wished. For some fractions, you could continue to add zeros forever and never complete the division. Such fractions are called **repeating decimals**.

EXAMPLE 8

Change $\frac{2}{3}$ to decimal form.

Solution

$$
\begin{array}{r}
.6\ 6\ 6 \\
3\overline{)2.0\ 0\ 0} \\
\underline{1\ 8} \\
2\ 0 \\
\underline{1\ 8} \\
2\ 0 \\
\underline{1\ 8} \\
2 \quad
\end{array}
$$

← The same remainder keeps coming up, so the process is never finished.

This is a *repeating decimal* and is indicated by three dots or by a bar over the repeating digits.

$$\frac{2}{3} = .666\ldots \qquad \text{or} \qquad \frac{2}{3} = .\overline{6} \qquad ■$$

ρ.   When you write repeating decimals, be careful to include the three dots or the bar. You **cannot** write

$$\frac{2}{3} = .6666 \qquad \text{or} \qquad \frac{2}{3} = .6667$$

because $3 \times .6666 = 1.9998$ and $3 \times .6667 = 2.0001$. All fractions have a decimal form that either terminates (as in Example 7) or repeats (as in Example 8).

You are often required to round decimals (for example, when you work with money). The following procedure should help you.

1. **Locate the rounding place digit.** This is identified by the column name.
2. **Determine the rounding place digit**
   a. It stays the same if the first digit to its right is a 0, 1, 2, 3, or 4.
   b. It increases by 1 if the digit to the right is a 5, 6, 7, 8, or 9. (If the rounding place digit is a 9 and 1 is added, there will be a carry in the usual fashion.)
3. **Change digits**
   a. All digits to the left of the rounding digit remain the same (unless there is a carry).
   b. All digits to the right of the rounding digit are changed to zeros.
4. **Dropping zeros**
   a. If the rounding place digit is to the left of the decimal point, drop all trailing zeros.
   b. If the rounding place digit is to the right of the decimal point, drop all trailing zeros to the right of the rounding place digit.

**Procedure for
Rounding a Decimal
Number**

EXAMPLE 9

Round each number as indicated.
a. Round 46.8217 to the nearest hundredth.

Step 1. Rounding place digit is in the hundredth column.

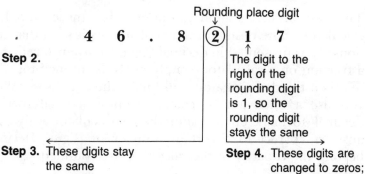

The rounded number is **46.82**.

b. Round 13.6992 to the nearest hundredth.

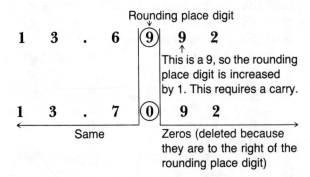

Rounding place digit

1   3   .   6   ⑨   9   2

↑ This is a 9, so the rounding place digit is increased by 1. This requires a carry.

1   3   .   7   ⓪   9   2

Same   Zeros (deleted because they are to the right of the rounding place digit)

The rounded number is **13.70**. Notice that the zero as the rounding place digit is *not* deleted.

c. Round 72,416.921 to the nearest hundred.

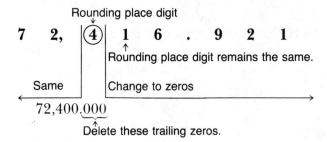

Rounding place digit

7   2,   ④   1   6   .   9   2   1

↑ Rounding place digit remains the same.

Same   Change to zeros

72,400.000

↑ Delete these trailing zeros.

The rounded number is **72,400**.   ∎

CALCULATOR COMMENT

The rest of this section is optional for those who have calculators. The design of most calculators is such that fractions are represented in decimal form. In order to represent a fraction on a calculator, simply press the numerator, then $\boxed{\div}$, then the denominator, and finally the equal sign. When you use a calculator to change from fractional form to decimal form, keep in mind that a calculator only gives answers correct to a certain number of digits. Decimals beyond the capacity of the calculator are rounded.

EXAMPLE 10

Represent the given fractions in decimal form by entering them into a calculator.

a. $\frac{4}{5}$ *Press:* 4 ÷ 5 =
  Display: 4. 4. 5. .8   Exact

b. $29\frac{7}{8}$ Remember that $29\frac{7}{8}$ means $29 + \frac{7}{8}$.

Enter the fractional part first.

Press: 7 ÷ 8 + 29 =
Display: 7. 7. 8. .875 29. 29.875

Exact

c. $\frac{1}{3}$ *Press:* 1 ÷ 3 =
  Display: 1. 1. 3. .33333333

Approximate

Now, $\frac{1}{3} \neq .33333333$, so when using a calculator, be sure to look for a pattern and write the exact answer, if possible:

$$\frac{1}{3} = .\overline{3}$$ ■

Problem Set 2.2

Write each of the improper fractions in Problems 1–6 as a mixed number.

1. a. $\frac{3}{2}$ **b.** $\frac{4}{3}$ **c.** $\frac{5}{4}$

2. a. $\frac{16}{3}$ **b.** $\frac{25}{4}$ **c.** $\frac{19}{2}$

3. a. $\frac{141}{10}$ **b.** $\frac{163}{10}$ **c.** $\frac{177}{10}$

4. a. $\frac{1,681}{10}$ **b.** $\frac{1,493}{10}$ **c.** $\frac{27}{16}$

5. a. $\frac{33}{16}$ **b.** $\frac{27}{7}$ **c.** $\frac{89}{21}$

6. a. $\frac{41}{12}$ **b.** $\frac{61}{5}$ **c.** $\frac{118}{15}$

Write each of the mixed numbers in Problems 7–14 as an improper fraction.

7. a. $2\frac{1}{2}$ **b.** $1\frac{2}{3}$ **c.** $5\frac{1}{4}$
8. a. $1\frac{3}{4}$ **b.** $3\frac{5}{8}$ **c.** $4\frac{3}{8}$

9.   **a.** $3\frac{2}{3}$ **b.** $5\frac{1}{5}$ **c.** $4\frac{2}{5}$

10.   **a.** $6\frac{1}{2}$ **b.** $3\frac{2}{5}$ **c.** $1\frac{3}{10}$

11.   **a.** $2\frac{2}{5}$ **b.** $3\frac{7}{10}$ **c.** $2\frac{9}{10}$

12.   **a.** $17\frac{2}{3}$ **b.** $12\frac{4}{5}$ **c.** $11\frac{3}{8}$

13.   **a.** $1\frac{14}{15}$ **b.** $2\frac{13}{17}$ **c.** $3\frac{11}{12}$

14.   **a.** $19\frac{3}{5}$ **b.** $17\frac{1}{8}$ **c.** $18\frac{7}{8}$

Change the fractions and mixed numbers in Problems 15–20 to decimal form.

15.   **a.** $\frac{1}{8}$ **b.** $\frac{3}{8}$ **c.** $\frac{7}{8}$

16.   **a.** $\frac{5}{6}$ **b.** $\frac{7}{6}$ **c.** $\frac{3}{5}$

17.   **a.** $2\frac{1}{2}$ **b.** $5\frac{1}{3}$ **c.** $6\frac{2}{3}$

18.   **a.** $\frac{3}{7}$ **b.** $3\frac{1}{6}$ **c.** $6\frac{1}{12}$

19.   **a.** $\frac{5}{9}$ **b.** $\frac{7}{9}$ **c.** $\frac{2}{7}$

20.   **a.** $7\frac{5}{6}$ **b.** $4\frac{1}{15}$ **c.** $3\frac{1}{30}$

Round the numbers in Problems 21–32 to the given degree of accuracy.

21.   2.312;   nearest tenth

22.   6,287.4513;   nearest hundredth

23.   5.291;   one decimal place

24.   6,287.4513;   nearest hundred

25.   12.8197;   two decimal places

26.   4.81792;   nearest thousandth

27. $12.993;   nearest cent

28.   4.81992;   nearest whole number

29. $14.998;   nearest cent

30. $6.9741;   nearest cent

31.   694.3814;   nearest ten

32. $4.018;   nearest cent

33.   A baseball player's batting average is found by dividing the number of hits by the number of times at bat. This number is then rounded to the nearest thousandth. Find each player's batting average.
    a.   Jack Foley was at bat 6 times with 2 hits.
    b.   Niels Sovndal was at bat 20 times with 7 hits.
    c.   Greg Kintzi had 5 hits in 12 times at bat.

34. If a person's annual salary is $15,000, what is the monthly salary?

35. If a person's annual tax is $512, how much is that tax per month?

36. If the sale of 150 shares of PERTEC grossed $1,818.75, how much was each share worth?

37. If an estate of $22,000 is to be divided equally among three children, what is each child's share?

38. If you must pay back $850 in 12 monthly payments, what is the amount of each payment?

39. If you must pay back $1,000 in 12 monthly payments, what is the amount of each payment?

40. A businesswoman bought a copy machine for her office. If it cost $674 and the useful life is six years, what is the cost per year?

41. A businessman bought a typewriter for his office. If it cost $890 and the useful life is seven years, what is the cost per year?

CALCULATOR PROBLEMS

Show the sequence of buttons to be pressed in order to enter each of the fractions in Problems 42 and 43 into a calculator.

42. a. $\dfrac{8}{13}$ **b.** $\dfrac{17}{21}$ **c.** $2\frac{1}{2}$

43. a. $14\frac{3}{4}$ **b.** $3\frac{2}{7}$ **c.** $4\frac{3}{7}$

Use a calculator to change each of Problems 44 and 45 to decimal form. Give the exact decimal value if possible. If you can't give the exact value, indicate that the answer you are giving is approximate.

44. a. $\dfrac{1}{11}$ **b.** $\dfrac{1}{12}$ **c.** $\dfrac{3}{17}$

45. a. $\dfrac{7}{15}$ **b.** $\dfrac{7}{22}$ **c.** $\dfrac{5}{19}$

Mind Bogglers

46. Insert appropriate operation signs ($+$, $-$, $\times$, or $\div$) between each digit so that the following becomes a true statement:

$$1\quad 2\quad 3\quad 4\quad 5\quad 6\quad 7\quad 8\quad 9 = 100$$

47. *B.C.* has a mental block against fours, as we can see in the cartoon. See if you can handle fours by writing the numbers from 1 to 10 using four 4s for each. Here are the first three completed for you:

$$\frac{4}{4} + 4 - 4 = 1 \qquad \frac{4}{4} + \frac{4}{4} = 2 \qquad \frac{4 + 4 + 4}{4} = 3$$

2.3 Common Fractions

One of the troublesome aspects of working with common fractions is that there are different representations for the same fraction. We say that a fraction is **reduced** if there is no counting number other than 1 that divides into both the numerator and denominator. The process of reducing a fraction relies on the **fundamental property of fractions** and on a process called **factoring**. When you multiply numbers, the numbers being multiplied are called **factors** and the process of taking a given number and writing it as the product of two or more other numbers is called **factoring**, with the result called a **factorization** of the given numbers.

EXAMPLE 1

Find the factors of the given numbers.

a. $1 = 1 \times 1$ The factor is: 1
b. $2 = 2 \times 1$ Factors: 1, 2
c. $3 = 3 \times 1$ Factors: 1, 3
d. $4 = 4 \times 1$ or 2×2 Factors: 1, 2, 4
e. $5 = 5 \times 1$ Factors: 1, 5
f. $6 = 6 \times 1$ or 2×3 Factors: 1, 2, 3, 6
g. $7 = 7 \times 1$ Factors: 1, 7
h. $8 = 8 \times 1$ or 4×2 or $2 \times 2 \times 2$ Factors: 1, 2, 4, 8 ∎

Let's categorize the numbers listed in Example 1.

Fewer than two factors	PRIMES: Exactly two factors	COMPOSITES: More than two factors
1	2	4
	3	6
	5	8
	7	

A **prime** number is a natural number with exactly two distinct factors, and a **composite** is a number with more than two factors. Will any number besides 1 have fewer than two factors? The primes smaller than 100 are given in the box.

2, 3, 5, 7, 11, 13, 17, 19, 23, 29, 31, 37, 41, 43, 47, 53, 59, 61, 67, 71, 73, 79, 83, 89, 97

List of Primes Smaller than 100

EXAMPLE 2

Find the prime factorization of the given numbers.

a. 36   **b.** 48

Solution

a. Step 1. From your knowledge of the basic multiplication facts, write down *any* two numbers whose product is the given numbers. Circle the prime factors.
Step 2. Repeat the process for uncircled numbers.
Step 3. When all the factors are circled, their product is the **prime factorization**.

$$36$$
$$② \times 18$$
$$② \times 9$$
$$③ \times ③$$

$$36 = 2 \times 2 \times 3 \times 3$$

b.

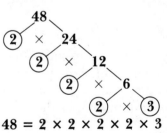

$$48 = 2 \times 2 \times 2 \times 2 \times 3$$

∎

**Procedure for
Reducing a Fraction**

> 1. Find all common factors (other than 1).
> 2. Divide out the common factors.

EXAMPLE 3

Reduce $\dfrac{36}{48}$.

Solution

Step 1. Completely factor both numerator and denominator. We did this in Example 2.

Step 2. Write out the numerator and denominator in factored form:

$$\frac{36}{48} = \frac{2 \times 2 \times 3 \times 3}{2 \times 2 \times 2 \times 2 \times 3}$$

Step 3. Use the fundamental property of fractions to eliminate the common factors. This process is sometimes called **canceling**.

$$\frac{36}{48} = \frac{\cancel{2} \times \cancel{2} \times 3 \times \cancel{3}}{\cancel{2} \times \cancel{2} \times 2 \times 2 \times \cancel{3}}$$

Step 4. Multiply the remaining factors. Treat the canceled factors as 1s.

$$
\begin{array}{c}
1 \times 1 \times 3 \times 1 = 3 \\[4pt]
\frac{36}{48} = \frac{\cancel{2} \times \cancel{2} \times 3 \times \cancel{3}}{\cancel{2} \times \cancel{2} \times 2 \times 2 \times \cancel{3}} = \frac{3}{4} \\[4pt]
1 \times 1 \times 2 \times 2 \times 1 = 4
\end{array}
$$

Notice how slashes are used as 1s.

∎

This process is rather lengthy and can sometimes be shortened by noticing common factors that are larger than prime factors. For example, you might have noticed that 12 is a common factor in Example 3, so that

$$\frac{36}{48} = \frac{3 \times \cancel{12}}{4 \times \cancel{12}}$$

$$= \frac{3}{4} \quad\longleftarrow\text{ Remember, this is 1.}$$

EXAMPLE 4

Reduce the fractions.

a. $\dfrac{4}{8} = \dfrac{1 \times \cancel{4}}{2 \times \cancel{4}}$ **b.** $\dfrac{75}{100} = \dfrac{3 \times \cancel{25}}{4 \times \cancel{25}}$ **c.** $\dfrac{35}{55} = \dfrac{7 \times \cancel{5}}{11 \times \cancel{5}}$

$\quad= \dfrac{1}{2}$ $\qquad\quad= \dfrac{3}{4}$ $\qquad\qquad= \dfrac{7}{11}$

If the fractions are complicated, you may reduce them in several steps.

d. $\dfrac{160}{180} = \dfrac{\cancel{10} \times 16}{\cancel{10} \times 18}$ **e.** $\dfrac{1,200}{9,000} = \dfrac{12 \times \cancel{100}}{90 \times \cancel{100}}$

$\quad= \dfrac{\cancel{2} \times 8}{\cancel{2} \times 9}$ $\qquad\qquad= \dfrac{\cancel{6} \times 2}{\cancel{6} \times 15}$

$\quad= \dfrac{8}{9}$ $\qquad\qquad\quad= \dfrac{2}{15}$ ∎

To multiply fractions, multiply numerators and multiply denominators.

Procedure for Multiplying Fractions

EXAMPLE 5

Multiply.

a. $\dfrac{1}{3} \times \dfrac{2}{5} = \boxed{\dfrac{1 \times 2}{3 \times 5}}$ ←——This step can often be done in your head.

$= \dfrac{2}{15}$

b. $\dfrac{2}{3} \times \dfrac{4}{7} = \dfrac{8}{21}$

c. $5 \times \dfrac{2}{3}$

When multiplying a whole number and a fraction, write the whole number as a fraction, and then multiply:

$$\frac{5}{1} \times \frac{2}{3} = \frac{10}{3}$$

d. $3\frac{1}{2} \times 2\frac{3}{5}$

When multiplying mixed numbers, write the mixed numbers as improper fractions, and then multiply:

$$\frac{7}{2} \times \frac{13}{5} = \frac{91}{10}$$ ∎

After the multiplication has been done, the product is often a fraction that can and should be reduced. Remember, a fraction is in reduced form if the only counting number that can divide evenly into both the numerator and the denominator is 1. In this book, all answers should be reduced unless you are otherwise directed.

EXAMPLE 6

Multiply.

a. $\dfrac{3}{4} \times \dfrac{2}{3} = \dfrac{6}{12}$ but $\dfrac{6}{12} = \dfrac{\cancel{3} \times \cancel{2}}{\cancel{4} \times \cancel{3}} = \dfrac{1}{2}$

Notice that the actual multiplication was a wasted step, because the answer was simply factored again in order to reduce it! Therefore, the proper procedure is to cancel common factors before you do the multiplication. Also, it is *not* necessary to write out prime factors but simply to find all the common factors. To reduce $\frac{6}{12}$, notice that 6 is a common factor and write

These little numbers mean that 6 divides into 6 one time, and 6 divides into 12 twice (6 is the common factor).

$$\frac{\cancel{6}^{1}}{\cancel{12}_{2}} = \frac{1}{2}$$

b. $\dfrac{2}{5} \times \dfrac{3}{4} = \dfrac{\cancel{2}^{1} \times 3}{5 \times \underbrace{\cancel{4}}_{2}} = \dfrac{3}{10}$

Notice that this is very similar to the original multiplication as stated to the left of the equal sign. You can save a step by canceling with the original product, as shown in the next example.

c. $\dfrac{\cancel{3}^{1}}{5} \times \dfrac{1}{\underset{1}{\cancel{3}}} = \dfrac{1}{5}$

d. $3\frac{2}{3} \times 2\frac{2}{5} = \dfrac{11}{\underset{1}{\cancel{3}}} \times \dfrac{\cancel{12}^{4}}{5} = \dfrac{44}{5}$

e. $\dfrac{4}{5} \times \dfrac{3}{8} \times \dfrac{2}{5}$

With a more lengthy problem, you may do your canceling in several steps. We recopy the problem at each step for the sake of clarity, but in your work the result would look like the last step only.

Step 1. $\dfrac{\cancel{4}^{1}}{5} \times \dfrac{3}{\underset{2}{\cancel{8}}} \times \dfrac{2}{5}$

Step 2. $\dfrac{\cancel{4}^{1}}{5} \times \dfrac{3}{\underset{1}{\underset{2}{\cancel{8}}}} \times \dfrac{\cancel{2}^{1}}{5}$

Step 3. $\dfrac{\cancel{4}^{1}}{5} \times \dfrac{3}{\underset{1}{\underset{2}{\cancel{8}}}} \times \dfrac{\cancel{2}^{1}}{5} = \dfrac{3}{25}$ ■

If the product of two numbers is 1, then those numbers are called **reciprocals**. To find the reciprocal of a given number, write the number in fractional form and then **invert**, as shown by Example 7.

EXAMPLE 7

Find the reciprocal of each number, and then prove it is the correct reciprocal by multiplication.

a. $\dfrac{5}{11}$ Invert for reciprocal: $\dfrac{\mathbf{11}}{\mathbf{5}}$ *Check:* $\frac{5}{11} \times \frac{11}{5} = 1$

b. 3   Write as a fraction: $\dfrac{3}{1}$

Invert for reciprocal: $\dfrac{\mathbf{1}}{\mathbf{3}}$ *Check:* $3 \times \frac{1}{3} = 1$

c. $2\frac{3}{4}$ Write as a fraction: $\dfrac{11}{4}$

Invert for reciprocal: $\dfrac{\mathbf{4}}{\mathbf{11}}$

Check: $2\frac{3}{4} \times \frac{4}{11} = \frac{11}{4} \times \frac{4}{11} = 1$

d. 0.2   Write as a fraction: $\dfrac{2}{10} = \dfrac{1}{5}$

Invert for reciprocal: $\dfrac{5}{1} = \mathbf{5}$ *Check:* $0.2 \times 5 = 1$

e. Zero is the only whole number that does not have a reciprocal, because any number multiplied by 0 is 0 (and not 1).   ■

The process of division is very similar to the process of multiplication. To divide fractions, you need to have an understanding of three ideas:

1. How to multiply fractions
2. Which term is called the divisor
3. How to find the reciprocal of a number

Procedure for Dividing Fractions

> To divide fractions, multiply by the reciprocal of the divisor. This is sometimes phrased as "invert and multiply."

EXAMPLE 8

Divide.

a. $\dfrac{7}{8} \div \dfrac{2}{3} = \dfrac{7}{8} \times \dfrac{3}{2}$

$\qquad = \dfrac{21}{16}$ or $1\frac{5}{16}$

c. $\dfrac{5}{8} \div 3 = \dfrac{5}{8} \times \dfrac{1}{3}$

$\qquad = \dfrac{5}{24}$

b. $\dfrac{3}{5} \div \dfrac{9}{20} = \dfrac{\overset{1}{\cancel{3}}}{\cancel{5}_{1}} \times \dfrac{\overset{4}{\cancel{20}}}{\cancel{9}_{3}}$

$\qquad = \dfrac{4}{3}$ or $1\frac{1}{3}$

d. $0 \div \dfrac{6}{7} = 0 \times \dfrac{7}{6}$

$\qquad = 0$ ∎

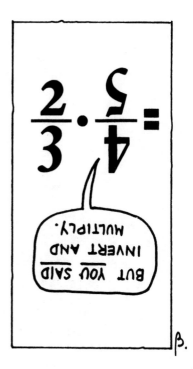

We have discussed how to change a fraction to a decimal by division. The reverse procedure, changing a terminating decimal to a fraction, can be viewed as a multiplication problem involving fractions. To do this you must:

1. Name the decimal position of the last digit:
  One place is tenth, or $\frac{1}{10}$
  Two places is hundredth, or $\frac{1}{100}$
  Three places is thousandth, or $\frac{1}{1000}$
  $\vdots$

2. Multiply the digits by the decimal name of the last digit.

EXAMPLE 9

a. .4; 4 tenths; decimal position is tenth

$$4 \times \frac{1}{10} = \frac{\overset{2}{\cancel{4}}}{1} \times \frac{1}{\underset{5}{\cancel{10}}} = \frac{2}{5}$$

b. .75; 75 hundredths; decimal position is hundredth

$$\overset{3}{\cancel{75}} \times \frac{1}{\underset{4}{\cancel{100}}} = \frac{3}{4}$$

c. .0014; 14 ten-thousandths; decimal position is ten-thousandth

$$\overset{7}{\cancel{14}} \times \frac{1}{\underset{5,000}{\cancel{10,000}}} = \frac{7}{5,000}$$

d. $.12\frac{1}{2};$ $12\frac{1}{2}$ hundredths

$$12\frac{1}{2} \times \frac{1}{100} = \frac{\overset{1}{\cancel{25}}}{2} \times \frac{1}{\underset{4}{\cancel{100}}} = \frac{1}{8}$$

e. $.3\frac{1}{3};$ $3\frac{1}{3}$ tenths

$$3\frac{1}{3} \times \frac{1}{10} = \frac{\overset{1}{\cancel{10}}}{3} \times \frac{1}{\underset{1}{\cancel{10}}} = \frac{1}{3}$$ ∎

Numbers in which decimal and fractional forms are mixed, as in Examples 9d and 9e, are called **complex decimals**.

Problem Set 2.3

Completely reduce the fractions in Problems 1–6.

1. a. $\dfrac{3}{9}$ **b.** $\dfrac{6}{9}$ **c.** $\dfrac{2}{10}$ **d.** $\dfrac{4}{10}$ **e.** $\dfrac{6}{10}$

2. a. $\dfrac{3}{12}$ **b.** $\dfrac{4}{12}$ **c.** $\dfrac{6}{12}$ **d.** $\dfrac{8}{12}$ **e.** $\dfrac{10}{12}$

3. a. $\dfrac{14}{7}$ **b.** $\dfrac{38}{19}$ **c.** $\dfrac{92}{2}$ **d.** $\dfrac{160}{8}$ **e.** $\dfrac{56}{8}$

4. a. $\dfrac{72}{15}$ **b.** $\dfrac{42}{14}$ **c.** $\dfrac{16}{24}$ **d.** $\dfrac{128}{256}$ **e.** $\dfrac{30}{120}$

5. a. $\dfrac{18}{30}$ **b.** $\dfrac{70}{105}$ **c.** $\dfrac{50}{400}$ **d.** $\dfrac{35}{21}$ **e.** $\dfrac{30}{18}$

6. a. $\dfrac{140}{420}$ **b.** $\dfrac{150}{1,000}$ **c.** $\dfrac{2,500}{10,000}$ **d.** $\dfrac{105}{120}$ **e.** $\dfrac{350}{720}$

Perform the operations in Problems 7–21. Give your answers in reduced form.

7. a. $\dfrac{1}{4} \times \dfrac{1}{6}$ **b.** $\dfrac{2}{3} \times \dfrac{4}{5}$ **c.** $\dfrac{3}{4} \times \dfrac{1}{6}$

   d. $\dfrac{4}{5} \times \dfrac{3}{8}$ **e.** $\dfrac{4}{7} \times \dfrac{14}{9}$

8. a. $\dfrac{5}{9} \times \dfrac{18}{25}$ **b.** $\dfrac{2}{3} \times \dfrac{3}{8}$ **c.** $\dfrac{4}{5} \times \dfrac{13}{16}$

   d. $\dfrac{5}{12} \times \dfrac{4}{15}$ **e.** $\dfrac{2}{5} \times \dfrac{15}{8}$

9. **a.** $\dfrac{7}{20} \times \dfrac{100}{14}$ **b.** $\dfrac{18}{25} \times \dfrac{5}{36}$ **c.** $\dfrac{9}{16} \times \dfrac{4}{27}$

    d. $\dfrac{5}{3} \times \dfrac{9}{15}$ **e.** $\dfrac{3}{5} \times \dfrac{20}{27}$

10. **a.** $\dfrac{1}{2} \div \dfrac{1}{3}$ **b.** $\dfrac{1}{3} \div \dfrac{1}{2}$ **c.** $\dfrac{2}{3} \div \dfrac{1}{2}$

    d. $\dfrac{2}{3} \div \dfrac{5}{6}$ **e.** $\dfrac{4}{5} \div \dfrac{3}{10}$

11. **a.** $\dfrac{3}{4} \div \dfrac{2}{3}$ **b.** $\dfrac{3}{8} \div \dfrac{15}{16}$ **c.** $\dfrac{5}{6} \div \dfrac{1}{3}$

    d. $\dfrac{4}{7} \div \dfrac{4}{5}$ **e.** $\dfrac{3}{5} \div \dfrac{3}{7}$

12. **a.** $\dfrac{4}{5} \div \dfrac{4}{5}$ **b.** $\dfrac{7}{9} \div \dfrac{7}{9}$ **c.** $\dfrac{8}{3} \div \dfrac{8}{3}$

    d. $\dfrac{6}{7} \div \dfrac{2}{3}$ **e.** $\dfrac{4}{9} \div \dfrac{3}{4}$

13. **a.** $\dfrac{4}{5} \times 5$ **b.** $\dfrac{2}{3} \times 3$ **c.** $\dfrac{6}{7} \times 7$

    d. $\dfrac{2}{5} \div 3$ **e.** $\dfrac{3}{8} \div 3$

14. **a.** $\dfrac{5}{6} \times 18$ **b.** $\dfrac{3}{8} \times 24$ **c.** $\dfrac{5}{8} \times 8$

    d. $\dfrac{3}{5} \div 5$ **e.** $3 \div \dfrac{1}{6}$

15. **a.** $2\frac{1}{2} \div 3$ **b.** $6\frac{1}{2} \div 3$ **c.** $3\frac{4}{5} \div 0$

    d. $7 \times \dfrac{9}{14}$ **e.** $52 \times \dfrac{5}{13}$

16. **a.** $5 \div 1\frac{1}{2}$ **b.** $4 \div 2\frac{2}{3}$ **c.** $6 \div 1\frac{5}{6}$

    d. $6\frac{1}{2} \times \dfrac{5}{6}$ **e.** $3\frac{4}{5} \times \dfrac{1}{2}$

17. **a.** $2\frac{2}{5} \times 1\frac{4}{5}$ **b.** $5\frac{1}{2} \times 3\frac{2}{3}$ **c.** $4\frac{1}{6} \times 2\frac{3}{8}$

    d. $1\frac{1}{6} \div 2\frac{1}{3}$ **e.** $2\frac{2}{3} \div 1\frac{1}{3}$

18. **a.** $2\frac{2}{5} \div 1\frac{2}{3}$ **b.** $5\frac{1}{2} \div 1\frac{4}{5}$ **c.** $4\frac{1}{2} \div 2\frac{3}{8}$

    d. $5 \times \dfrac{3}{5}$ **e.** $2\frac{1}{2} \times \dfrac{3}{4}$

19. **a.** $\dfrac{1}{2} \times \dfrac{8}{9} \times \dfrac{3}{16}$ **b.** $\dfrac{2}{3} \times \dfrac{5}{8} \times \dfrac{16}{100}$ **c.** $\dfrac{2}{3} \times \dfrac{4}{5} \times \dfrac{15}{16}$

20.   **a.** $\left(\dfrac{1}{2} \div \dfrac{1}{3}\right) \div \dfrac{1}{4}$ **b.** $\dfrac{1}{2} \div \left(\dfrac{1}{3} \div \dfrac{1}{4}\right)$ **c.** $\dfrac{3}{8} \times \dfrac{4}{5} \times \dfrac{15}{9}$

21.   **a.** $2\frac{1}{2} \times 3\frac{1}{6} \times 1\frac{1}{5}$ **b.** $\dfrac{2}{3} \div \left(\dfrac{1}{2} \div \dfrac{3}{4}\right)$ **c.** $\left(\dfrac{2}{3} \div \dfrac{1}{2}\right) \div \dfrac{3}{4}$

Change the decimals in Problems 22–25 to fractional form.

22.   **a.** .7   **b.** .9   **c.** .8   **d.** .18   **e.** .48
23.   **a.** .78   **b.** .85   **c.** .246   **d.** .505   **e.** .015
24.   **a.** $.66\frac{2}{3}$ **b.** $.87\frac{1}{2}$ **c.** $.16\frac{2}{3}$ **d.** $.1\frac{1}{9}$ **e.** $.5\frac{5}{9}$
25.   **a.** $.37\frac{1}{2}$ **b.** $.8\frac{8}{9}$ **c.** $.000\frac{1}{3}$ **d.** $.08\frac{1}{3}$ **e.** $.03\frac{1}{4}$

26.   If Karl owns $\frac{3}{16}$ of a mutual water system, and a new well is installed at a cost of \$12,512, how much does Karl owe for his share?

27.   A recipe calls for $\frac{5}{8}$ cup of sugar, one egg, and $\frac{7}{8}$ cup of flour. How much of each ingredient is needed to double this mixture?

28.   Stock prices are quoted in eighths of dollars. If Sears stock is selling for $20\frac{1}{8}$ per share, how much would 100 shares cost?

29.   What would 100 shares of Xerox cost when it is selling for $56\frac{5}{8}$ per share?

30.   If you received \$277.50 from selling 20 shares of Brunswick stock, what is the price per share (stated as a mixed fraction)?

31.   If $4\frac{2}{5}$ acres sell for \$44,000, what is the price per acre?

32.   If Shannon drives $5\frac{3}{4}$ miles to work and he drives the same distance home every day five days each week, how many miles does he drive to and from work each week?

33.   If two-thirds of a person's body weight is water, what is the weight of water in a person who weighs 180 pounds?

2.4 Adding and Subtracting Fractions

If the fractions you are adding or subtracting are similar (all halves, thirds, fourths, fifths, sixths, and so on), then the procedure is straightforward: Add or subtract the numerators. In this case, we say that the fractions have **common denominators**. If the fractions are not similar, then you *cannot* add or subtract them directly. You must change the form of the fractions so that they are similar. This process is called **finding common denominators**.

To add or subtract fractions with common denominators, add or subtract the numerators. The denominator of the sum or difference is the same as the common denominator.

Procedure for Adding or Subtracting Fractions with Common Denominators

EXAMPLE 1

Perform the indicated operations.

a. $\dfrac{2}{7} + \dfrac{3}{7} = \dfrac{5}{7}$ **b.** $\dfrac{11}{20} - \dfrac{9}{20} = \dfrac{2}{20}$

$= \dfrac{1}{10}$ Don't forget to reduce your answers.

c. $\dfrac{3}{2} + \dfrac{7}{2} = \dfrac{10}{2}$

$= 5$

This example could also be added in the form of mixed numbers:

┌── Add fraction part first.

$$\begin{array}{r} 1\frac{1}{2} \\ + 3\frac{1}{2} \\ \hline 4\frac{2}{2} \end{array}$$

└── Next, add whole number part.

Since $\frac{2}{2} = 1$,

$$4\tfrac{2}{2} = 4 + \tfrac{2}{2} = 4 + 1 = 5$$

d.
$$\begin{array}{r} 3\frac{2}{3} \\ 1\frac{2}{3} \\ \frac{1}{3} \\ + 4\frac{2}{3} \\ \hline 8\frac{7}{3} = 10\frac{1}{3} \end{array}$$

Notice that the carry may be more than 1. In this example, $\frac{7}{3} = 2\frac{1}{3}$.   ∎

Sometimes when doing subtraction, it is necessary to borrow from the units column in order to have enough fractional parts to carry out the subtraction. Remember,

$$1 = \frac{2}{2} = \frac{3}{3} = \frac{4}{4} = \frac{5}{5} = \frac{6}{6} = \frac{7}{7} = \frac{8}{8} = \frac{9}{9} = \frac{10}{10} = \cdots$$

EXAMPLE 2

Borrow 1 from the units column and combine with the fractional part.

a. $3\frac{1}{3}$ You would do the following steps in your head and write down only the answer:

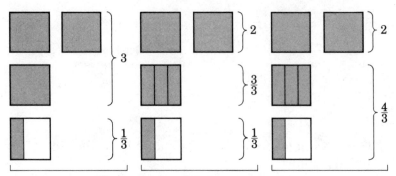

These all represent the same number

Thus, $3\frac{1}{3} = \mathbf{2\frac{4}{3}}$.

b. $5\frac{3}{5} = 4\frac{8}{5}$ **c.** $3\frac{1}{2} = 2\frac{3}{2}$ **d.** $1\frac{1}{3} = \frac{4}{3}$ **e.** $14\frac{4}{5} = 13\frac{9}{5}$ ∎

Using the idea shown in Example 2, we can carry out some subtractions with mixed numbers.

EXAMPLE 3

Find the differences.

a.
$$\begin{array}{r} 3\frac{1}{3} = \quad 2\frac{4}{3} \\ - 1\frac{2}{3} = - 1\frac{2}{3} \\ \hline 1\frac{2}{3} \end{array}$$

b.
$$\begin{array}{r} 4\frac{3}{5} = \quad 3\frac{8}{5} \\ - 2\frac{4}{5} = - 2\frac{4}{5} \\ \hline 1\frac{4}{5} \end{array}$$

c.
$$\begin{array}{r} 1\frac{1}{3} = \quad \frac{4}{3} \\ - \frac{2}{3} = - \frac{2}{3} \\ \hline \frac{2}{3} \end{array}$$ ∎

If the fractions to be added or subtracted do not have common denominators, we need to use the fundamental property of fractions to change the form of one or more of the fractions so that they do have the same denominators. Use the following guidelines to find the best common denominator.

First: The common denominator should be a number that all the given denominators divide into evenly. This means that the product of the given denominators will always be a common denominator. Many students learn this and *always* find the common denominator by multiplication. Even though this works for all numbers, it is inefficient except for small numbers. Therefore, we have a second condition to find the best common denominator.

Second: The common denominator should be as small as possible. This number is called the **lowest common denominator**, denoted by **LCD**.

---

1. Factor all given denominators into prime factors.
2. Circle common factors.
3. List all the prime factors; circled factors are counted as one factor.
4. Every factor listed is a factor of the LCD; circled factors are counted as one factor of the LCD.

Procedure for Finding the Lowest Common Denominator (LCD)

EXAMPLE 4

Find the LCD for the given denominators.

a. 6; 8

$$6 = 2 \times 3$$
$$8 = 2 \times 2 \times 2$$
$$\text{LCD} = 2 \times 3 \times 2 \times 2 = \mathbf{24}$$

b. 8; 12

$$8 = 2 \times 2 \times 2$$
$$12 = 2 \times 2 \times 3$$
$$\text{LCD} = 2 \times 2 \times 2 \times 3 = \mathbf{24}$$

c. 24;   30

$$24 = 2 \times 2 \times 2 \times 3$$
$$30 = 2 \times 3 \times 5$$
$$LCD = 2 \times 2 \times 2 \times 3 \times 5 = \mathbf{120}$$

d. 8;   24;   60   This same procedure works no matter how many denominators are given.

$$8 = 2 \times 2 \times 2$$
$$24 = 2 \times 2 \times 2 \times 3$$
$$60 = 2 \times 2 \times 3 \times 5$$
$$LCD = 2 \times 2 \times 2 \times 3 \times 5 = \mathbf{120}$$

e. 300;   144   If the numbers are larger, you may need factor trees:

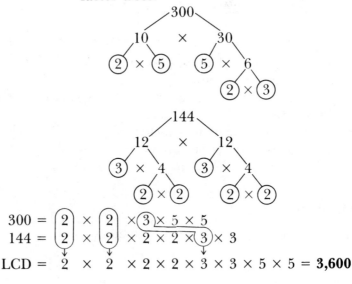

$$300 = 2 \times 2 \times 3 \times 5 \times 5$$
$$144 = 2 \times 2 \times 2 \times 2 \times 3 \times 3$$
$$LCD = 2 \times 2 \times 2 \times 2 \times 3 \times 3 \times 5 \times 5 = \mathbf{3,600}$$

■

Procedure for Adding or Subtracting Fractions that Do Not Have Common Denominators	**1.** Find the LCD. **2.** Change the forms of the fractions to obtain forms with common denominators. **3.** Add or subtract the numerators of the fractions with common denominators.

EXAMPLE 5

Perform the given operations.

a. $\dfrac{1}{5} + \dfrac{3}{4}$ Write in column form:

$$\dfrac{1}{5} = \dfrac{}{20}$$

$$+\dfrac{3}{4} = \dfrac{}{20}$$

Find the LCD: 4 and 5 have no common factors, so
LCD = 4 × 5 = 20
Next, change form:

$$\dfrac{1}{5} = \dfrac{4}{20}$$

$$+\dfrac{3}{4} = \dfrac{15}{20}$$

Finally, add:

$$\dfrac{\mathbf{19}}{\mathbf{20}}$$

↑
Be sure to reduce your answers, if possible.

b. $\dfrac{4}{9} - \dfrac{1}{4}$

Combining the steps outlined in Example 5a, you'll obtain:

$$\dfrac{4}{9} = \dfrac{16}{36}$$

$$-\dfrac{1}{4} = \dfrac{9}{36}$$

$$\dfrac{\mathbf{7}}{\mathbf{36}}$$

c.
$$\begin{aligned} 2\tfrac{1}{6} &= 2\tfrac{4}{24} \\ +\,5\tfrac{3}{8} &= 5\tfrac{9}{24} \\ \hline &\ \ 7\tfrac{13}{24} \end{aligned}$$

From Example 4a, the LCD of 6 and 8 is 24; also,

$$\dfrac{1}{6} = \dfrac{4}{24} \quad \text{and} \quad \dfrac{3}{8} = \dfrac{9}{24}$$

d.
$$12\tfrac{7}{24} = 12\tfrac{35}{120}$$

From Example 4c, the LCD is 120;

$$\dfrac{7}{24} = \dfrac{35}{120}$$

120 ÷ 24 = 5   and   5 × 7 = 35

$$\begin{aligned} -\ \ 5\tfrac{4}{30} &= \ \ 5\tfrac{16}{120} \\ \hline &\ \ 7\tfrac{19}{120} \end{aligned}$$

$$\dfrac{4}{30} = \dfrac{16}{120}$$

120 ÷ 30 = 4   and   4 × 4 = 16

e. $\dfrac{3}{8} + \dfrac{11}{12} + \dfrac{3}{20}$

LCD: $8 = \boxed{2} \times \boxed{2} \times 2$
$12 = \boxed{2} \times \boxed{2} \times 3$
$20 = \boxed{2} \times \boxed{2} \times 5$

LCD $= 2 \times 2 \times 2 \times 3 \times 5 = 120$

$\dfrac{3}{8} = \dfrac{45}{120}$ $120 \div 8 = 15$ and $15 \times 3 = 45$

$\dfrac{11}{12} = \dfrac{110}{120}$ $120 \div 12 = 10$ and $10 \times 11 = 110$

$+\dfrac{3}{20} = \dfrac{18}{120}$ $120 \div 20 = 6$ and $6 \times 3 = 18$

$\dfrac{173}{120} = 1\dfrac{53}{120}$

$\left\{\begin{array}{l}\text{Borrow so that you can}\\\text{complete the subtraction.}\end{array}\right.$

f. $16\frac{1}{3} = 16\frac{2}{6} = 15\frac{8}{6}$
$-4\frac{1}{2} = \ \ 4\frac{3}{6} = \ \ 4\frac{3}{6}$
$11\frac{5}{6}$

The order of operations is the same for fractions as it is for whole numbers.

Order of Operations

1. Operations within parentheses should be performed first.
2. Then do multiplications and divisions, working from left to right.
3. Finally, do additions and subtractions, working from left to right.

EXAMPLE 6

Perform the indicated operations.

a. $\dfrac{3}{4} \times \dfrac{1}{3} + \dfrac{3}{4} \times \dfrac{2}{3} = \dfrac{1}{4} + \dfrac{1}{2}$ **b.** $\dfrac{3}{4} \times \left(\dfrac{1}{3} + \dfrac{2}{3}\right) = \dfrac{3}{4} \times 1$

$$= \dfrac{1}{4} + \dfrac{2}{4} \qquad\qquad\qquad = \dfrac{3}{4}$$

$$= \dfrac{3}{4}$$ ■

Sometimes in mathematics a number is multiplied by a quantity within parentheses and no operation symbol is shown. This is called **juxtaposition** and means multiplication. Example 6b might have been written

$$\frac{3}{4}\left(\frac{1}{3} + \frac{2}{3}\right)$$

EXAMPLE 7

Perform the indicated operations.

a. $\dfrac{5}{6}\left(\dfrac{3}{4} + \dfrac{1}{2}\right) + \dfrac{2}{3} \times \dfrac{9}{10} = \dfrac{5}{6}\left(\dfrac{3}{4} + \dfrac{2}{4}\right) + \dfrac{2}{3} \times \dfrac{9}{10}$

$$= \frac{5}{6}\left(\frac{5}{4}\right) + \frac{\overset{1}{\cancel{2}}}{\underset{1}{\cancel{3}}} \times \frac{\overset{3}{\cancel{9}}}{\underset{5}{\cancel{10}}}$$

$$= \frac{25}{24} + \frac{3}{5}$$

$$= \frac{125}{120} + \frac{72}{120}$$

$$= \frac{197}{120} \quad \text{or} \quad 1\frac{77}{120}$$

b. $\dfrac{1}{2} \div \dfrac{1}{3} \div \left(\dfrac{3}{4} + \dfrac{3}{5}\right) = \dfrac{1}{2} \div \dfrac{1}{3} \div \left(\dfrac{15}{20} + \dfrac{12}{20}\right)$

$$= \frac{1}{2} \div \frac{1}{3} \div \frac{27}{20}$$

$$= \frac{1}{2} \times \frac{3}{1} \div \frac{27}{20}$$

$$= \frac{3}{2} \div \frac{27}{20}$$

$$= \frac{\overset{1}{\cancel{3}}}{\underset{1}{\cancel{2}}} \times \frac{\overset{10}{\cancel{20}}}{\underset{9}{\cancel{27}}}$$

$$= \frac{10}{9} \quad \text{or} \quad 1\frac{1}{9}$$ ■

CALCULATOR COMMENT

The rest of this section is optional for those who have calculators. It is not necessary to find common denominators when you add or subtract decimal fractions and therefore it is not necessary to find common denominators when you add or subtract common fractions on a calculator, because the work is done in decimal form. Suppose you want your answer in fractional form but still wish to use a calculator? (There are some fractional-form calculators on the market, but they are more like novelty calculators than practical ones.) When using calculators you usually operate on the fractions in decimal form and then change your answer to fractional form. However, you can do this only for certain fractions for two reasons:

1. Round-off error may occur because the calculator does not work with exact fractions, but rather with decimal approximations.
2. Repeating decimals may be difficult to change back into fractional form.

Instead of finding common denominators, we'll add and subtract on a calculator according to the following procedures:

Addition

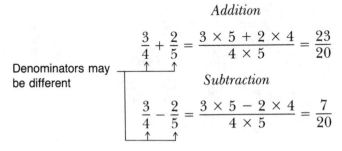

$$\frac{3}{4} + \frac{2}{5} = \frac{3 \times 5 + 2 \times 4}{4 \times 5} = \frac{23}{20}$$

Denominators may be different

Subtraction

$$\frac{3}{4} - \frac{2}{5} = \frac{3 \times 5 - 2 \times 4}{4 \times 5} = \frac{7}{20}$$

This process may be summarized by the numbered steps:

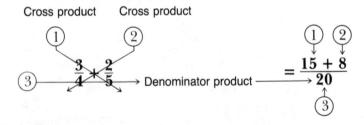

Cross product   Cross product

$$= \frac{15 + 8}{20}$$

EXAMPLE 8

Add or subtract by using this procedure instead of finding common denominators.

a. $\dfrac{1}{2} + \dfrac{1}{3} = \dfrac{1 \times 3 + 2 \times 1}{2 \times 3}$ **b.** $\dfrac{4}{5} - \dfrac{2}{3} = \dfrac{4 \times 3 - 5 \times 2}{5 \times 3}$

       $= \dfrac{3 + 2}{6}$ $= \dfrac{12 - 10}{15}$

       $= \dfrac{5}{6}$ $= \dfrac{2}{15}$ ■

This procedure is cumbersome when working with large fractions, but when you have a calculator, it is easy to carry out.

EXAMPLE 9

a. $\dfrac{11}{12} + \dfrac{3}{20}$

$$\boxed{11} \; \boxed{\times} \; \boxed{20} \; \boxed{+} \; \boxed{12} \; \boxed{\times} \; \boxed{3} \; \boxed{=} \; 256.$$

$$\boxed{12} \; \boxed{\times} \; \boxed{20} \; \boxed{=} \; 240.$$

Reduce $\dfrac{256}{240} = \dfrac{16}{15},$ or as a decimal,

$$\boxed{256} \; \boxed{\div} \; \boxed{240} \; \boxed{=} \; 1.0666667$$

b. $\dfrac{55}{144} - \dfrac{25}{108}$ (Answer in decimal form.)

$$\boxed{55} \; \boxed{\times} \; \boxed{108} \; \boxed{-} \; \boxed{144} \; \boxed{\times} \; \boxed{25} \; \boxed{=} \; \boxed{\div}$$
$$\boxed{(} \; \boxed{144} \; \boxed{\times} \; \boxed{108} \; \boxed{)} \; \boxed{=} \; .15046296 \quad ■$$

Note: If your calculator is not algebraic (if it does not do multiplication before addition), be careful to insert parentheses or to work the operations in the proper order.

Problem Set 2.4

Perform the indicated operations in Problems 1–6. Give your answers in reduced form.

1. **a.** $\dfrac{2}{5} + \dfrac{1}{5}$ **b.** $\dfrac{3}{7} + \dfrac{5}{7}$ **c.** $\dfrac{5}{11} + \dfrac{3}{11}$

2. **a.** $\dfrac{9}{13} - \dfrac{5}{13}$ **b.** $\dfrac{6}{23} - \dfrac{5}{23}$ **c.** $\dfrac{9}{7} - \dfrac{2}{7}$

3. **a.** $\dfrac{13}{15} + \dfrac{2}{15}$ **b.** $\dfrac{7}{12} - \dfrac{1}{12}$ **c.** $\dfrac{5}{8} - \dfrac{3}{8}$

4. **a.** $\dfrac{3}{2} + \dfrac{5}{2}$ **b.** $\dfrac{7}{3} - \dfrac{4}{3}$ **c.** $\dfrac{5}{9} + \dfrac{1}{9}$

5. **a.** $\begin{array}{r} 2\frac{2}{3} \\ + 1\frac{1}{3} \\ \hline \end{array}$ **b.** $\begin{array}{r} 6\frac{4}{5} \\ + 2\frac{1}{5} \\ \hline \end{array}$ **c.** $\begin{array}{r} 8\frac{3}{8} \\ + 4\frac{5}{8} \\ \hline \end{array}$

6. **a.** $\begin{array}{r} 5\frac{1}{3} \\ - 3\frac{2}{3} \\ \hline \end{array}$ **b.** $\begin{array}{r} 14\frac{1}{8} \\ - 8\frac{5}{8} \\ \hline \end{array}$ **c.** $\begin{array}{r} 6\frac{1}{4} \\ - 5\frac{3}{4} \\ \hline \end{array}$

Find the LCD for the denominators given in Problems 7–10.

7. **a.** 4; 8   **b.** 2; 6   **c.** 2; 5

8. **a.** 5; 12   **b.** 4; 12   **c.** 12; 90

9. **a.** 12; 336   **b.** 90; 210   **c.** 60; 72

10. **a.** 12; 20; 36   **b.** 6; 8; 10   **c.** 9; 12; 14

Perform the indicated operations in Problems 11–25. Give your answers in reduced form.

11. **a.** $\dfrac{1}{2} + \dfrac{2}{3}$ **b.** $\dfrac{1}{2} + \dfrac{3}{8}$ **c.** $\dfrac{1}{2} - \dfrac{1}{6}$

12. **a.** $\dfrac{1}{2} + \dfrac{2}{5}$ **b.** $\dfrac{1}{2} - \dfrac{2}{5}$ **c.** $\dfrac{5}{6} + \dfrac{2}{3}$

13. **a.** $\dfrac{5}{6} - \dfrac{1}{3}$ **b.** $\dfrac{5}{6} + \dfrac{5}{8}$ **c.** $\dfrac{5}{8} - \dfrac{1}{3}$

14. **a.** $\dfrac{3}{4} - \dfrac{5}{12}$ **b.** $\dfrac{3}{5} + \dfrac{1}{6}$ **c.** $\dfrac{3}{4} + \dfrac{1}{12}$

15. **a.** $\dfrac{2}{45} + \dfrac{1}{6}$ **b.** $\dfrac{41}{45} - \dfrac{5}{6}$ **c.** $\dfrac{4}{9} - \dfrac{5}{12}$

16. **a.** $\dfrac{3}{5} + \dfrac{1}{12}$ **b.** $\dfrac{5}{24} - \dfrac{4}{30}$ **c.** $\dfrac{5}{27} + \dfrac{1}{90}$

17. **a.** $\begin{array}{r} 2\frac{1}{2} \\ +\ 4\frac{3}{4} \\ \hline \end{array}$ **b.** $\begin{array}{r} 1\frac{2}{3} \\ +\ 5\frac{1}{2} \\ \hline \end{array}$ **c.** $\begin{array}{r} 3\frac{3}{8} \\ +\ 5\frac{1}{2} \\ \hline \end{array}$

18. **a.** $\begin{array}{r} 5\frac{1}{8} \\ -\ 3\frac{3}{4} \\ \hline \end{array}$ **b.** $\begin{array}{r} 17\frac{1}{2} \\ -\ 6\frac{2}{3} \\ \hline \end{array}$ **c.** $\begin{array}{r} 12\frac{1}{3} \\ -\ 4\frac{1}{2} \\ \hline \end{array}$

19. **a.** $\begin{array}{r} 6\frac{3}{8} \\ -\ 5\frac{1}{6} \\ \hline \end{array}$ **b.** $\begin{array}{r} 2\frac{1}{3} \\ +\ 1\frac{5}{12} \\ \hline \end{array}$ **c.** $\begin{array}{r} 6\frac{1}{16} \\ -\ 5\frac{3}{8} \\ \hline \end{array}$

20. **a.** $5 - \dfrac{3}{8}$ **b.** $7 - \dfrac{2}{3}$ **c.** $3\frac{3}{4} + 2\frac{1}{6}$

21. **a.** $\dfrac{1}{8} + 2\frac{2}{3} + \dfrac{1}{6}$ **b.** $\dfrac{4}{5} + \dfrac{3}{7} + \dfrac{3}{10}$ **c.** $6\frac{1}{8} + 3\frac{2}{5} + 1\frac{1}{4}$

22. **a.** $\begin{array}{r} 5\frac{1}{2} \\ 6\frac{3}{4} \\ +\ 4\frac{1}{8} \\ \hline \end{array}$ **b.** $\begin{array}{r} 7\frac{2}{3} \\ 5\frac{1}{2} \\ +\ 12\frac{1}{6} \\ \hline \end{array}$ **c.** $\begin{array}{r} 12\frac{3}{8} \\ 2\frac{1}{2} \\ +\ 5\frac{1}{8} \\ \hline \end{array}$

23. **a.** $\dfrac{1}{2} \times \dfrac{2}{3} + \dfrac{1}{5}$ **b.** $\dfrac{1}{2} + \dfrac{2}{3} \times \dfrac{1}{5}$ **c.** $\dfrac{1}{5} \div \dfrac{1}{3} \div \dfrac{1}{4}$

24. **a.** $\dfrac{1}{2} \div \left(\dfrac{1}{3} \div \dfrac{1}{4}\right)$ **b.** $\dfrac{3}{4}\left(\dfrac{9}{13} + \dfrac{4}{13}\right)$ **c.** $\dfrac{4}{5}\left(\dfrac{5}{16} + \dfrac{11}{16}\right)$

25. **a.** $\dfrac{3 \times 3 + 5 \times 2}{5 \times 3}$ **b.** $\dfrac{3 \times 5 + 7 \times 4}{7 \times 5}$ **c.** $\dfrac{1 \times 8 + 2 \times 5}{2 \times 8}$

26. A recipe calls for $\frac{2}{3}$ cup milk and $\frac{1}{2}$ cup water.
   a. What is the total amount of liquid?
   b. If you wish to make $\frac{1}{4}$ of this recipe, how much of each ingredient is needed, and what is the total amount?

27. Suppose you have items to mail that weigh $1\frac{1}{4}$ lb, $2\frac{2}{3}$ lb, and $3\frac{1}{2}$ lb. What is the total weight of these packages?

28. Loretta received three boxes of candy for her birthday: a $1\frac{1}{2}$ lb box, a $\frac{3}{4}$ lb box, and a $2\frac{15}{16}$ lb box. What is the total weight of the candy she received?

29. Suppose you are installing molding around a table and you need pieces $5\frac{1}{4}$ in., $7\frac{1}{2}$ in., and $5\frac{3}{16}$ in. long. If each time a cut is made the saw chews up $\frac{1}{16}$ in. of material, what is the smallest single length of molding that can be used to do this job?

30. If the outside diameter of a piece of tubing is $\frac{15}{16}$ in. and the wall is $\frac{3}{16}$ in. thick, what is the size of the inside diameter?

CALCULATOR PROBLEMS

Perform the indicated operations in Problems 31–36.

31. $\dfrac{3}{4} \times \dfrac{119}{200} + \dfrac{3}{4} \times \dfrac{81}{200}$ (fractional answer)

32. $\dfrac{4}{5} \times \dfrac{17}{95} + \dfrac{4}{5} \times \dfrac{78}{95}$ (fractional answer)

33. $\dfrac{7}{8} \times \dfrac{107}{147} + \dfrac{7}{8} \times \dfrac{40}{147}$ (decimal answer)

34. $\left(\dfrac{4}{5} + \dfrac{2}{3} \right) \div \dfrac{1}{5} + 2$ (decimal answer)

35. $\dfrac{19}{300} + \dfrac{55}{144} + \dfrac{25}{108}$ (decimal answer)

36. $\dfrac{19}{300} + \dfrac{55}{144} + \dfrac{25}{108}$ (reduced fraction answer)

Mind Bogglers

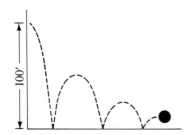

37. A rubber ball is known to rebound half the height it drops. If the ball is dropped from a height of 100 feet, how far will it have traveled by the time it hits the ground for:
  a. The first time?
  b. The second time?
  c. The third time?
  d. The fourth time?
  e. The fifth time?
  f. Look for a pattern, and decide if there is a maximum distance the ball will travel if we assume that it will bounce indefinitely.

38. Look for a pattern in the following problem. Verify by division (show your work):
  a. $\dfrac{1}{9} = .111\ldots$ **b.** $\dfrac{2}{9} = .222\ldots$ **c.** $\dfrac{3}{9} = .333\ldots$

  d. $\dfrac{4}{9} = .444\ldots$ $\ldots$ **e.** $\dfrac{8}{9} = .888\ldots$

  f. What is $\frac{9}{9}$ according to the pattern?

2.5 Review Problems

1. Perform the indicated operations.   Section 2.1
   a. $12 + 20 \div 2$ **b.** $(12 + 20) \div 2$

2. Perform the indicated operations.   Section 2.1
   a. $3 \times (200 + 50 + 6)$ **b.** $3 \times 200 + 3 \times 50 + 3 \times 6$

3. **a.** Write $\dfrac{114}{7}$ as a mixed number.   Section 2.2

   b. Write $2\frac{4}{5}$ as a decimal fraction.

4. **a.** Write $4\frac{2}{3}$ as an improper fraction.   Section 2.2

   b. Write $\dfrac{5}{6}$ as a decimal fraction.

5. **a.** Reduce $\dfrac{192}{240}$ to lowest terms.   Section 2.3

   b. Change $.4\frac{5}{11}$ to fractional form.

Perform the indicated operations in Problems 6–10.

6. **a.** $4\frac{1}{6} \times 3\frac{2}{5}$ **b.** $6\frac{1}{2} \div 3\frac{3}{4}$ Section 2.3

7. **a.** $\dfrac{5}{7} + \dfrac{3}{7}$ **b.** $\dfrac{3}{8} + \dfrac{5}{12}$ Section 2.4

8. **a.** $\dfrac{7}{12} - \dfrac{2}{15}$ **b.** $7\frac{2}{15} - 3\frac{7}{12}$ Section 2.4

9. $\dfrac{4}{10} + \dfrac{7}{15} - \dfrac{5}{6}$ Section 2.4

10. $\dfrac{2}{3} + \dfrac{4}{5} \times \dfrac{1}{2}$ Section 2.4

The World of Numbers

3.1 Signed Numbers

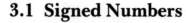

You have probably seen the need for numbers to represent quantities less than zero many times. Some common examples are temperatures below zero, card scores that are "in the hole," business debts, and altitudes below sea level. To represent these ideas, we introduce **signed numbers**.

1. With each counting number, we associate a **positive sign**:

$$1 = +1; \quad 2 = +2; \quad 3 = +3; \quad \dots$$

Although the symbolism is the same as that of an addition sign, remember that addition requires two numbers—for example, $6 + 5$. Using positive signs, this addition problem would look like this: $(+6) + (+5)$.

2. For each counting number, we define a *new number*, called its **opposite**, by using a **negative sign**.

The opposite of 1 is denoted by -1.
The opposite of 2 is denoted by -2.
The opposite of 3 is denoted by -3.
And so on.

The **opposite** of a given number is defined to be that number, which when added to the given number, gives a sum of zero:

Given number		*Opposite*	
1	+	(-1)	$= 0$
2	+	(-2)	$= 0$
3	+	(-3)	$= 0$
⋮			

Also:

Given number		*Opposite*	
0	+	0	$= 0$
(-1)	+	1	$= 0$
(-2)	+	2	$= 0$
⋮			

3. The counting numbers, along with their opposites, and the number 0, form a set called the **integers**.

> The set of **integers** is the set
>
> $$\{\dots, -4, -3, -2, -1, 0, 1, 2, 3, 4, \dots\}$$

Integers

Historically, the negative integers were developed quite late. There are indications that the Chinese had some knowledge of negative numbers as early as 200 B.C., and in the 7th century A.D. the Hindu Brahmagupta stated the rules for operations with positive and negative numbers. The Chinese represented negative integers by putting them in red (compare with the present-day accountant), and the Hindus represented them by putting a circle or a dot over the number. However, as late as the 16th century, some European scholars were calling numbers such as -1 absurd. In 1545, Cardano (1501–1576), an Italian scholar who presented the elementary properties of negative numbers, called the positive numbers "true" numbers and the negative numbers "fictitious" numbers. However, they did become universally accepted, and as a matter of fact, the word *integer* that we use to describe this set is derived from "numbers with integrity."

The easiest way to understand the operations and properties of integers is to represent them on a **number line**, as shown in Figure 3.1.

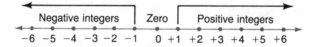

FIGURE 3.1   A number line

To draw a number line, locate any two convenient points and label the one on the left 0 and the one on the right $+1$. The distance between these points is called a **unit scale** and can be used to mark off equal distances in both directions. These points correspond to the integers, as shown in Figure 3.1. A number line can be used to order numbers:

> **Less than** (symbol, $<$) means *to the left* on a number line.
> **Greater than** (symbol, $>$) means *to the right* on a number line.
> **Equal to** (symbol, $=$) means *the same point* on a number line.

Order Symbols

EXAMPLE 1

Fill in $<$, $>$, or $=$ in the blank.

a. 2 ____ 5   Fill in $<$, since 2 is to the left of 5 on a
                      number line.
b. 6 ____ 3   Fill in $>$.
c. 4 ____ 4   Fill in $=$.
d. -2 ____ -3 Fill in $>$, since -2 is to the right of -3.
e. 0 ____ -2 Fill in $>$.   ∎

With the introduction of negative values, we also need to introduce a symbol to represent distance, because distances are nonnegative.

Absolute Value

> The **absolute value** of a number is the distance of that number from 0 on the number line.

EXAMPLE 2

a. The absolute value of 5, symbolized by $|5|$, is 5 because 5 is 5 units from 0.
b. The absolute value of -5, symbolized by $|-5|$, is 5 because -5 is 5 units from 0.
c. $|-3| = 3$ **d.** $|0| = 0$ **e.** $|-349| = 349$ ∎

EXAMPLE 3

Tell which number is larger. Then show the larger absolute value.

	Larger number	Larger absolute value
a. 2; 5	5	5
b. -2; -5	-2	5
c. -8; 10	10	10
d. 8; -10	8	10

∎

The number line is also used to illustrate addition of integers. It is agreed that adding a positive number means moving to the right, and adding a negative number means moving to the left. Adding zero means no move at all. This process is illustrated by Example 4.

EXAMPLE 4

Add on a number line.

a. Positive + Positive: $5 + 6$

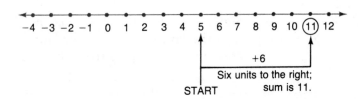

b. Positive + Negative: $5 + (-6)$ Notice: To avoid confusion, negative numbers are often enclosed in parentheses when combined with other operations.

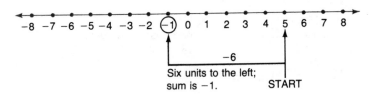

c. Negative + Positive: $(-5) + 6$

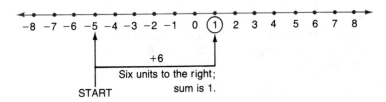

d. Negative + Negative: $(-5) + (-6)$

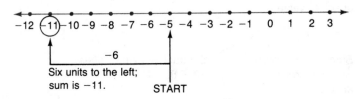

The method of adding on a number line shown in Example 4 is not practical for continued use, so we use it to lead us to the following rules for the addition of integers:

**Procedure for
Adding Integers**

	Sign of sum	Absolute values
1. Pos + Pos	+	Added
2. Neg + Neg	−	Added
3. Pos + Neg ⎫ **4.** Neg + Pos ⎭	Sign of number with the larger absolute value	Subtract the smaller number from the larger number

EXAMPLE 5

Find the sums.

a. $45 + 27 = 72$ Positives are understood.

b. $\overset{\uparrow}{-}18 + (\overset{\uparrow}{-}21) = \overset{}{-}39$

Neg + Neg
is negative ————⌐ Add absolute values:
$18 + 21 = 39$

c. $-12 + (-35) = -47$

d. $-128 + (-19) = -147$

e. $\overset{\uparrow}{13} + (\overset{\uparrow}{-5}) = \overset{\uparrow}{+8}$

Pos + Neg
Use the sign of the   Subtract absolute values:
number with the larger $13 - 5 = 8$
absolute value.

f. $-13 + 5 = \overset{\uparrow}{-8}$

Sign of the number   Subtract absolute values:
with the larger $13 - 5 = 8$
absolute value

g. $-8 + 12 = \overset{\uparrow}{+4}$

Sign of the number   Subtract absolute values:
with the larger $12 - 8 = 4$
absolute value

h. $8 + (-12) = -4$

i. $-48 + 53 = 5$

          └─ Plus sign understood

j. $127 + (-127) = 0$ Remember, adding opposites gives 0. ■

Subtraction is related to addition in the following way:

To subtract, add the opposite of the number being subtracted.

Procedure for Subtracting Integers

EXAMPLE 6

Find the differences.

a. $7 - 10 = 7 + (-10)$

          Add the opposite.

       $= -3$ ◄── Use the rules of **addition** to obtain this result.

b. $7 - (-10) = 7 + (+10)$
$$= 17$$

c. $4 - (-8) = 4 + 8$
$$= 12$$

d. $-2 - (-9) = -2 + 9$
$$= 7$$

e. $2 - (-9) = 2 + 9$
$$= 11$$

f. $-2 - 9 = -2 + (-9)$
$$= -11$$

g. $-345 - 527 = -345 + (-527)$
$$= -872$$ ■

For the whole numbers, multiplication can be defined as repeated addition, since we say that 3×4 means

$$\underbrace{3 + 3 + 3 + 3}$$

4 addends

However, this *cannot* be done for all integers, since $3 \times (-4)$ or

$$\underbrace{3 + 3 + \cdots + 3}_{-4 \text{ addends}}$$

doesn't "make sense." Thus, it is necessary to generalize the definition of multiplication to include all integers. There are four cases to consider.

Positive integers are the same as the natural numbers, so the above definition of repeated addition applies: **The product of two positive numbers is a positive number.**

Now, consider $(+3) \times (-4)$ by looking at a pattern.

$$(+3) \times (+4) = +12$$
$$(+3) \times (+3) = +9$$
$$(+3) \times (+2) = +6$$

Would you know what to write next? Here it is:

$$(+3) \times (+1) = +3$$
$$(+3) \times (\ 0\ ) = \ \ 0$$

What comes next? Do you see the pattern? As the factor *decreases by 1, the product decreases by 3:*

$$(+3) \times (-1) = \ \ -3$$
$$(+3) \times (-2) = \ \ -6$$
$$(+3) \times (-3) = \ \ -9$$
$$(+3) \times (-4) = -12$$

Do you see how to continue? Try building a few more such patterns using different numbers. What do you see about the product of a positive and a negative number? **The product of a positive number and a negative number is a negative number.**

The order in which two numbers are multiplied has no effect on the product, so

$$(-3) \times (+4) = (+4) \times (-3)$$
$$= -12$$

The product of a negative number and a positive number is a negative number.

Consider the final example—the product of two negative integers:

$$(-3) \times (-4)$$

Let's build another pattern.

$$(-3) \times (+4) = -12$$
$$(-3) \times (+3) = -9$$
$$(-3) \times (+2) = -6$$

What would you write next? Here it is:

$$(-3) \times (+1) = -3$$
$$(-3) \times (\ 0\ ) = \ \ 0$$

What comes next? You should notice that as the factor *decreases by 1, the product increases by 3:*

$$(-3) \times (-1) = \ +3$$
$$(-3) \times (-2) = \ +6$$
$$(-3) \times (-3) = \ +9$$
$$(-3) \times (-4) = +12$$

The product of two negative numbers is a positive number.

	Sign of product	Absolute values
1. Pos × Pos	+	Multiplied
2. Pos × Neg	−	Multiplied
3. Neg × Pos	−	Multiplied
4. Neg × Neg	+	Multiplied

Procedure for Multiplying Integers

EXAMPLE 7

Find the products.

a. $6 \times 9 = 54$ **b.** $6 \times (-2) = -12$

   Positives understood

c. $(-3)7 = -21$ **d.** $(-8)(-9) = 72$

   Remember that numbers in parentheses written right next to each other mean multiplication.

e. $-1(-1)(-1) = -1$ **f.** $-2(-3)5 = 30$ ■

You know that division can be thought of as multiplying by the reciprocal of a number, so division may be written as a multiplication. For example,

$$(-10) \div 2 = (-10) \times \tfrac{1}{2}$$
$$= -5$$

Negative **times** positive is negative

This fact means that the rules for division of integers are identical to those for multiplication.

Procedure for Dividing Integers

	Sign of quotient	Absolute values
1. Pos ÷ Pos	+	Divided
2. Pos ÷ Neg	−	Divided
3. Neg ÷ Pos	−	Divided
4. Neg ÷ Neg	+	Divided

EXAMPLE 8

Find the quotients.

a. $10 \div 5 = 2$ **b.** $10 \div (-5) = -2$

c. $(-10) \div 5 = -2$ **d.** $(-10) \div (-5) = 2$

e. $90 \div (-9) = -10$ **f.** $-282 \div (-6) = 47$ ■

CALCULATOR COMMENT

The rest of this section is optional for those who have calculators. When using a calculator, you must distinguish between a negative sign, as in -5, and a subtraction sign, as in $8 - 5$. For entering negative numbers into a calculator you'll find a key marked

$$\boxed{+/-} \qquad \text{or} \qquad \boxed{\text{CHS}}$$

These keys change the sign of a number. For example, $5 + (-6)$ is entered as:

$$\boxed{5}\ \boxed{+}\ \boxed{6}\ \boxed{+/-}\ \boxed{=}$$

EXAMPLE 9

Indicate the sequence of keys to enter $(-8) - (-5)$ into a calculator.

Solution

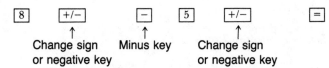

| 8 | +/− | | − | 5 | +/− | | = |

Change sign   Minus key   Change sign
or negative key   or negative key   ∎

Problem Set 3.1

Find the sums in Problems 1–6.

1. a. $5 + 9$ **b.** $5 + (-9)$
    c. $-5 + 9$ **d.** $-5 + (-9)$
    e. $-7 + 6$ **f.** $7 + 6$

2. a. $-7 + (-6)$ **b.** $7 + (-6)$
    c. $-7 + (-6)$ **d.** $-9 + 4$
    e. $9 + 4$ **f.** $9 + (-4)$

3. a. $8 + (-6)$ **b.** $-8 + (-6)$
    c. $-8 + 6$ **d.** $8 + 6$
    e. $42 + (-60)$ **f.** $-43 + 80$

4. a. $-76 + (-65)$ **b.** $-56 + 20$
    c. $28 + (-76)$ **d.** $-72 + 56$
    e. $-98 + (-84)$ **f.** $-62 + 79$

5. a. $7 + (-3)$ **b.** $-9 + 15$
    c. $-10 + (-4)$ **d.** $162 + (-27)$
    e. $-18 + (-4 + 3)$ **f.** $-5 + (-6 + 10)$

6. a. $-15 + 8$ **b.** $(-7 + 5) + (-4 + 5)$
    c. $-14 + 27$ **d.** $42 + (-121)$
    e. $62 + (-62)$ **f.** $(-8 + 6) + (6 + 8)$

Find the differences in Problems 7–12. Show your solutions as two steps: First rewrite the subtraction as addition; then carry out the addition to find the answer.

7. a. $9 - 5$ **b.** $9 - (-5)$ **c.** $-9 - 5$
    d. $-9 - (-5)$

8. a. $17 - (-8)$ **b.** $-17 - 8$ **c.** $17 - 8$
    d. $-17 - (-8)$

9. a. $-21 - 7$ **b.** $-21 - (-7)$ **c.** $21 - (-7)$
    d. $21 - 7$

10. **a.** $-13 - (-6)$ **b.** $13 - 6$ **c.** $13 - (-6)$
    d. $-13 - 6$

11. **a.** $8 - 23$ **b.** $-8 - (-23)$ **c.** $-8 - 23$
    d. $8 - (-23)$

12. **a.** $-46 - (-46)$ **b.** $-7 - (-18)$ **c.** $62 - (-112)$
    d. $5 - (-416)$

Find the product or quotient in Problems 13–18.

13. **a.** $6 \times (-9)$ **b.** $3 \times (-15)$ **c.** $(-2) \times 47$
    d. $4 \times (-12)$

14. **a.** $4(-5)$ **b.** $8(-8)$ **c.** $-6(-11)$
    d. $-7(-4)$

15. **a.** $100 \div (-5)$ **b.** $-56 \div 8$ **c.** $-88 \div (-8)$
    d. $48 \div (-4)$

16. **a.** $\dfrac{92}{-2}$ **b.** $\dfrac{-104}{4}$ **c.** $\dfrac{-63}{-9}$

    d. $\dfrac{-528}{-4}$

17. **a.** $0 \div (-8)$ **b.** $-4(18)$ **c.** $(-8) \div 0$
    d. $(-3)(-6)$

18. **a.** $(-121)(-11)$ **b.** $\dfrac{-300}{-30}$ **c.** $(-1)(2)(-2)(3)(4)$
    d. $(-19) \div 0$

Perform the operations indicated in Problems 19–24.

19. **a.** $3(2 + 8)$ **b.** $-3(3 + 7)$
    c. $-5(2 - 10)$

20. **a.** $6(12 - 20)$ **b.** $(-5)(3 - 10)$
    c. $(-5 \times 8) - 12$

21. **a.** $2(-3)(-3)(-4)$ **b.** $2 + 3 - 5(4 - 10)$
    c. $(-4 + 7)(-4 + 7)$

22. **a.** $3 + 10 \div 2 - 3(4 - 6)$ **b.** $13 - (-2)(-5)$
    c. $10 + 2(-5)$

23. **a.** $3 + 12 \div (-3)$ **b.** $(3 + 12) \div (-3)$
    c. $\dfrac{3 + 12}{-3}$

24. **a.** $\dfrac{2 + (-8)}{2}$ **b.** $\dfrac{-15 - 21}{6}$

    c. $\dfrac{12 - (-3)}{5}$

25. Suppose you are given the following number line:

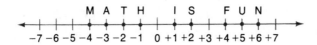

    a. Explain why a move from *H* to *N* is described by +7.
    b. Describe a move from *T* to *F*.
    c. Describe a move from *U* to *A*.

26. Add the following integers on a number line:
    a. $(+5) + (+3)$ **b.** $(-5) + (+3)$ **c.** $(+4) + (-7)$

27. Add the following integers on a number line:
    a. $(+3) + (+5)$ **b.** $(-3) + (+5)$ **c.** $(-2) + (-4)$

28. In a game of rummy, a player's scores for five hands were 25, −120, 45, −10, and 60. What is the player's final score?

29. What is the final temperature in a freezer if at 9:00 A.M. it is −5°C and then it goes up 10°, drops 15°, drops 6°, and finally goes up 7°?

30. In Death Valley, California, the temperature can vary from 134°F above zero to 25°F below zero. What is the difference between these temperature extremes?

31. What is the difference in elevation between the top of a mountain 8,520 ft above sea level and a point in a valley 253 ft below sea level?

32. What is the difference in elevation between a plane flying at 25,400 ft above sea level and a submarine traveling 450 ft below sea level?

33. IBM stock had the following changes during a certain week: +1, +3, −2, −1, and −3. What is the net change for the week?

34. If the Dow Jones Average stock prices for a particular week have the following changes, what is the total change for the week?

    Monday, +9;   Tuesday, −15;   Wednesday, −29;
    Thursday, +7;   Friday, +6

CALCULATOR PROBLEMS

Perform the indicated operations in Problems 35–38 on your calculator.

35. a. $487 - 843$ **b.** $-381 - (-843)$
36. a. $919 + (-483)$ **b.** $-1,439 - 816$

37. a. $-4,567 + (-3,891) + 458 - 7,912$
   b. $-982 - (-458) + (-402) - 4,921$
38. a. $4,987 + (-4,583) - 478 + (-5,670)$
   b. $-8,211 - (-10,209) + 4,511 - (-4,529)$

Mind Bogglers

39. Explain, in your own words, the difference between $0 \div 5$ and $5 \div 0$.

40. Perform the indicated operations.

   a. $(42 \div -7) \div -2$ **b.** $\dfrac{\frac{42}{-7}}{-2}$

   c. $-56 \div (14 \div 2)$ **d.** $\dfrac{-56}{\frac{14}{2}}$

3.2 Exponents and Roots

We often encounter numbers that comprise repeated multiplication of the same numbers. For example,

$$10 \times 10 \times 10 \qquad \text{or} \qquad 6 \times 6 \times 6 \times 6 \times 6 \qquad \text{or}$$

$$15 \times 15 \times 15 \times 15 \times 15 \times 15 \times 15 \times 15 \times 15 \times 15$$
$$\times 15 \times 15 \times 15$$

These numbers can be written more simply by inventing a new notation:

$$10^3 = \underbrace{10 \times 10 \times 10}_{\text{3 factors}}$$

$$6^5 = \underbrace{6 \times 6 \times 6 \times 6 \times 6}_{\text{5 factors}}$$

$$15^{13} = \underbrace{15 \times 15 \times \cdots \times 15}_{\text{13 factors}}$$

We call this **power** or **exponential notation**. The number that is multiplied as a repeated factor is called the **base**, and the number of times it is repeated is called the **exponent**.

EXAMPLE 1

a. $10^3 = 10 \times 10 \times 10$ or 1,000   The base is 10; the exponent is 3.

b. $6^2 = 6 \times 6$ or 36   The base is 6; the exponent is 2.

c. $7^5 = 7 \times 7 \times 7 \times 7 \times 7$ or 16,807   The base is 7; the exponent is 5.

d. $2^{63} = \underbrace{2 \times 2 \times 2 \times \cdots \times 2 \times 2}_{63\ \text{factors}}$ or

9,223,372,036,854,775,808   I used a computer to multiply this out—I don't expect you to do that, but I thought you might like to see it. ∎

We now use this notation to observe a pattern for **powers of 10**.

$$10^1 = 10$$
$$10^2 = 10 \times 10 = 100$$
$$10^3 = 10 \times 10 \times 10 = 1,000$$
$$10^4 = 10 \times 10 \times 10 \times 10 = 10,000$$

Do you see a relationship between the exponent and the number?

$$10^5 = \underbrace{100,000}_{5\ \text{zeros}}$$

Exponent is a 5

Notice that the exponent and the number of zeros are the same.

Could you write 10^{12} without actually multiplying?

There is a similar pattern for multiplication of any number by a power of 10. Consider the following examples, and notice what happens to the decimal point:

$$9.42 \times 10^1 = 94.2$$
$$9.42 \times 10^2 = 942.$$
$$9.42 \times 10^3 = 9,420.$$
$$9.42 \times 10^4 = 94,200.$$

We find these answers by direct multiplication.

Do you see a pattern? If we multiply 9.42×10^5, how many places to the right will the decimal point be moved? See the next page.

$$9.42 \times 10^5 = 942,000.$$

This answer is found by observing the pattern, not by directly multiplying.

5 places $\longrightarrow$

Using this pattern, can you multiply the following *without direct calculation*?

$$9.42 \times 10^{12}$$

We will investigate one final pattern of 10s, this time looking at smaller values in this pattern.

$$\vdots$$
$$100,000 = 10^5$$
$$10,000 = 10^4$$
$$1,000 = 10^3$$
$$100 = 10^2$$
$$10 = 10^1$$

Continuing with the same pattern:

$$1 = 10^0$$
$$.1 = 10^{-1}$$
$$.01 = 10^{-2}$$
$$.001 = 10^{-3}$$
$$.0001 = 10^{-4}$$
$$.00001 = 10^{-5}$$
$$\vdots$$

When we multiply a power of 10 by some number, a pattern emerges:

$$9.42 \times 10^2 = 942.$$
$$9.42 \times 10^1 = 94.2$$
$$9.42 \times 10^0 = 9.42$$

We interpret the zero exponent as "decimal point moves 0 places."

To continue, direct calculation extends the pattern:

$$9.42 \times 10^{-1} = .942$$
$$9.42 \times 10^{-2} = .0942$$
$$9.42 \times 10^{-3} = .00942$$

These numbers are found by direct multiplication.

Do you see that the same pattern for multiplying by a negative exponent also holds? Can you multiply 9.42×10^{-6} *without direct calculation*?

The solution is as follows:

$$9.42 \times 10^{-6} = .00000942$$

Moved 6 places to the left

These patterns lead to a useful way for writing large and small numbers, called **scientific notation**.

The **scientific notation** of a number is that number written as a power of 10 or as a decimal number between 1 and 10 times a power of 10.

Scientific Notation

EXAMPLE 2

Write the given numbers in scientific notation.

a. $123,600 = 1.236 \times 10^?$

Step 1. Fix decimal point after the first nonzero digit.
Step 2. From this number, count the number of decimal places to restore the number to its given form:

$$123,600 = 1.23600 \times 10^?$$

5 places to the right

Step 3. The exponent is the same as the number of decimal places needed to restore scientific notation to the original given number:

$$123,600 = 1.236 \times 10^5$$

b. $.000035 = 3.5 \times 10^?$

Step 1

Step 2. $.000035 = 00003.5 \times 10^?$

5 places to the left;
this is −5

Step 3. $.000035 = 3.5 \times 10^{-5}$

c. $48,300 = 4.83 \times 10^4$
d. $.0821 = 8.21 \times 10^{-2}$
e. $1,000,000,000,000 = 10^{12}$
f. $7.35 = 7.35 \times 10^0$ or just 7.35 ∎

You may have noticed that the elementary operations we've discussed can be paired as **opposite operations**. Two operations are classified as opposites if one operation followed by the other has the same result as if neither operation had been performed. For example, start with the number 6. Then perform an operation (say, add 5; the result is 11). Next, perform the opposite operation (subtract 5; $11 - 5 = 6$). The result (6) is the same as the number with which you started.

EXAMPLE 3

	Operation	Opposite operation	
a.	Open a window	Close a window	Everyday operations
b.	Take off your shoe	Put on your shoe	
c.	Raise a flag	Lower a flag	
d.	Addition	Subtraction	Elementary operations
e.	Subtraction	Addition	
f.	Multiplication	Division	
g.	Division	Multiplication	∎

Now consider the operation of raising a number to the second power, called **squaring a number**. The result is called a **perfect square** if the number you are squaring is an integer.

EXAMPLE 4

Square each of the given numbers.

a. 7; $7^2 = 49$; 7^2 is pronounced **"seven squared,"** and the number **49** is called a **perfect square**

b. -6; $(-6)^2 = 36$ **c.** 6; $6^2 = 36$
d. 13; $13^2 = 169$ **e.** 0; $0^2 = 0$ ∎

The opposite operation of squaring a number is called **finding a square root**. This means finding a number whose square is equal to the given number. For example, by studying Example 4, you can see that a square root of 49 is 7, of 169 is 13, and of 0 is 0. Also notice from Examples 4b and 4c that 36 has two square roots, 6 and -6. To avoid the possible confusion of having two answers for a particular operation, we define the symbol $\sqrt{}$ to mean the **positive square root**.

EXAMPLE 5

Find the positive square root of each of the given numbers.

a. 4; $\sqrt{4} = 2$, since $2 \times 2 = 4$

b. 49; $\sqrt{49} = 7$, since $7 \times 7 = 49$

c. $\frac{1}{4}$; $\sqrt{\frac{1}{4}} = \frac{1}{2}$, since $\frac{1}{2} \times \frac{1}{2} = \frac{1}{4}$

d. -9; $\sqrt{-9}$ doesn't exist, because no number squared can be negative:

$$\text{Pos} \times \text{Pos} = \text{Pos}$$
$$\text{Neg} \times \text{Neg} = \text{Pos}$$

e. 324; $\sqrt{324}$ If the number is not one you recognize from the times table, you can sometimes find it by trial-and-error multiplication:

$$10^2 = 100$$
$$20^2 = 400, \quad \text{so } \sqrt{324} \text{ is between } 10 \text{ and } 20$$
$$15^2 = 225, \quad \text{so it is between } 15 \text{ and } 20$$
$$17^2 = 289, \quad \text{so it is between } 17 \text{ and } 20$$
$$19^2 = 361, \quad \text{so it is between } 17 \text{ and } 19$$
$$18^2 = 324$$

$$\sqrt{324} = 18 \qquad \blacksquare$$

Example 5 illustrates finding square roots of perfect squares. What if the number is not a perfect square? To answer this question, we need to talk about a type of number called a **rational number**. If we include all possible fractions (positive and negative) with the set of integers, the resulting set is called the set of **rational numbers**. A number is rational if it can be written in a decimal form that terminates or eventually repeats.

EXAMPLE 6

Show that each number is rational by representing it as a terminating or a repeating decimal.

a. $3 = 3.$ Terminating decimal

b. $\dfrac{1}{2} = .5$ Terminating decimal

c. $\dfrac{1}{3} = .\overline{3}$ Repeating decimal

d. $\dfrac{1}{7} = .\overline{142857}$ Repeating decimal

e. $-6\frac{1}{18} = -6.0\overline{5}$ Repeating decimal

f. $\dfrac{4}{-70} = -.0\overline{571428}$ Repeating decimal   ∎

Now consider the square root of a number that is not a perfect square, say $\sqrt{2}$. You might try to represent this square root as a decimal, but you would not be successful:

$$1^2 = 1$$
$$2^2 = 4, \qquad \text{so it is between 1 and 2}$$
$$1.5^2 = 2.25, \quad \text{so it is between 1 and 1.5}$$
$$1.4^2 = 1.96, \quad \text{so it is between 1.4 and 1.5}$$
$$\vdots$$

You may even turn to a calculator:

$$\boxed{2}\ \boxed{\sqrt{x}}, \quad \text{which displays } 1.414213562$$

but $1.414213562^2 \neq 2$ so even this number is not equal to a square root of 2. It is proved in more advanced courses that $\sqrt{2}$ **is not rational** and cannot be represented as a repeating or terminating decimal! Such a number is called an **irrational number**, and is written as $\sqrt{2}$. Notice that the symbol $\sqrt{}$ was first used to indicate a process, but now it is also used to indicate a *number*. Square roots of counting numbers that are not perfect squares are irrational numbers. A few irrational numbers are

$$\sqrt{2}, \ \ \sqrt{3}, \ \ \sqrt{5}, \ \ \sqrt{6}, \ \ \sqrt{7}, \ \ \sqrt{8}, \ \ \sqrt{10}, \ \ \sqrt{11}, \ \ \sqrt{12},$$
$$\sqrt{13}, \ \ \sqrt{14}, \ \ \sqrt{15}, \ \ \sqrt{17}$$

EXAMPLE 7

Classify each given number as rational or irrational. If it is irrational, place it between two integers.

a. $\sqrt{64}$ is rational because $8^2 = 64$

b. $\sqrt{65}$ is irrational because 65 is not a perfect square; since $8^2 = 64$ and $9^2 = 81$, $\sqrt{65}$ is between 8 and 9

c. $\sqrt{441}$; $20^2 = 400$
$30^2 = 900$, so it is between 20 and 30
$25^2 = 625$, so it is between 20 and 25
$22^2 = 484$, so it is between 20 and 22
$21^2 = 441$

Rational; $\sqrt{441} = 21$

d. $\sqrt{965}$; $30^2 = 900$
$40^2 = 1,600$, so it is between 30 and 40
$32^2 = 1,024$, so it is between 30 and 32
$31^2 = 961$, so it is between 31 and 32

Consecutive counting numbers, so 965 is not a perfect square

Irrational; $\sqrt{965}$ is between 31 and 32 and cannot be represented as a repeating or terminating decimal ∎

If the square root of a number is not a whole number, you can determine whether it is rational or irrational by writing the number as a fraction and then handling the numerator and denominator separately.

EXAMPLE 8

Classify each number as rational or irrational.

a. $\sqrt{.25} = \sqrt{\dfrac{25}{100}}$

$= \dfrac{5}{10} = \dfrac{1}{2}$ Rational

b. $\sqrt{1.5625} = \sqrt{\dfrac{15,625}{10,000}}$ 15,625: $100^2 = 10,000$
$200^2 = 40,000$
$125^2 = 15,625$

$= \dfrac{125}{100}$ Rational ∎

CALCULATOR COMMENT

The remainder of this section is optional for those who have calculators. Many of the processes of this section can be simplified with a calculator. Remember that finding powers is nothing more than repeated multiplication, so even a four-function calculator can be used for exponents.

EXAMPLE 9

Find 7^5 on a calculator.

$$\boxed{7}\ \boxed{\times}\ \boxed{7}\ \boxed{\times}\ \boxed{7}\ \boxed{\times}\ \boxed{7}\ \boxed{\times}\ \boxed{7}\ \boxed{=}$$
Display: 16807.   ∎

There are two keys on some calculators that can simplify the process of finding powers. Some calculators have a constant button, $\boxed{K}$. Press $\boxed{K}$, the base, $\boxed{\times}$, and then press $\boxed{=}$ one less time than the number indicated by the exponent.

Press:	$\boxed{K}$	$\boxed{7}$	$\boxed{\times}$	$\boxed{=}$	$\boxed{=}$	$\boxed{=}$	$\boxed{=}$
Display:	0.	7.	7.	49.	343.	2401.	16807.

The most efficient way to find powers is with an exponent key, $\boxed{Y^x}$. If your calculator has this button, Example 9 can be worked by pressing

$$\boxed{7}\ \boxed{Y^x}\ \boxed{5}\ \boxed{=}$$

For finding square roots, you can use the trial-and-error procedure illustrated by Example 7 if you have a four-function calculator. However, if your calculator has a square root button, $\boxed{\sqrt{x}}$, you can find square roots quite easily.

EXAMPLE 10

Classify each number as rational or irrational. You may use a calculator.

a. $\sqrt{20.4304} = \sqrt{\dfrac{204,304}{10,000}}$ $\boxed{204304}\ \boxed{\sqrt{x}}$ *Display* 452.

$= \dfrac{452}{100} = 4.52$ Rational

b. $\sqrt{3.525} = \sqrt{\dfrac{3,525}{1,000}}$

$\boxed{3525}\ \boxed{\sqrt{x}}$ *Display*

$\boxed{59}\ \boxed{\times}\ \boxed{59}\ \boxed{=}$ 59.37171044

$\boxed{60}\ \boxed{\times}\ \boxed{60}\ \boxed{=}$ 3481.

3600.

$\sqrt{3,525}$ is between consecutive integers (59 and 60)

$\sqrt{3.525}$ is irrational.   ∎

After studying Example 10, you might wonder why not simply press

$\boxed{20.4304}\ \boxed{\sqrt{x}}$ 4.52

and

$\boxed{3.525}\ \boxed{\sqrt{x}}$ 1.87749834

You must keep in mind the directions. You were not asked to *perform the operation* of square root, but rather to *classify the numbers* as rational or irrational. Calculators *always* make irrational numbers look like rational numbers because they represent them as terminating decimals. If the numerator and denominator can be written as perfect squares, then the number is rational; if they can be placed between two *consecutive* counting numbers, then the number is irrational.

However, sometimes it is necessary to limit your work to the set of rational numbers; in these cases you'll need to approximate an irrational number with a rational number. You can do this by:

1. Successive approximations (trial multiplication)
2. Table (see Table 12.3 on page 443)
3. Calculator

EXAMPLE 11

Approximate $\sqrt{18}$ with a rational number of two decimal places.

Solution

Method I.

$$4^2 = 16$$
$$5^2 = 25,$$ so it is between 4 and 5

$$4.2^2 = 17.64$$
$$4.3^2 = 18.49,$$ so it is between 4.2 and 4.3

$$4.24^2 = 17.9776$$
$$4.25^2 = 18.0625,$$ so it is between 4.24 and 4.25

$$4.245^2 = 18.020025,$$ so it is closer to 4.24

Method II. Use Table 12.3 on page 443.
$\sqrt{18}$ is 4.243 in the table; round answer to two decimal places: 4.24

Method III. Use a calculator.

$$\boxed{18}\ \boxed{\sqrt{x}} \qquad \textit{Display:}\quad 4.242640687$$

Round to two decimal places: 4.24

Method I can be used if you have a calculator without a square root button.   ∎

In algebra, we group together all the rational numbers and all the irrational numbers into a set called the set of **real numbers**.

Problem Set 3.2

1. Consider the number 10^6.
    a. What is the common name for this number?
    b. What is the base?
    c. What is the exponent?
    d. According to the definition of exponential notation, what does the number mean?

2. Consider the number 10^{-1}.
    a. What is the common name for this number?
    b. What is the base?
    c. What is the exponent?
    d. According to the definition of exponential notation, what does the number mean?

Write each of the numbers in Problems 3–11 in scientific notation.

3. a. 3,200   b. 25,000
    c. 18,000,000   d. 640

4. a. .004   b. .02
    c. .0035   d. .00000 045

5. a. 5,624   b. 23.79
    c. 24.006   d. .00081 7

6. a. 35,000,000,000   b. 63,000,000
    c. .00001   d. .00000 00000 00000 00003 5

7. a. .00008 61   b. 249,000,000
    c. 100   d. $11\frac{1}{2}$

8. A light year is the distance that light travels in one year; this is about 5,869,713,600,000 miles.

9. The world's largest library, the Library of Congress, has approximately 59,000,000 items.

10. In 1979, Americans spent $212.2 billion on health care.

11. A thermochemical calorie is about 41,840,000 ergs.

Write each of the numbers in Problems 12–20 without using exponents.

12. a. 7.2×10^{10} b. 4.5×10^{3} c. 3.1×10^{2}
    d. 6.8×10^{8}

13. a. 2.1×10^{-3} b. 4.6×10^{-7} c. 2.05×10^{-1}
    d. 3.013×10^{-2}

14. a. 3.2×10^{0} b. 8.03×10^{-4} c. 5.06×10^{3}
    d. 6.81×10^{0}

15. a. 7^{2} b. 5^{2} c. 2^{6} d. 6^{3}

16. a. 4^{3} b. 2^{5} c. 10^{4} d. 8^{2}

17. The estimated age of the earth is about 5×10^{9} years.

18. Saturn is about 8.86×10^{8} miles from the sun.

19. The mass of the sun is about 3.33×10^{5} times the mass of the earth.

20. The sun develops about 5×10^{23} horsepower per second.

WHAT IS A TRILLION?

How MUCH IS A TRILLION? It is 1,000,000,000,000, or a million millions. A trillion inches is more than 15.8 million miles or half the distance to Venus. The moon is about 230,000 miles, so a trillion *inches* is about 68 round trips to the moon. A trillion seconds is about 81,700 years!

21. A **million**, 1,000,000, is a large number. A million days ago, Christ walked on the earth. At 15% interest, a million dollars would bring about $411 per day income forever! Write a million in scientific notation.

22. A **billion**, 1,000,000,000, is a large number. A billion seconds ago, we were in World War II. At 15% interest, a billion dollars would bring about $17,123 per hour income forever! Write a billion in scientific notation.

23. A **trillion** is described in the news article. Write a trillion in scientific notation.

24. A **googol** is a very large number which is defined in the cartoon. Write a googol in scientific notation.

Find the positive square root of each of the numbers given in Problems 25–30.

25. **a.** 9   **b.** 1   **c.** 0   **d.** −9

26. **a.** 81   **b.** −25   **c.** 169   **d.** 196

27. **a.** 225   **b.** $\dfrac{25}{36}$ **c.** $\dfrac{100}{144}$ **d.** $\dfrac{9}{49}$

28. **a.** 36   **b.** 25   **c.** −16   **d.** 625

29. **a.** 1,225   **b.** 2,025   **c.** 9,604   **d.** 6,084

30. **a.** 10,000   **b.** 1,089   **c.** 10,404   **d.** 3,364

Classify each of the numbers given in Problems 31–36 as rational or irrational. If the number is rational, write it as a terminating or repeating decimal. If it is irrational, place it between two integers.

31. **a.** 5   **b.** $\sqrt{5}$ **c.** $\sqrt{25}$ **d.** $\dfrac{1}{4}$

32. **a.** $\sqrt{\dfrac{1}{4}}$ **b.** $\sqrt{\dfrac{1}{9}}$ **c.** $-2\tfrac{7}{10}$ **d.** $\sqrt{.49}$

33. **a.** $\sqrt{10}$ **b.** $\sqrt{15}$ **c.** $\sqrt{16}$ **d.** $\sqrt{17}$

34. **a.** $\dfrac{1}{36}$ **b.** $\sqrt{784}$ **c.** $\sqrt{580}$ **d.** $\sqrt{2,400}$

35. **a.** $\sqrt{2,401}$ **b.** $\sqrt{2,402}$ **c.** $\sqrt{18.49}$ **d.** $\sqrt{12.4}$
36. **a.** $\sqrt{12.3904}$ **b.** $\sqrt{1.2321}$ **c.** $\sqrt{1.2}$ **d.** $\sqrt{\dfrac{1}{10}}$

CALCULATOR PROBLEMS

Approximate each irrational number in Problems 37–41 with a rational number of two decimal places.

37. **a.** $\sqrt{15}$ **b.** $\sqrt{17}$ **c.** $\sqrt{20}$
38. **a.** $\sqrt{30}$ **b.** $\sqrt{40}$ **c.** $\sqrt{80}$
39. **a.** $\sqrt{123}$ **b.** $\sqrt{119}$ **c.** $\sqrt{150}$
40. **a.** $\sqrt{190}$ **b.** $\sqrt{1,000}$ **c.** $\sqrt{2,000}$
41. **a.** $\sqrt{250}$ **b.** $\sqrt{875}$ **c.** $\sqrt{4,210}$

Mind Bogglers

42. If it took one second to write down each digit, how long would it take to write all the numbers from 1 to 1,000,000?

43. Imagine that you have written down the numbers from 1 to 1,000,000. What is the total number of zeros you have recorded?

3.3 Large Numbers

Scientific notation is only one method for handling large and small numbers. In the 17th century a Scottish mathematician named John Napier discovered a property of numbers that simplified the multiplication process and allowed him to handle very large numbers. We'll see what he did by considering an old story.

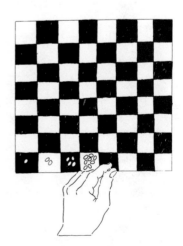

> The story is about a king who, being under obligation to one of his subjects, offered to reward him in any way the subject desired. Being of mathematical mind and modest tastes, the subject simply asked for a chessboard with one grain of wheat on the first square, two on the next square, four on the following square, and so forth. The old king was delighted with this request because he had a beautiful daughter and had feared the subject would ask for her hand in marriage. However, the king was soon sorry he had granted the request.

Suppose we number the squares from 0 to 63.

FIRST SEQUENCE																		
Number of squares on chessboard:	0	1	2	3	4	5	6	7	8	9	10	11	12	13	14	15	16	...
SECOND SEQUENCE																		
Number of grains:	1	2	4	8	16	32	64	128	256	512	1,024	2,048	4,096	8,192	16,384	32,768	65,536	...

You will recognize the first sequence as an arithmetic sequence with a common difference of 1, and the second sequence as a geometric sequence with a common ratio of 2. Notice, as did Napier, that if you multiply two numbers in the second sequence, such as

$$8 \times 32 = 256$$

the result is someplace else in that same sequence. Moreover, the location can be found by **adding the corresponding numbers** in the first sequence:

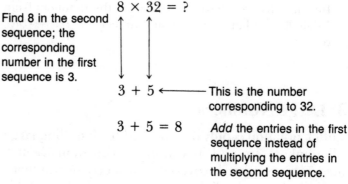

Find 8 in the second sequence; the corresponding number in the first sequence is 3.

$8 \times 32 = ?$

$3 + 5 \leftarrow$ ——— This is the number corresponding to 32.

$3 + 5 = 8$ *Add* the entries in the first sequence instead of multiplying the entries in the second sequence.

The answer is the number that corresponds to 8, which is 256. Thus,

$$8 \times 32 = 256 \qquad \text{Found by addition, not multiplication}$$

Napier called the numbers in the first sequence the **logarithms** of the numbers in the second sequence. Since the common ratio is 2, these are called **logarithms to the base two**.

EXAMPLE 1

a. The logarithm to the base two of 8 is 3.
b. The logarithm to the base two of 32 is 5.

c. The logarithm to the base two of 256 is 8.
d. The logarithm to the base two of 4,096 is 12.
e. The logarithm to the base two of 32,768 is 15. ■

We now can use logarithms to reduce multiplication to addition by using Napier's addition property of logarithms, as shown in Example 2.

EXAMPLE 2

a. 32×128

The logarithm to the base two of 32 is: $\quad 5$
The logarithm to the base two of 128 is: $\quad \underline{7}$
$\qquad\qquad\qquad\qquad Add$ logarithms: $\quad \mathbf{12}$
$\qquad\qquad\qquad\qquad\qquad\qquad\qquad \uparrow$

Now find this number in the first sequence. The answer is the corresponding number in the second sequence.

The number whose logarithm to the base two is 12 is 4,096.

b. $2{,}048 \times 16$

$\quad \downarrow \qquad \downarrow$
$\quad 11 \; + \; 4 \; = \mathbf{15}$ $\qquad$ Find the corresponding number in the
$\qquad \uparrow$ $\qquad\qquad\qquad\qquad$ second sequence.
Add logarithms

Thus,

$$2{,}048 \times 16 = 32{,}768$$

c. $32 \times 64 \times 8 = \mathbf{16{,}384}$
$\quad \downarrow \quad\;\; \downarrow \quad\; \downarrow \quad\;\; \uparrow$ $\qquad$ Look up corresponding entry.
$\quad 5 \; + \; 6 \; + \; 3 \; = \mathbf{14}$

d. $(256)^2 = 256 \times 256 = \mathbf{65{,}536}$
$\qquad\qquad\;\; \downarrow \qquad\;\; \downarrow \quad\;\; \uparrow$
$\qquad\qquad\;\; 8 \; + \; 8 \; = \mathbf{16}$ ■

Before we leave logarithms to the base two, let's find out the number of grains of wheat on the last square of the chessboard problem. Look at the pattern—since a chessboard has 64 squares, the last square alone will have 2^{63} grains. Do you have any idea how big this is? I used a computer to find 2^{63}, which I thought you might like to see multiplied out:

$$2^{63} = 9{,}223{,}372{,}036{,}854{,}775{,}808$$

and I did some calculations just to give you an idea about the size of this big number. I purchased some raw wheat at a health food store and counted out 250 grains per cubic inch. Next, I calculated that there are 2,150 cubic inches per bushel, or 537,500 grains of wheat in a bushel. This means 2^{63} grains of wheat equal about 17,383,000,000,000 bushels. For the last several years, the annual output of wheat in the United States has been about 1.25 billion bushels. At this rate of production, it would take the United States almost 14,000 years to satisfy the requirements for the last square of the chessboard alone (let alone the rest of the chessboard)! The story goes no further. The chances are that the king lost his temper and the subject lost his head before the 64th square of the chessboard was reached.

The primary difficulty with working with base two is the fact that it is difficult to see what the additional numbers in the sequence are. For this reason, it is much more common to use **base ten**. When working with base ten, we simply write **log** to mean **logarithm to the base 10**. Let's see how the sequences for base 10 begin:

Number:	1	10	100	1,000	10,000	100,000	1,000,000	10,000,000	100,000,000	1,000,000,000	...
Logarithm:	0	1	2	3	4	5	6	7	8	9	...

EXAMPLE 3

a. $\log 1{,}000 = 3$
b. $\log 100 = 2$
c. $\log 100{,}000 = 5$
d. $\log 10^6 = 6$ *Note:* $10^6 = 1{,}000{,}000$
e. $\log 10^9 = 9$ ■

Notice an additional pattern. The logarithm to the base 10 is the same as the exponent on the number 10. Logarithms to the base 10 are called **common logarithms**.

> A **common logarithm** is a logarithm to the base 10 and   **Common Logarithm**
> is defined as the exponent on 10. For example, the
> logarithm of 3 is written as **log 3** and is defined as the
> exponent of 10 that gives 3.

In arithmetic you learned shortcuts for multiplying numbers with a lot of zeros. The reason those shortcuts work is because of the pattern of logarithms.

EXAMPLE 4

Find $1,000 \times 10,000$ using logarithms.

Solution

$$\log 1,000 = 3 \qquad \text{and} \qquad \log 10,000 = 4$$

so $1,000 \times 10,000$ is the number whose log is
$3 + 4 = 7$: $10,000,000$.   ∎

About now an anxiety attack may have set in because you are probably wondering why all the bother for multiplying easy numbers like $1,000 \times 10,000$. How about some hard numbers? Or even how about multiplying 2×3 using logarithms? We first need to find the logarithms of 2 and 3.

Number:	1	2	3	4	5	6	7	8	9	10
Logarithm:	0									1

It looks like the logarithms of these numbers should be between 0 and 1. Mathematicians have worked these out, as shown below.

Number:	1	2	3	4	5	6	7	8	9	10
Logarithm:	0.000	0.301	0.477	0.602	0.699	0.778	0.845	0.903	0.954	1.000

Thus, $\log 2 = 0.301$ and $\log 3 = 0.477$, so 2×3 is the number whose log is $0.301 + 0.477 = 0.778$; you can see from the table above that this is 6.

TABLE 3.1   Logarithm Table

No.	Log	No.	Log	No.	Log	No.	Log	No.	Log
1.0	0.000	3.0	0.477	5.0	0.699	7.0	0.845	9.0	0.954
1.1	0.041	3.1	0.491	5.1	0.708	7.1	0.851	9.1	0.959
1.2	0.079	3.2	0.505	5.2	0.716	7.2	0.857	9.2	0.964
1.3	0.114	3.3	0.519	5.3	0.724	7.3	0.863	9.3	0.968
1.4	0.146	3.4	0.531	5.4	0.732	7.4	0.869	9.4	0.973
1.5	0.176	3.5	0.544	5.5	0.740	7.5	0.875	9.5	0.978
1.6	0.204	3.6	0.556	5.6	0.748	7.6	0.881	9.6	0.982
1.7	0.230	3.7	0.568	5.7	0.756	7.7	0.886	9.7	0.987
1.8	0.255	3.8	0.580	5.8	0.763	7.8	0.892	9.8	0.991
1.9	0.279	3.9	0.591	5.9	0.771	7.9	0.898	9.9	0.996
2.0	0.301	4.0	0.602	6.0	0.778	8.0	0.903	10.0	1.000
2.1	0.322	4.1	0.613	6.1	0.785	8.1	0.908		
2.2	0.342	4.2	0.623	6.2	0.792	8.2	0.914		
2.3	0.362	4.3	0.633	6.3	0.799	8.3	0.919		
2.4	0.380	4.4	0.643	6.4	0.806	8.4	0.924		
2.5	0.398	4.5	0.653	6.5	0.813	8.5	0.929		
2.6	0.415	4.6	0.663	6.6	0.820	8.6	0.934		
2.7	0.431	4.7	0.672	6.7	0.826	8.7	0.940		
2.8	0.447	4.8	0.681	6.8	0.833	8.8	0.944		
2.9	0.462	4.9	0.690	6.9	0.839	8.9	0.949		

Tables for logarithms to the base 10 for numbers between 1 and 10 are readily available. A simple one is shown in Table 3.1.

EXAMPLE 5

Find:

a. log 5   **b.** log 2.4   **c.** log 1.7   **d.** log 9.3
e. log 1.4873

Solution

a. log 5 = 0.699   Look at Table 3.1 to find these logs.
b. log 2.4 = 0.380
c. log 1.7 = 0.230
d. log 9.3 = 0.968

e. To use Table 3.1, you will need to round the given number to one decimal place: $1.4873 \approx 1.5$. (Notice that the symbol $\approx$ is used when rounding because the numbers 1.4873 and 1.5 are not exactly the same. That is, $\approx$ means "approximately equal to.") Thus,

$$\log 1.4873 \approx \log 1.5 = .176 \qquad \blacksquare$$

Now, to find the logarithms of really big numbers (in the case of Table 3.1, any number larger than 10), we combine the ideas of scientific notation, Napier's addition property of logarithms, and Table 3.1.

EXAMPLE 6

Find:

a. $\log 38,000$ **b.** $\log 293,612$ **c.** $\log 93,000,000$
d. $\log 2^{63}$

Solution

a. Step 1. Write in scientific notation:
$$38,000 = 3.8 \times 10^4$$
Step 2. Use Table 3.1:
$$\log 3.8 = 0.580$$
$$\underline{\log 10^4 = 4}$$
Step 3. Add: 4.580
Thus, $\log 38,000 = 4.580$.

b. Step 1. $293,612 = 2.93612 \times 10^5$
Step 2. $\log 2.93612 \approx \log 3.0 = 0.477$
$$\underline{\log 10^5 = 5}$$
Step 3. Add: 5.477
Thus, $\log 293,612 = 5.477$.

c. The steps are often combined as shown below.

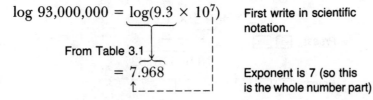

$$\log 93,000,000 = \log(9.3 \times 10^7)$$ First write in scientific notation.

$$= 7.968$$ Exponent is 7 (so this is the whole number part)

d. $\log 2^{63} \approx \log(9.2 \times 10^{18})$ This was worked out for you earlier in this section—you would not be expected to do this step.

From Table 3.1

$= 18.964$

Exponent is 18   ∎

Do you see why logarithms are used to represent big numbers? A number whose logarithm is 18 is a very large number. However, logarithms are not only used to represent large numbers but are important in a variety of applications. Some of these applications are considered in the next section.

CALCULATOR COMMENT

Logarithmic tables are used less often today because of the widespread availability of calculators. If you don't have a calculator, then skip to the problem set now. If you do have a calculator with a key marked $\boxed{\text{LOG}}$, then you don't need to use scientific notation or log tables—you can simply press a single calculator button. If you use a calculator, your answers will be more accurate than if you use Table 3.1.

EXAMPLE 7

Find the logs of Example 6 using a calculator.

a. $\boxed{38,000}\ \boxed{\text{LOG}}$ 4.57978360
b. $\boxed{293,612}\ \boxed{\text{LOG}}$ 5.46777380
c. $\boxed{93,000,000}\ \boxed{\text{LOG}}$ 7.96848295
d. Most calculators will not accept

$$9,223,372,036,854,775,808$$

directly, so you can write

$$2^{63} \approx 9.2 \times 10^{18}$$

Press: $\boxed{9.2}\ \boxed{\text{LOG}}$

≈ 18.963787827

Press: $\boxed{+}\ \boxed{18}\ \boxed{=}$

If your calculator has an $\boxed{Y^x}$ key, you can calculate this log directly:

$\boxed{2}$ $\boxed{Y^x}$ $\boxed{63}$ $\boxed{=}$ $\boxed{LOG}$ 18.96488973   ∎

Problem Set 3.3

1. In the chessboard problem, we showed the number of grains on the first sixteen squares. Continue this table for the next four squares.

Use logarithms to the base two to find the products in Problems 2–13 without direct multiplication.

2. 16×32 3. 16×64 4. 128×512

5. $1,024 \times 32$ 6. $1,024 \times 64$ 7. $1,024 \times 256$

8. $1,024 \times 512$ 9. 16^2 10. 64^2

11. 256^2 12. 64^3 13. 32^4

Find the logarithms requested in Problems 14–33 by using Table 3.1 or a calculator.

14. $\log 10^2$ 15. $\log 10^5$
16. $\log 10^9$ 17. $\log 10^8$
18. $\log 10,000$ 19. $\log 10$
20. $\log 1,000$ 21. $\log 10,000,000$
22. $\log 4$ 23. $\log 2$
24. $\log 3.5$ 25. $\log 7.7$
26. $\log 9.62$ 27. $\log 1.38$
28. $\log 2.7183$ 29. $\log 3.1416$
30. $\log 24,000$ 31. $\log 84,000,000$
32. $\log 128,000,000$ 33. $\log 143,000$

34. A googol was defined in the *Peanuts* cartoon on page 114. What is the logarithm to the base 10 of a googol?

35. Use the table on page 116 to find $\sqrt{4,096}$.

36. Use the table on page 116 to find $\sqrt{16,384}$.

37. Build a logarithmic table using base 3 by filling in the blanks:

Number:	3	9	27	a.	b.	c.	d.	e.	19,683	59,049
Logarithm:	1	2	3	4	5	6	7	8	9	10

38. Use the table in Problem 37 to find 81×27.

39. Use the table in Problem 37 to find $(243)^2$.

40. Name the largest of the following numbers:

48,385,426,712,790; 5.14×10^{13};   the number whose log is 13

41. Name the largest of the following numbers:

198,345,285,612,000,000,000; 5.8×10^{18};   the number
whose log is 21

Mind Bogglers

42. Approximately how high would a stack of one million $1 bills be? There are about 233 new $1 bills per inch.

43. Estimate how many pennies it would take to make a stack one inch high. A handful of pennies and a ruler will help you. Approximately how high would a stack of one million pennies be?

44. A sheet of notebook paper is approximately .003 inch thick. Tear a sheet of paper in half and place the two halves in a pile. Repeat, so that there are 4 sheets. Repeat again, so that there is a pile of 8 sheets. If it were possible to continue in this fashion until the paper has been halved 50 times, how high would you guess the final pile would be? Having guessed, *compute* the height.   [*Hint:* $2^{50} = 1,125,899,906,842,624$]

3.4 Earthquakes, Rock Concerts, and Altitudes

The variety of applications for the logarithm of a number is almost limitless. In this section we'll see this special type of number applied to three very different applications—the Richter scale, the intensity of sound, and altitude as measured by atmospheric pressure.

The strength of an earthquake is measured on a logarithmic scale called the **Richter scale**. It was developed by Gutenberg and Richter in 1935 and relates the energy, E, released by an earthquake in terms of a specific magnitude. These **Richter numbers** are shown in Table 3.2.

QUAKES

Waverly Person, a geophysicist at the U.S. Geological Survey's Earthquake Information Center in Colorado, said a quake is considered "significant if it is a magnitude 6.5 or greater on the Richter Scale, or if it causes extensive damage, fatalities, or injuries."

The quake that hit Italy Nov. 23 officially registered 6.8 on the open-ended, logarithmic scale and is considered significant on all counts.

So far this year, there have been 65 significant quakes, compared to 58 last year and a long-term yearly average of 50 to 60, Person said.

However, there have been only 12 readings of 7 or greater on the Richter Scale—the threshold for a major quake. That compares to a long-term yearly average of about 19 major quakes and the even dozen recorded last year, he said.

The year's biggest quake—reading 7.3 with a 6.2 aftershock—struck Algeria Oct. 10, killing more than 3,000 persons and devastating a city. The quake in southern Italy has killed at least 3,000 persons, with 2,000 more missing and also presumed dead.

This year's earthquake death toll thus is well ahead of 1979, when 1,479 persons died, Person said. But the long-term quake toll has been about 10,000 deaths a year.

For instance, the 8.0 jolt that leveled Tang-shan city in northeast China in 1976 killed an estimated 250,000 persons. The 1972 Managua, Nicaragua, quake, which registered 6.2, took an estimated 5,000 lives.

TABLE 3.2 Richter Scale*

Richter number	Effects of earthquake
1	Only detectable by a seismograph
2	Felt by a few persons at rest
3	Felt by many, especially those indoors
4	Felt by all; buildings shake, glass breaks
5	Furniture falls; chimneys break
6	Damage to wooden houses; ground cracked
7	Buildings fall; damage to dams and bridges
8	Catastrophic damage over wide area

*The Richter scale, used by seismologists to measure the magnitude of earthquakes, operates on a logarithmic scale so that each Richter number measures an earthquake that is 10 times more powerful than an earthquake with the next smaller number. A magnitude of 6 is 10 times as great as an earthquake of magnitude 5 and 100 times greater than an earthquake of magnitude 4.

In physics, energy is measured using a unit called an **erg**. If the amount of energy, E, released by an earthquake is known, then the Richter number, M, can be found by using the following formula:

$$M = 0.67 \log E - 7.9$$

This formula enables you to find the magnitude by following these steps:

1. Find the logarithm of the amount of energy, measured in ergs.
2. Multiply by 0.67.
3. Subtract 7.9.

EXAMPLE 1

A small earthquake is one that releases about 10^{17} ergs of energy. What is the Richter number of such an earthquake?

Solution

Step 1. $\log 10^{17} = 17$
Step 2. $0.67 \times 17 = 11.39$
Step 3. $11.39 - 7.9 = 3.49$

This earthquake would have a magnitude of about 3.5 on the Richter scale. ∎

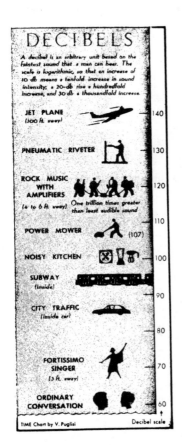

Another application of logarithms involves measuring the intensity of sound. This scale, like the Richter scale, is logarithmic. The unit of intensity of sound is called the **decibel**, and is named after Alexander Graham Bell, the inventor of the telephone. An increase of 10 decibels means that the sound is 10 times louder, and an increase of 20 decibels means that the sound is $10 \times 10 = 100$ times louder. The clipping from *Time* magazine explains this scale and gives some examples.

In physics, the intensity of sound is measured in watts per cubic centimeter. If the intensity of a sound, I, is known, then the number of decibels, D, is found according to the following formula:

$$D = 160 + 10 \log I$$

To use this formula, follow these steps:

1. Find the logarithm of the intensity of sound, measured in watts per cubic centimeter.
2. Multiply by 10.
3. Add 160.

EXAMPLE 2

Suppose that it is found that the intensity of the sound of a certain vacuum cleaner is 8.3×10^{-9} watts per cubic centimeter. What is the intensity of sound measured in decibels?

Solution

Step 1. $\log(8.3 \times 10^{-9}) = \log 8.3 + \log 10^{-9}$

Addition property of logarithms

$= 0.919 + (-9)$

Exponent on 10

From Table 3.1

$= -8.081$

Step 2. $10 \times -8.081 = -80.81$ Multiply by 10.
Step 3. $-80.81 + 160 = 79.19$ Add 160.

The intensity of sound for the vacuum cleaner is about 79 decibels. ∎

EXAMPLE 3

If the intensity of sound of a rock concert is 1.1×10^{-5} watts per cubic centimeter, what is the decibel rating, and how many times louder is it than the vacuum cleaner of Example 2?

Solution

Step 1. $\log(1.1 \times 10^{-5}) = \log 1.1 + \log 10^{-5}$

$= 0.041 + (-5)$

Look up 1.1 in Table 3.1.

$= -4.959$

Step 2. $10 \times (-4.959) = -49.59$
Step 3. $-49.59 + 160 = 110.41$

The intensity of sound for this rock concert is about 110 decibels. To compare this with the vacuum cleaner, compare 80 decibels with 110 decibels:

80 to 90;   90 to 100;   100 to 110

10   ×   10   ×   10   = 1,000

The rock concert is about 1,000 times louder than a vacuum cleaner. ∎

The last application of logarithms we will discuss here involves finding the altitude above sea level. The altitude, a, measured in miles above sea level, is related to the atmospheric pressure, P, according to the following formula:

$$a = -11 \log P + 12.8$$

To use this formula, carry out the following steps:

1. Find the logarithm of the atmospheric pressure, measured in pounds per square inch.
2. Multiply by -11.
3. Add 12.8.

EXAMPLE 4

If the atmospheric pressure of Denver is 11.92 pounds per square inch, what is the approximate altitude of this city?

Solution

Step 1. $\log 11.92 = \log(1.192 \times 10)$ Write in scientific
$\qquad\qquad = \log 1.192 + \log 10$ notation.
$\qquad\qquad = 0.079 + 1$
$\qquad\qquad\qquad\quad \uparrow$ Look up log 1.2
$\qquad\qquad = 1.079$ in Table 3.1.

Step 2. $-11 \times 1.079 = -11.869$
Step 3. $-11.869 + 12.8 = 0.931$

The altitude is about 0.93 mile. Do you see why Denver is called the mile-high city? ■

CALCULATOR COMMENT

EXAMPLE 5

If the atmospheric pressure outside an airplane is 7.83 pounds per square inch, what is the approximate altitude in feet?

Solution

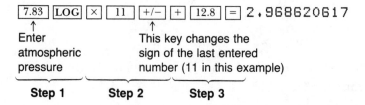

| Step 1 | Step 2 | Step 3 |

To change to feet, multiply by 5,280:

$$\boxed{\times}\ \boxed{5280}\ \boxed{=}\ \ 15674.31686$$

The altitude of the plane is about 16,000 ft.   ■

Problem Set 3.4

Earthquakes

1. A large earthquake is one that releases about 10^{25} ergs of energy. What is the magnitude of such an earthquake on the Richter scale?

2. A medium-sized earthquake is one that releases about 10^{20} ergs of energy. What is the magnitude of such an earthquake on the Richter scale?

Find the magnitude (rounded to the nearest tenth) of the earthquakes described in Problems 3–9.

3. The worst European earthquake since 1915 hit Naples, Italy, in 1980. The energy released was estimated to be 10^{22} ergs.

4. The 1980 Eureka earthquake released 2×10^{22} ergs of energy. It did very little damage because it was centered in the ocean.

5. The 1971 San Fernando earthquake released 5×10^{21} ergs of energy. The death toll was 64, and the damage was estimated at $1 billion.

6. One of the most famous earthquakes is the 1906 San Francisco earthquake. It released 1.78×10^{24} ergs of energy and killed over 500 people.

7. The 1964 Alaska earthquake was the largest ever to strike North America. The energy released was estimated to be 3.5×10^{24} ergs.

8. A 1952 Bakersfield earthquake released 2.24×10^{23} ergs of energy.

9. A 1981 Santa Rosa earthquake released 3.98×10^{16} ergs of energy.

Intensity of Sound

Find the intensity (in decibels) of the sounds described in Problems 10–16.

10. The intensity of a whisper is 10^{-13} watts per cubic centimeter.

11. The intensity of the sound inside a new car is 10^{-8} watts per cubic centimeter.

12. The intensity of the world's loudest shout was 10^{-5} watts per cubic centimeter. This record was set by Skipper Kenny Leader.

13. The intensity of the sound of normal conversation is 3.16×10^{-10} watts per cubic centimeter.

14. The intensity of the sound of a rototiller is 2.51×10^{-6} watts per cubic centimeter.

15. The intensity of the sound of a rock concert is 5.23×10^{-6} watts per cubic centimeter.

16. The intensity of sounds during the 1981 Rolling Stones tour of the United States was measured as 3.16×10^{-4} watts per cubic centimeter.

ALTITUDE ABOVE SEA LEVEL— CALCULATOR APPLICATIONS

17. What is the altitude of a city with an atmospheric pressure of 13 pounds per square inch?

18. What is the altitude of an airplane with an atmospheric pressure of 4 pounds per square inch?

19. What is the altitude of a city with an atmospheric pressure of 14.58 pounds per square inch?

20. What is the altitude of a city with an atmospheric pressure of 10.2 pounds per square inch?

21. What is the altitude of an airplane with an atmospheric pressure of 6.5 pounds per square inch?

22. What is the altitude of an airplane with an atmospheric pressure of 3.82 pounds per square inch?

23. What is the altitude of a village with an atmospheric pressure of 9.85 pounds per square inch?

pH of a Substance

Many shampoos and even one new acne treatment advertise that they have the "proper pH." What does the pH of a substance mean? The pH of a substance measures the amount of acidity or alkalinity. It is found by the formula

$$pH = -\log H^+$$

where H^+ is the concentration of hydrogen ions in a certain amount of the liquid. The hydrogen ion concentration can be found from a chemical analysis. Given the H^+ values in Problems 24–30, find the corresponding pH values.

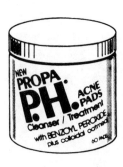

24. A shampoo has a concentration of hydrogen ions of 10^{-7}; what is the pH?

25. A soft drink has a concentration of hydrogen ions of 10^{-3}; what is the pH?

26. Milk has a concentration of hydrogen ions of 3.98×10^{-7}; what is the pH?

27. Rainwater has a concentration of hydrogen ions of 6.31×10^{-7}; what is the pH?

28. Seawater has a concentration of hydrogen ions of 3.16×10^{-9}; what is the pH?

29. Vinegar has a concentration of hydrogen ions of 1.26×10^{-3}; what is the pH?

30. Tomatoes have a concentration of hydrogen ions of 6.31×10^{-5}; what is the pH?

Mind Bogglers

POPULATION GROWTH—CALCULATOR APPLICATIONS

You can find the growth rate for a population in a certain geographical area if you have two population figures and carry out the following steps:

Step 1. *Divide the initial (or beginning) population into the final (or ending) population.*

Step 2. *Find the log of the quotient.*

Step 3. *Multiply by 2.3.*

Step 4. *Divide by the length of time between the initial and the ending population figures.*

Step 5. *Multiply by 100.*

The result is the growth rate as a percent.

31. The world population reached 3 billion in 1930 and 4 billion in 1976. What is the growth rate for this period of time?

32. If San Jose, California, grew from 459,913 in 1970 to 625,763 in 1980, what was the growth rate for this period?

33. If Boston, Massachusetts, declined from 641,071 in 1970 to 562,118 in 1980, what was the growth rate for this period? [*Note:* A negative growth rate means the population declined.]

34. From your local Chamber of Commerce, obtain the population figures for your city or state for 1960, 1970, and 1980. Find the growth rate for each period of time.

3.5 Review Problems

Section 3.1   **1.** Perform the indicated operations.
   a. $2 + 3 - 7 + 2 - 5$ **b.** $6 - 8 - 10 + 2 - 5$
   c. $-15 + 15 \div (-5)$ **d.** $4(8 - 12)$

Section 3.2   **2.** Write in scientific notation.
   a. .0034   **b.** 4,000,300   **c.** 17,400   **d.** 5

Section 3.2   **3.** Write without exponents.
   a. 4^3 **b.** 9^2 **c.** 5.79×10^{-4} **d.** 4.01×10^5

Section 3.2   **4.** Find the positive square root of each number.
   a. 100   **b.** -16 **c.** 0   **d.** 7,569

Section 3.2   **5.** Classify each number as rational or irrational. If the number is rational, write it as a terminating or a repeating decimal. If it is irrational, place it between two consecutive integers.
   a. $\sqrt{529}$ **b.** $\sqrt{1,000}$ **c.** $\sqrt{\dfrac{4}{9}}$ **d.** $\sqrt{6.76}$

Section 3.3   *Find the logarithms in Problems 6 and 7. You may use Table 3.1 on page 120 or a calculator.*

   6. **a.** $\log 10^{12}$ **b.** $\log 10,000$ **c.** $\log 8$
   d. $\log 3.5$ **e.** $\log 1.6483$

   7. **a.** $\log 43,000$ **b.** $\log 418,000,000$
   c. $\log(1.28 \times 10^9)$ **d.** $\log(4.5 \times 10^{-3})$

Section 3.4   **8.** In 1933, there was an earthquake in Long Beach, California, that released 2.2×10^{17} ergs of energy. What was the Richter number for this earthquake?

Section 3.4   **9.** If the intensity of sound from a jet takeoff is 10^{-2} watts per cubic centimeter, what is the decibel rating?

Section 3.4   **10.** **Calculator problem.** What is the altitude (to the nearest 100 ft) of a city with an atmospheric pressure of 11.9 pounds per square inch?

CHAPTER 4

Algebra and
Problem Solving

4.1 Symbol Shock

In arithmetic, you learned about numbers and operations with numbers. Elementary algebra is also a study of numbers and their properties, but it begins where arithmetic leaves off. In algebra, extensive use of symbols is made, and that is where many of us go into a type of "symbol shock." However, our society is filled with symbols that we've learned to use intuitively. Books have been written to help us interpret the symbols of body language, we can learn to interpret our dreams, and the very language we use is a symbolic representation. For example, the letters FACE can stand for a variety of ideas, depending on the context in which they are used.

FACE

FACE

F A C E

The cartoon in the margin uses letters to represent words; can you read it? In algebra, we use letters of the alphabet to represent numbers with unknown values. A letter used in this way is called a **variable**.

I M A U-M B-N.

U R N N-M-L.

EXAMPLE 1

Think of a counting number from one to ten. Add five. Multiply the result by two. Subtract six. Divide by two. Subtract the original number. The result is two. Why does this number trick work?

Solution

Without variables
Let □ represent a box containing the number you have chosen (its value is unknown to us). Let stars (☆ ☆ · · ·) represent the numbers stated in the question (known numbers).

With variables
Let n = UNKNOWN NUMBER

Think of a number:

$$\square \qquad\qquad\qquad n$$

Add five:

$$\square\;\boxed{\star\star\star\star\star} \qquad\qquad n + 5$$

Multiply by two:

$$\square\;\star\star\star\star\star$$
$$\boxed{\square\;\star\star\star\star\star}$$

$$2(n + 5) = 2n + 10$$
Distributive property

Subtract six:

$$\square\;\star\star\boxed{\star\star\star} = \square\;\star\star$$
$$\square\;\star\star\boxed{\star\star\star} = \square\;\star\star$$

$$2n + 10 - 6 = 2n + 4$$
$$\qquad\qquad\qquad = 2(n + 2)$$
Distributive property

Divide by two:

$$\dfrac{\square\;\star\star}{\boxed{\square\;\star\star}} = \square\;\star\star$$

$$\dfrac{2(n + 2)}{2} = n + 2$$

Subtract original number:

$$\boxed{\square}\;\star\star = \star\star$$

$$n + 2 - n = 2 \qquad \blacksquare$$

As you look at Example 1, which solution seems easier, the one with or without variables? At this point, I would expect that the one without variables is easier for you to understand. Our study of algebra is a study in becoming familiar enough with the variables to feel comfortable with manipulations like those shown at the right in Example 1.

A variable represents a number from a given set of numbers called the **domain** of the variable. To express the idea "The sum of a number and two" you can write

$$n + 2$$

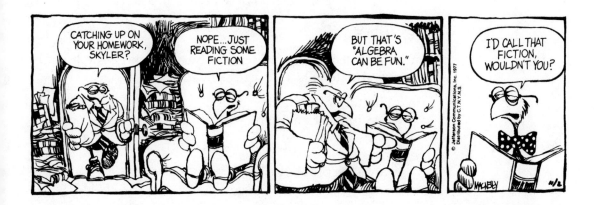

where n represents the unknown number. If an expression contains at least one variable and at least one defined operation, it is called a **variable expression**. The domain for this variable expression is the set of all numbers because there is no reason to restrict the possible replacements of n. (That is, n can be 6, -3, $\frac{1}{3}$, or any other number that comes to mind.) On the other hand, if n represents the number of people on an elevator, then $n + 2$ would represent the idea that two additional people boarded the elevator, and the domain for n is the set of whole numbers up to some maximum capacity of the elevator. That is, if the domain $D = \{0, 1, 2, 3, \ldots, 30\}$, then the possible number of people on the elevator after the two additional people boarded is 2, 3, 4, 5, $\ldots$, 32.

EXAMPLE 2

Let x be the variable and let $D = \{-5, 0, 5, 10\}$ be the domain. Then the variable expression

$$x + 7$$

represents

$-5 + 7 = 2$ if $x = -5$		$0 + 7 = 7$ if $x = 0$
$5 + 7 = 12$ if $x = 5$		$10 + 7 = 17$ if $x = 10$

It *cannot* represent

$$9 + 7 = 16$$

because 9 is not in the domain.   ■

There are some terms involving the operations with which you should be familiar. These are summarized in Table 4.1.

EXAMPLE 3

Choose a letter to represent the variable and write a mathematical statement to express the idea.

a. The sum of a number and 13.
   Sum indicates addition, so the statement is

$$(\text{NUMBER}) + (\text{THIRTEEN})$$

Now select some variable (your choice), say $N = \text{NUMBER}$.

TABLE 4.1 Translating to Algebra

Symbol	Verbal description
=	Is equal to; equals; are equal to; is the same as; is; was; becomes; will be; results in
+	The numbers being added are called **terms** and the result is called the **sum**. Plus; the sum of; added to; more than; greater than; increased by
−	The numbers being subtracted are called **terms** and the result is called the **difference**. Minus; the difference of; the difference between; is subtracted from; less than; smaller than; decreased by; is diminished by
× · ( )	A times sign is used primarily in arithmetic. A raised dot is used to indicate multiplication in algebra, as in $11 \cdot 7$ or $x \cdot 1$. Parentheses are used to indicate multiplication in algebra, as in $3(4 + 10)$ or $6(x + 2)$ or $5(w)$. Juxtaposition (no symbol) is used to indicate multiplication in algebra, especially with variables, as in xy or $5x$, but not with numerals (for example, 24 still means twenty-four—not two times four). The numbers being multiplied are called **factors** and the result is called the **product**. Times; product; is multiplied by
÷	The division symbol is used primarily in arithmetic. The fractional bar, as in $\frac{3}{4}$ (meaning 3 divided by 4) or $\frac{x}{y}$ meaning x divided by y) or $\frac{x + 2}{3}$ (meaning $x + 2$ divided by 3) is frequently used in algebra. In $\frac{x}{y}$, x is called the **dividend**, y is called the **divisor**, and the result is called the **quotient**. Divided by; quotient of

Then the symbolic statement is

$$N + 13$$

b. The difference of a number subtracted from 10.

$$(\text{TEN}) - (\text{NUMBER})$$

Let $x = \text{NUMBER}$; then the variable expression is

$$10 - x$$

c. The quotient of two numbers.

$$\frac{\text{NUMBER}}{\text{ANOTHER NUMBER}}$$

If there is more than one unknown in a problem, then more than one variable may be needed. Let $m = $ NUMBER and $n = $ ANOTHER NUMBER; then

$$\frac{m}{n}$$

d. The product of two consecutive numbers.

$$(\text{NUMBER})(\text{NEXT CONSECUTIVE NUMBER})$$

If there is more than one unknown in a problem, but a relationship between those unknowns is given, then *do not choose more variables than you need* for the problem. In this problem, "a consecutive number" means one more than the first number:

$$(\text{NUMBER})(\text{NUMBER} + 1)$$

Let $y = $ NUMBER; then the variable expression is

$$y(y + 1) \qquad\blacksquare$$

In the last chapter you simplified numerical expressions. If $x = a$, then x and a name the same number; x may be replaced by a in any expression and the value of that expression remains unchanged. When you replace variables by given numerical values and then simplify the resulting numerical expression, the process is called **evaluating an expression**.

EXAMPLE 4

Evaluate the given expressions where $a = -2$, $b = -1$, and $c = 3$.

a. $a + cb$ Remember, cb means c **times** b.

Step 1. Replace each variable with the corresponding numerical value. Additional parentheses may be necessary to make sure you don't change the order of operations.

$$\begin{matrix} a & + & c & b \\ \downarrow & & \downarrow & \downarrow \\ -2 & + & 3(-1) & \end{matrix}$$ ←— Parentheses are necessary so that the product cb is not changed to a subtraction, $3 - 1$.

Step 2. Simplify.

$$-2 + 3(-1) = -2 + (-3)$$
$$= -5$$

b. $a^2 - b^2 = (-2)^2 - (-1)^2$ Parentheses are necessary
$$= 4 - 1$$ because -2^2 is not
$$= 3$$ the same as $(-2)^2$.   ∎

Remember that a particular variable is replaced by a single value when evaluating an expression. You should also be careful to write capital letters differently from lowercase letters because they often represent different values. This means that just because $a = -2$ in Example 4, you should not assume $A = -2$.

EXAMPLE 5

Let $a = -1, b = 3, c = -2$, and $d = -3$. Find the value of the given capital letters. After you have found the value of a capital letter, fill it into the box that corresponds with its numerical value. This exercise will help you check your work to see if you have it right.

28	-9	-5	0	3

a. $G = bc - a$
$$= 3(-2) - (-1)$$
$$= -6 + 1$$
$$= -5$$

b. $H = 3c - 2d$
$$= 3(-2) - 2(-3)$$
$$= -6 - (-6)$$
$$= -6 + 6$$
$$= 0$$

c. $I = 3a - 2b$
$$= 3(-1) - 2(3)$$
$$= -3 + (-6)$$
$$= -9$$

d. $R = a^2 - b^2d$
$$= (-1)^2 - (3)^2(-3)$$
$$= 1 - (-27)$$
$$= 1 + 27$$
$$= 28$$

e. $S = \dfrac{2(b - d)}{2c}$

$= \dfrac{2[3 - (-3)]}{2(-2)}$

$= \dfrac{2[3 + 3]}{-4}$

$= \dfrac{12}{-4}$

$= -3$

f. $T = \dfrac{3a + bc + b}{c}$

$= \dfrac{3(-1) + 3(-2) + 3}{-2}$

$= \dfrac{-3 + (-6) + 3}{-2}$

$= \dfrac{-6}{-2}$

$= 3$

After you have filled in the appropriate boxes the result is:

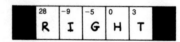

■

Problem Set 4.1

Let x be the variable and let D = {0, 1, 3, 7} be the domain. Find the values for the variable expressions in Problems 1–3.

1. $x + 5$ **2.** $x - 10$ **3.** $2x - 1$

Let y be the variable and let D = {−3, 0, 3, 10} be the domain. Find the values for the variable expressions in Problems 4–6.

4. $y + 8$ **5.** $y - 6$ **6.** $3y + 2$

Let z be the variable and let D = {−10, 0, 10} be the domain. Find the values for the variable expressions in Problems 7–9.

7. $|z|$ **8.** z^2 **9.** $10 - z^2$

In Problems 10–33, let w = −2, x = −1, y = 2, and z = −4. Find the values of the given capital letters.

10. $A = x + z$ **11.** $B = |x| + y - z$

12. $C = -w$ **13.** $D = 3z$

14. $E = -y^2$ **15.** $F = w(y - x + wz)$

16. $G = 5x + 3z$ **17.** $H = 3x + 2w$

18. $I = 3y - 2z$ **19.** $J = 2w - 3z$

20. $K = wxy$ **21.** $L = x^2 - y^2$

22. $M = (x - y)^2$ **23.** $N = x^2 + 2xy + y^2$

24. $P = x^2 + y^2 + z^2$ **25.** $Q = w(x + y)$

26. $R = x^2 - y^2 - z^2$ **27.** $S = (x - y + z)^2$

28. $T = x^2 - y^2 z$　　**29.** $U = \dfrac{w + y}{z}$

30. $V = \dfrac{3wyz}{x}$　　**31.** $W = \dfrac{3w + 6z}{xy}$

32. $X = (x^2 z + x)^2 z$　　**33.** $Y = wy^2 + w^2 y - 3x$

Mind Bogglers

34. This problem will help you check your work in Problems 10–33. Fill in the capital letters from Problems 10–33 to correspond with their numerical values (the letter O has been filled in for you). Some letters may not appear in the boxes. When you are finished, darken all the blank spaces to separate the words in the secret message. Notice that one of the blank spaces has also been filled in to help you.

−5	5	21	−19	O	−22	−4	49	49	O	−19
14	49	−1	O	1	−4	−6	15	−7	O	6
−71	17	−5	−3	4	49	■		14	1	−8	−8
10	10	49	O	9	−4	O		1	−4	11	12
−4	−3	49	−4	49	−9	49	−3	−4	−4	21

35. Think of a counting number less than 20. Add six. Multiply by two. Subtract eight. Divide by two. Subtract your original number. The answer is 2. Explain why this trick works.

36. Think of a counting number less than 100. Add five. Multiply by two. Add ten. Subtract twenty. Divide by two. The answer is your original number. Explain why this trick works.

37. The pictures below describe a number trick. Describe it in words.

(1) □　　(2) □☆☆☆　　(3) □☆☆☆
　　　　　　　　　　　　　　　□☆☆☆

(4) □☆☆　(5) ☆☆
　　□☆☆　　　☆☆

38. The algebraic steps below describe a number trick. Describe it in words.

(1) x (2) $x + 7$ (3) $2(x + 7) = 2x + 14$

(4) $2x + 14 - 4 = 2x + 10 = 2(x + 5)$

(5) $\dfrac{2(x + 5)}{2} = x + 5$ (6) $x + 5 - x = 5$

39. My favorite number is 7. Make up a number trick in which you ask someone to think of a number and carry on some operations, with the final answer always 7.

40. Bill Leonard's favorite number is 23. Make up a number trick in which you ask someone to think of a number and carry on some operations, with the final answer always 23.

41. Translate the following symbolic message:

    EZ4NE12CYU $\dfrac{R}{WEIGHT}$

42. Translate the following symbolic message:

    U8N8N8N8NR $\dfrac{2}{ACTIVE}$

4.2 Equations

Have you ever been asked to find the unknown?

Can You Tell Who's Who by Their Hair?

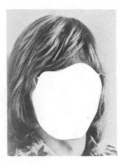 

THE ANSWERS ARE GIVEN ON PAGE 144

An **equation** is a statement of equality that can be true, false, or depend on the values of the variable. If the equation depends on the values of the variable, it is called a **conditional equation**. The values that make the conditional equation true are said to **satisfy** the equation and are called the **solution** or **roots** of the equation.

Just as you needed to look for visual clues to find the identities of the celebrities pictured above, you must also check the clues of the given equation to find the solution. You must look for things to do in order to make the solution or roots more obvious. The key ideas are those of **opposites** (for addition and subtraction) and **reciprocals** (for multiplication and division).

Numbers whose sum is 0 are called **opposites**.
Numbers whose product is 1 are called **reciprocals**.

Opposites
Reciprocals

EXAMPLE 1

Use the idea of opposites to find the sums.

a. $5 + (-5) = 0$ **b.** $-18 + 18 = 0$
c. $x + (-x) = 0$ **d.** $x - x = x + (-x) = 0$
e. $175 + 7 + (-7) = 175$ **f.** $14 + 23 - 23 = 14$
g. $x - 18 + 18 = x$ **h.** $y + 10 - 10 = y$ ■

The goal when solving an equation is to **isolate the variable** on one side of the equation. To begin our study, we state the first of four properties about equation solving.

The solution of an equation is unchanged by **adding the same number to both sides of the equation.**

Addition Property of Equations

You will use this property to add some number to both sides of an equation so that, after it is simplified, the only expression on one side of the equation is the variable.

The celebrities in the photographs on the preceding page are Henry "The Fonz" Winkler, Barbra Streisand, John Davidson, John Denver, and Lee Majors.

EXAMPLE 2

Solve the given equations.

a. $x - 7 = 12$

The opposite of subtracting 7 is adding 7, so you **add 7 to both sides** of the equation.

$x - 7 + \mathbf{7} = 12 + \mathbf{7}$

$x = 19$ Simplify

This says, if $x = 19$, the original equation is true.

Check: $19 - 7 = 12$

Replace x by 19 to check.

b. $y - 10 = 5$

$y - 10 + \mathbf{10} = 5 + \mathbf{10}$ Add 10 to both sides.

$y = 15$ Simplify

Can you check?

c. $z - 12 = -25$ Notice that when solving

$z - 12 + \mathbf{12} = -25 + \mathbf{12}$ equations you write one

$z = -13$ equation under the other with
                                                    the equal signs aligned.

d. $-108 = k - 92$

$-108 + \mathbf{92} = k - 92 + \mathbf{92}$ It doesn't matter whether the

$-16 = k$ variable ends up on the left or
                                                    the right side.   ∎

The addition property is used whenever some number is being subtracted from the variable. **If a number is being added to the variable, then the opposite is subtraction,** and you can use the following property.

The solution of an equation is unchanged by **subtracting the same number from both sides of the equation.**

Subtraction Property of Equations

EXAMPLE 3

Solve the given equations.

a. $x + 5 = 13$

—The opposite of adding 5 is subtracting 5, so you **subtract 5 from both sides** of the equation.

$$x + 5 - 5 = 13 - 5$$
$$x = 8 \qquad \text{Simplify}$$
$$\qquad\qquad \textit{Check: } \ 8 + 5 = 13$$

b. $y + 8 = 5$
$$y + 8 - 8 = 5 - 8 \qquad \text{Subtract 8 from both sides.}$$
$$y = -3 \qquad \text{Simplify}$$

Can you check?

c. $17 = a + 12$
$$17 - 12 = a + 12 - 12$$
$$5 = a$$

d. $19 = b + 48$
$$19 - 48 = b + 48 - 48$$
$$-29 = b \qquad\qquad\qquad\qquad\qquad ■$$

The idea of opposite operations extends to examples in which the variable is multiplied or divided by some number. That is, multiplication and division are opposite operations. Remember that when reciprocals are multiplied, the product is 1.

EXAMPLE 4

Simplify each expression.

a. $\dfrac{3x}{3} = x$ since $\dfrac{3x}{3}$ means $3x \div 3 = 3x\left(\dfrac{1}{3}\right)$

$$= x(3)\left(\dfrac{1}{3}\right)$$

$$= x(1)$$

$$= x$$

In practice, you don't go through these steps, but think: "If some number, x, is multiplied by 3 and then the result is divided by 3, the result is the original number, x." As you are thinking this, you write:

$$\dfrac{3x}{3} = x$$

b. $\dfrac{5x}{5} = x$ **c.** $\dfrac{19p}{19} = p$ **d.** $\dfrac{-12q}{-12} = q$

e. $\left(\dfrac{x}{2}\right)(2) = x$ since $\left(\dfrac{x}{2}\right)(2)$ means $(x \div 2)(2) = x\left(\dfrac{1}{2}\right)(2)$

$$= x(1)$$

$$= x$$

Think: "If some number, x, is divided by 2 and then the result is multiplied by 2, the result is the original number, x." As you are thinking this, you write:

$$\left(\dfrac{x}{2}\right)(2) = x$$

f. $\left(\dfrac{x}{7}\right)(7) = x$ **g.** $\left(\dfrac{t}{-3}\right)(-3) = t$ ■

To solve an equation in which some number is dividing the variable, you'll use the following property.

Multiplication Property of Equations

The solution of an equation is unchanged by **multiplying both sides of the equation by the same nonzero number.**

EXAMPLE 5

Solve each equation.

a. $\dfrac{x}{2} = 9$

L The opposite of dividing by 2 is multiplying by 2, so you **multiply both sides** of the equation **by 2**.

$$\left(\dfrac{x}{2}\right)(2) = 9(2)$$

$$x = 18 \qquad \text{Simplify} \quad \textit{Check:} \ \dfrac{18}{2} = 9$$

b. $\dfrac{w}{8} = 7$ **c.** $\dfrac{r}{-5} = 12$

$$\left(\dfrac{w}{8}\right)(8) = 7(8) \qquad\qquad \left(\dfrac{r}{-5}\right)(-5) = 12(-5)$$

$$w = 56 \qquad\qquad\qquad\qquad r = -60$$

d. $-3 = \dfrac{m}{6}$ **e.** $0 = \dfrac{n}{4}$

$$-3(6) = \left(\dfrac{m}{6}\right)(6) \qquad\qquad 0(4) = \left(\dfrac{n}{4}\right)(4)$$

$$-18 = m \qquad\qquad\qquad 0 = n \qquad\qquad \blacksquare$$

The last property of equations is used if the variable is multiplied by some number.

The solution of an equation is unchanged by **dividing both sides of the equation by the same nonzero number.**

Division Property of Equations

EXAMPLE 6

Solve each equation.

a. $5a = 20$

L The opposite of multiplying by 5 is dividing by 5, so you **divide both sides** of the equation **by 5.**

$$\dfrac{5a}{5} = \dfrac{20}{5}$$

$$a = 4 \qquad \text{Simplify} \quad \textit{Check:} \ 5(4) = 20$$

b. $3b = 39$

$\dfrac{3b}{3} = \dfrac{39}{3}$

$b = 13$

c. $-7c = 42$

$\dfrac{-7c}{-7} = \dfrac{42}{-7}$

$c = -6$

d. $-104 = 8d$

$\dfrac{-104}{8} = \dfrac{8d}{8}$

$-13 = d$

e. $-70 = -14e$

$\dfrac{-70}{-14} = \dfrac{-14e}{-14}$

$5 = e$ ∎

Problem Set 4.2

Solve each equation in Problems 1–48. Show each step of the process until the variable is isolated. Be sure to keep your equal signs aligned.

1. $a - 5 = 10$

2. $b - 8 = 14$

3. $c - 2 = 0$

4. $d - 9 = -7$

5. $d - 7 = -10$

6. $e - 1 = -35$

7. $f - 3 = -12$

8. $6 = g - 2$

9. $-5 = h - 4$

10. $-40 = i - 92$

11. $-137 = j - 49$

12. $0 = k - 112$

13. $m + 2 = 7$

14. $n + 8 = 12$

15. $n + 19 = 20$

16. $p + 8 = 2$

17. $q + 7 = -15$

18. $r + 6 = -14$

19. $8 + s = 4$

20. $12 + t = 15$

21. $36 + u = 40$

22. $18 + v = 10$

23. $-10 + w = 14$

24. $-12 + x = -6$

25. $\dfrac{y}{4} = 8$

26. $\dfrac{z}{2} = 18$

27. $\dfrac{A}{3} = 7$

28. $\dfrac{B}{12} = 8$

29. $\dfrac{C}{5} = -12$

30. $\dfrac{D}{8} = -15$

31. $\dfrac{E}{-4} = 11$

32. $\dfrac{F}{-12} = 16$

33. $\dfrac{G}{-13} = -3$

34. $7 = \dfrac{H}{-8}$

35. $-4 = \dfrac{I}{-10}$

36. $-18 = \dfrac{J}{-4}$

37. $\dfrac{K}{19} = 0$

38. $4L = 12$

39. $3M = 33$

40. $4N = -48$

41. $-8P = -96$

42. $-Q = 5$

43. $-R = 14$

44. $-S = 19$

45. $-12T = 168$

46. $13U = -234$

47. $-5V = 225$

48. $13W = 0$

Mind Bogglers

49. What is wrong with the following "proof"?

i.	12 eggs = 1 dozen	Multiply both sides of Step i by 2.
ii.	24 eggs = 2 dozen	Divide both sides of Step i by 2.
iii.	6 eggs = $\frac{1}{2}$ dozen	Multiply Step iii times Step ii
iv.	144 eggs = 1 dozen	(equals times equals are equal).
v.	12 dozen = 1 dozen	Substitute, since
		144 eggs = 12 dozen.

50. A hippopotamus and a little bird want to play on a teeter-

WHAT DO YOU SAY WE PLAY SOMETHING DIFFERENT?

totter. The bird says that it's impossible, but the hippopotamus assures the little bird that it will work out and that she
will prove it, since she has had a little algebra. She presents
the following argument:

Let H = WEIGHT OF HIPPOPOTAMUS
b = WEIGHT OF BIRD

Now, there must be some weight, w (probably very large), so
that

$$H = b + w$$

Multiply both sides by $H - b$:

$$H(\boldsymbol{H - b}) = (b + w)(\boldsymbol{H - b})$$

Using the distributive property:

$$H^2 - Hb = bH + wH - b^2 - wb$$

Subtract wH from both sides:

$$H^2 - Hb - wH = bH - b^2 - wb$$

Use the distributive property again:

$$H(H - b - w) = b(H - b - w)$$

Divide both sides by $H - b - w$:

$$H = b$$

Thus, the weight of the hippopotamus is the same as the weight of the bird. "Now," says the hippopotamus, "since our weights are the same, we'll have no problem on the teeter-totter."

"Wait!" hollers the bird. "Obviously this is false." But where is the error in the reasoning?

4.3 Solving Equations

One of the most important tools in problem solving is solving equations. In this section, we'll take the fundamentals you learned in the last section and apply them to more complicated equations. The first step is to learn to solve the equations from the last section *mentally*.

EXAMPLE 1

Solve each equation.

a. $7h = -56$
   $h = -8$

   Mentally divide both sides by 7.

b. $j + 7 = -56$
   $j = -63$

   Mentally subtract 7 from both sides.

c. $\frac{r}{7} = -56$
   $r = -392$

   Mentally multiply both sides by 7.

d. $s - 7 = -56$
   $s = -49$

   Mentally add 7 to both sides. ■

Sometimes it is necessary to add or subtract variables that are not opposites. If the variables are different, then the sum or difference is in *simplified form*:

$$\left. \begin{array}{l} a + b \\ 3x + 4y \\ s - 2t \\ x^2 + 2x \end{array} \right\} \quad \text{These are all simplified.}$$

If the variables are the *same* (including exponents), then the terms are called **similar** or **like** terms and can be added or subtracted by using the distributive property.

Similar terms				
$5x,$	$2x,$	$\frac{5}{2}x,$	$-3x,$	5^2x
$4,$	$12,$	$5,$	0	
$a,$	$5a,$	$-2a$		

Not similar terms			
$5x,$	$5y,$	$3xy,$	$5x^2$
$4x,$	$12,$	$6p,$	$3p^2$
$a,$	$3,$	$a^2,$	b

EXAMPLE 2

Simplify each expression.

a. $3x + 5x = (3 + 5)x$
$\quad\quad\quad\quad = 8x$

b. $9y - 3y = (9 - 3)y$
$\quad\quad\quad\quad = 6y$

c. $12b - 8b = 4b$

d. $t + t + t + t = 4t$

e. $3x + 15 - 15 + x = 4x$

f. $x - 36 + 36 + x = 2x$

g. $2y + 14 + y - 14 = 3y$

h. $3x + 5 - 3x + 2 = 7$

i. $5x + 3 - 4x + 5 = x + 8$

j. $6x + 3 - 5x + 7 = x + 10$ ∎

Equations can now be solved by first simplifying the left and right sides *before* using one of the equation properties.

EXAMPLE 3

Solve each equation.

a. $3x - 2 - 2x = 5 + 2x - 2x$

$\quad\quad\quad\quad x - 2 = 5$ First simplify

$\quad\quad\quad\quad\quad\quad x = 7$ Add 2 to both sides (done mentally).

b. $4 + 5x - 5x = 7 + 6x - 5x$

$\quad\quad\quad\quad\quad 4 = 7 + x$ Simplify

$\quad\quad\quad\quad -3 = x$ Subtract 7 from both sides (mentally).

c. $\frac{x}{2} + 4 - 4 = 5 - 4$

$\quad\quad\quad\quad \frac{x}{2} = 1$ Simplify

$\quad\quad\quad\quad x = 2$ Multiply both sides by 2.

d. $4 - 10 = 10 - 3x - 10$

$\quad\quad\quad -6 = -3x$ Simplify

$\quad\quad\quad\quad 2 = x$ Divide both sides by -3. ∎

Sometimes it is necessary to use more than one of the equation properties when solving a simple equation.

**Procedure for
Solving Simple
Equations**

> 1. Simplify the left and right sides.
> 2. Use equation properties to isolate the variable on one side:
>   a. First, use the addition or subtraction properties.
>   b. Next, use the multiplication or division properties.

EXAMPLE 4

Solve each equation.

a. $4x + 3 = 7$

$\qquad 4x = 4$ Subtract 3 from both sides (addition or subtraction first).

$\qquad\quad x = 1$ Divide both sides by 4.

b. $2x - 7 = 13$

$\qquad 2x = 20$ Add 7 to both sides.

$\qquad\quad x = 10$ Divide both sides by 2.

c. $4x + 3 = 3x + 8$

$\qquad x + 3 = 8$ Subtract 3x from both sides (isolate the variable on one side).

$\qquad\quad x = 5$ Subtract 3 from both sides.

d. $2x - 4 = 5x + 2$

$\qquad -3x - 4 = 2$ Subtract 5x from both sides.

$\qquad\quad -3x = 6$ Add 4 to both sides.

$\qquad\qquad x = -2$ Divide both sides by −3.

e. $41 = \dfrac{x}{2} - 7$

$\qquad 48 = \dfrac{x}{2}$ Add 7 to both sides.

$\qquad 96 = x$ Multiply both sides by 2.

f. $-4 = \dfrac{x}{3} + 2$

$\qquad -6 = \dfrac{x}{3}$ Subtract 2 from both sides.

$\qquad -18 = x$ Multiply both sides by 3. ∎

Problem Set 4.3

Solve the equations in Problems 1–48. Show all your work and be sure to keep the equal signs aligned.

1. $2x + 1 = 9$

2. $3x + 4 = 19$

3. $5x + 7 = 47$

4. $4y - 3 = 77$

5. $6y - 9 = 21$

6. $9y - 23 = 49$

7. $\dfrac{z}{5} + 3 = 4$

8. $\dfrac{z}{3} + 1 = 0$

9. $\dfrac{z}{9} + 12 = 6$

10. $\dfrac{w}{2} - 5 = 4$

11. $\dfrac{w}{-4} - 7 = 6$

12. $\dfrac{w}{3} - 19 = 47$

13. $6 + 2t = 4$

14. $4 + 7t = 53$

15. $9 + 3t = 45$

16. $-s + 2 = 9$

17. $-s - 5 = 14$

18. $-s - 7 = 10$

19. $4 + (-u) = 17$

20. $3 + (-s) = 15$

21. $17 + (-s) = 21$

22. $4 - s = 17$

23. $3 - s = 15$

24. $18 - s = 31$

25. $3A + 7 = 49$

26. $2B + 5 = 65$

27. $3C - 8 = 115$

28. $2D - 10 = 52$

29. $\dfrac{E}{7} + 12 = 12$

30. $\dfrac{F}{5} - 12 = 53$

31. $9 = \dfrac{G}{2} + 5$

32. $-8 = \dfrac{H}{6} - 3$

33. $4 + 3I = 67$

34. $2 - 5J = 62$

35. $\dfrac{K}{7} - 10 = 0$

36. $\dfrac{L}{12} + 4 = -3$

37. $-M = 14$

38. $6 + \dfrac{N}{2} = -10$

39. $-P + 5 = -12$

40. $7 + (-Q) = -15$

41. $8 - R = 5$

42. $5 - S = 10$

43. $10 - T = -5$

44. $4 - 2U = 8$

45. $7 - 3V = 10$

46. $6 - 5W = 26$

47. $1 - 5X = 126$

48. $12 - 21Y = 75$

Mind Bogglers

49. This problem will help you check your work in Problems 25–48. Fill in the capital letters from Problems 25–48 to correspond with their numerical values shown in the blanks below. (The letter O has been filled in for you.) Some letters may not appear in the boxes. When you finish putting letters in the boxes, darken in all the blank spaces to separate the words in the message.

2	12	21	-8	-30	O	17	0	1	-3	O	-2	4	14	3	0	2	4
325	21	-32	31	21	-32	8	35	15	-30	21	-5	5	0	14	-5	-3	-15
14	-32	31	7	14	3	0	24	-84	0	14	3	-32	21	-32	8	-20	-15
-2	-5	0	325	-2	-84	35	-14	14	15	-30	0	-14	14	15	21	41	-5

50. A turkey weighs 10 lb plus one-fourth of its weight. How much does it weigh?

51. Answer the question in the *Peanuts* cartoon strip.

4.4 Problem Solving with Algebra

Our goal in applying algebra to problem solving is to find solutions to a wide variety of problems. In the last two sections you've seen techniques for solving equations. In this section, we'll introduce a *technique for solving problems.* Eventually, you want to be able to solve applied problems to obtain *answers,* but when beginning, we must necessarily start with simple problems—ones where the answers are obvious or trivial. You must, therefore, keep in mind that we are not seeking answers in this section, but a strategy for attacking word problems.

When confronted with a word problem, many people begin by looking at the last sentence to see what is being asked for, declare that this quantity is the unknown, and then proceed to read the problem backwards in order to come up with an equation. This procedure complicates the thinking process and can be frustrating. Instead, we begin with the problem as stated, using as many unknowns as we wish with our initial statement. We then let this relationship *evolve* until we have a statement with a single unknown. At *this* point a variable is chosen as a natural result of the thinking process, and the resulting equation is solved.

1. **Read the problem.** Note what it is all about. Focus on processes rather than numbers. You can't work a problem you don't understand.
2. **Restate the problem.** Write down a verbal description of the problem using operations signs and an equal sign. Look for equality. If you can't find equal quantities, you will never formulate an equation.
3. **Evolve the equation.** Don't rush this step; "evolve" is the key word. The discovery of this final equation with a single unknown should be the result of your understanding of the problem, not your algebraic skill.
4. **Choose a variable.** This choice should be the natural result of the process in Step 3. Don't force your choice of variable to be the same as what is asked for in the question.
5. **Solve the equation.** This is the easy step. Be sure your answer makes sense by checking it with the original question in the problem.
6. **State an answer.** There were no variables defined when you started, so $x = 3$ is not an answer. Pay attention to units of measure and other details of the problem. Remember to answer the question that was asked.

Procedure for Problem Solving

EXAMPLE 1

An advertisement for an Oldsmobile had the information shown in the margin. What is the size of the tank?

Solution

Step 1. Do you understand the question? Can you rephrase it in your own words?

336
Miles per tankful.
Estimated city
driving range.
EPA estimated
MPG ⑯ CITY.

441
Miles per tankful.
Estimated highway
driving range.
EPA estimated
21 HIGHWAY.

Step 2. $\begin{pmatrix} \text{MILES PER} \\ \text{GALLON} \end{pmatrix} \times \begin{pmatrix} \text{NUMBER OF} \\ \text{GALLONS} \end{pmatrix} = \begin{pmatrix} \text{DISTANCE} \\ \text{TRAVELED} \end{pmatrix}$

Step 3. $\quad\quad 16 \quad \times \begin{pmatrix} \text{NUMBER OF} \\ \text{GALLONS} \end{pmatrix} = \quad 336$

Step 4. Let g = NUMBER OF GALLONS.

Step 5.
$$16g = 336$$
$$g = 21 \quad \text{Divide both}$$
sides by 16.

Step 6. The car has a 21 gallon tank. ∎

EXAMPLE 2

Three prices for dog food are mentioned in the cartoon.

What are the prices for the three types of dog food (assuming that all the cans are the same size)? If the boy bought 10 cans, what is his savings by buying the least expensive as compared with the most expensive?

Solution

$$\begin{pmatrix} \text{NUMBER OF} \\ \text{CANS} \end{pmatrix} \begin{pmatrix} \text{PRICE PER} \\ \text{CAN} \end{pmatrix} = \begin{pmatrix} \text{TOTAL} \\ \text{COST} \end{pmatrix}$$

Apply this general formula for each of the dog foods. Let a, b, and c be the prices for the three brands, respectively.

$$\text{First:}$$
$$3\left(\begin{array}{c}\text{PRICE PER}\\\text{CAN}\end{array}\right) = 78$$
$$3a \quad\ = 78$$
$$a \quad\ = 26$$

$$\text{Second:}$$
$$2\left(\begin{array}{c}\text{PRICE PER}\\\text{CAN}\end{array}\right) = 68$$
$$2b \quad\ = 68$$
$$b \quad\ = 34$$

$$\text{Third:}$$
$$10\left(\begin{array}{c}\text{PRICE PER}\\\text{CAN}\end{array}\right) = 254$$
$$10c \quad\ = 254$$
$$c \quad\ = 25.4$$

Notice that we worked in pennies (this is often easier than working in dollars). The third dog food is the cheapest.

$$\text{SAVINGS} = \left(\begin{array}{c}\text{TOTAL COST}\\\text{OF MOST}\\\text{EXPENSIVE}\end{array}\right) - \left(\begin{array}{c}\text{TOTAL COST}\\\text{OF LEAST}\\\text{EXPENSIVE}\end{array}\right)$$

$$= \left(\begin{array}{c}\text{NO. OF CANS}\\\text{OF MOST}\\\text{EXPENSIVE}\end{array}\right)\left(\begin{array}{c}\text{COST PER}\\\text{CAN OF MOST}\\\text{EXPENSIVE}\end{array}\right) - \left(\begin{array}{c}\text{NO. OF CANS}\\\text{OF LEAST}\\\text{EXPENSIVE}\end{array}\right)\left(\begin{array}{c}\text{COST PER}\\\text{CAN OF LEAST}\\\text{EXPENSIVE}\end{array}\right)$$

$$= \quad (10) \quad\quad (34) \quad\quad - \quad (10) \quad\quad (25.4)$$

$$= 340 - 254$$

$$= 86$$

The savings is \$0.86. ■

EXAMPLE 3

A dispatcher must see that 75,000 condensers are delivered immediately. He has two sizes of trucks; one will carry 15,000 condensers and the other will carry 12,000 condensers. If the dispatcher has only two of the larger trucks available, how many smaller trucks are necessary?

MIDTOWN ELECTRIC

"Great scott! You got the order backwards! We wanted twenty 75,000 micro-fared condensers!"

Solution

$$\left(\begin{matrix}\text{AMOUNT DELIVERED} \\ \text{BY LARGER TRUCKS}\end{matrix}\right) \quad + \quad \left(\begin{matrix}\text{AMOUNT DELIVERED} \\ \text{BY SMALLER TRUCKS}\end{matrix}\right) \quad = \quad \left(\begin{matrix}\text{TOTAL} \\ \text{AMOUNT} \\ \text{DELIVERED}\end{matrix}\right)$$

$$\left(\begin{matrix}\text{NO. OF} \\ \text{LG. TRUCKS}\end{matrix}\right)\left(\begin{matrix}\text{CAPACITY OF} \\ \text{LG. TRUCKS}\end{matrix}\right) + \left(\begin{matrix}\text{NO. OF SM.} \\ \text{TRUCKS}\end{matrix}\right)\left(\begin{matrix}\text{CAPACITY OF} \\ \text{SM. TRUCKS}\end{matrix}\right) = \quad 75,000$$

$$2 \qquad\qquad (15,000) \quad + \left(\begin{matrix}\text{NO. OF SM.} \\ \text{TRUCKS}\end{matrix}\right) \quad (12,000) \quad = \quad 75,000$$

Let n = NUMBER OF SMALLER TRUCKS.

$$2(15,000) + n(12,000) = 75,000$$
$$30,000 + 12,000n = 75,000$$
$$12,000n = 45,000$$
$$n = 3.75$$

Notice that the solution of the equation is not necessarily the answer to the problem. You must interpret this solution to answer the question that was asked. **The dispatcher will need to send 4 smaller trucks.** ∎

One formula used in many everyday applications is the distance-rate-time formula. Suppose you drove for 2 hours at 55 mph. How far did you go? The formula is

$$\text{DISTANCE} = (\text{RATE})(\text{TIME}) \qquad \text{or} \qquad d = rt$$

Example 4 makes use of this formula.

EXAMPLE 4

Once upon a time (about 450 B.C.), a Greek named Zeno made up a word problem about a race between Achilles and a tortoise. The tortoise has a 100 yard head start. Achilles runs at a rate of 10 yards per second, while the tortoise runs 1 yard per second (it is an extraordinarily swift tortoise). How long does it take Achilles to catch up with the tortoise?

Solution

Step 1. Reread the problem.
Step 2. Write down a verbal description of the problem. We use the diagram to help us.

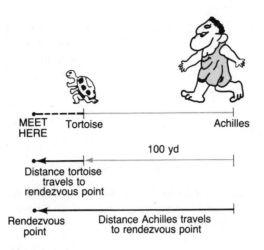

Step 3. Evolve the equation.

$$\left(\begin{array}{c}\text{ACHILLES' DISTANCE}\\\text{TO RENDEZVOUS}\end{array}\right) = \left(\begin{array}{c}\text{TORTOISE'S DISTANCE}\\\text{TO RENDEZVOUS}\end{array}\right) + 100$$

$$\left(\begin{array}{c}\text{RATE OF}\\\text{ACHILLES}\end{array}\right)\left(\begin{array}{c}\text{TIME TO}\\\text{RENDEZVOUS}\end{array}\right) = \left(\begin{array}{c}\text{RATE OF}\\\text{TORTOISE}\end{array}\right)\left(\begin{array}{c}\text{TIME TO}\\\text{RENDEZVOUS}\end{array}\right) + 100$$

$$10 \left(\begin{array}{c}\text{TIME TO}\\\text{RENDEZVOUS}\end{array}\right) = 1 \left(\begin{array}{c}\text{TIME TO}\\\text{RENDEZVOUS}\end{array}\right) + 100$$

Step 4. Let t = TIME TO RENDEZVOUS. t must represent a number; don't let t = rendezvous

Step 5. $10t = t + 100$

$9t = 100$

$t = \dfrac{100}{9}$

$t = 11\frac{1}{9}$

Step 6. It takes Achilles $11\frac{1}{9}$ seconds to catch up with the tortoise. ■

Problem Set 4.4

*Fill in the blanks in Problems 1–6. Notice that the answers to the questions have been given. What you are practicing here is a **procedure**. The lowercase letters designate the answer that is to be filled in for that location.*

1. The Toyota Corolla has an estimated MPG of 22, with a cruising range of 286 miles. What is the size of the tank to the nearest tenth gallon?

Solution: $\begin{pmatrix} \text{MILES PER} \\ \text{GALLON} \end{pmatrix} \begin{pmatrix} \text{NUMBER OF} \\ \text{GALLONS} \end{pmatrix} = \begin{pmatrix} \text{DISTANCE} \\ \text{TRAVELED} \end{pmatrix}$

$\underline{\quad\text{a.}\quad} \begin{pmatrix} \text{NUMBER OF} \\ \text{GALLONS} \end{pmatrix} = \underline{\quad\text{b.}\quad}$

Let $G = \underline{\quad\quad\text{c.}\quad\quad}$

$22G = \underline{\quad\text{d.}\quad}$

$G = \underline{\quad\text{e.}\quad}$

The car has a 13 gallon tank.

2. The Rolls-Royce Silver Shadow I has an estimated MPG of 9, with a cruising range of 234 miles. What is the size of the tank?

Solution: $\underline{\quad\text{a.}\quad} \begin{pmatrix} \text{NUMBER OF} \\ \text{GALLONS} \end{pmatrix} = \underline{\quad\text{b.}\quad}$

$\underline{\quad\text{c.}\quad} \quad \underline{\quad\text{d.}\quad} = \underline{\quad\text{e.}\quad}$

Let $G = $ NUMBER OF GALLONS.

$\underline{\quad\text{f.}\quad} = 234$

$G = \underline{\quad\text{g.}\quad}$

The car has a 26 gallon tank.

3. Del Monte asparagus spears sell for 78¢ for 8 oz and Green Giant asparagus spears are 99¢ for 12 oz. Which is the better buy, and by how much, rounded to the nearest $\frac{1}{10}$ cent?

Solution: $\begin{pmatrix} \text{NUMBER OF} \\ \text{OUNCES} \end{pmatrix} \begin{pmatrix} \text{PRICE PER} \\ \text{OUNCE} \end{pmatrix} = \begin{pmatrix} \text{TOTAL} \\ \text{PRICE} \end{pmatrix}$

Del Monte: $\underline{\quad\text{a.}\quad} \begin{pmatrix} \text{PRICE PER} \\ \text{OUNCE} \end{pmatrix} = \underline{\quad\text{b.}\quad}$

Let $D = $ PRICE PER OUNCE for Del Monte.

$8D = 78$

$D = \underline{\quad\text{c.}\quad}$

Green Giant: $\underline{\quad\quad\text{d.}\quad\quad} = \begin{pmatrix} \text{TOTAL} \\ \text{PRICE} \end{pmatrix}$

Let $G = $ PRICE PER OUNCE for Green Giant. [*Note:* Different variables in the same problem must be denoted by different letters.]

$\underline{\quad\text{e.}\quad} = 99$

$G = \underline{\quad\text{f.}\quad}$

$$\text{SAVINGS} = \left(\begin{array}{c}\text{COST PER OUNCE OF} \\ \text{MORE EXPENSIVE SPEARS}\end{array}\right) - \left(\begin{array}{c}\text{COST PER OUNCE OF} \\ \text{LESS EXPENSIVE SPEARS}\end{array}\right)$$

$$= \underline{\hspace{2cm}\text{g.}\hspace{2cm}} - \underline{\hspace{2cm}\text{h.}\hspace{2cm}}$$

$$= \underline{\hspace{3cm}\text{i.}\hspace{3cm}}$$

The Green Giant asparagus spears are $1\frac{1}{2}$¢ per ounce less expensive than Del Monte's.

4. Libby's green beans cost 48¢ for 15 ounces and Diet Delight green beans cost 26¢ for 6 ounces. Which is the better buy, and by how much, rounded to the nearest $\frac{1}{10}$ cent?

Solution: $\underline{\hspace{1.5cm}\text{a.}\hspace{1.5cm}}\left(\begin{array}{c}\text{PRICE PER} \\ \text{OUNCE}\end{array}\right) = \left(\begin{array}{c}\text{TOTAL} \\ \text{PRICE}\end{array}\right)$

Libby: $\underline{\hspace{1.5cm}\text{b.}\hspace{1.5cm}}\left(\begin{array}{c}\text{PRICE PER} \\ \text{OUNCE}\end{array}\right) = \underline{\hspace{1.5cm}\text{c.}\hspace{1.5cm}}$

Let L = PRICE PER OUNCE for Libby.

$$\underline{\hspace{2cm}\text{d.}\hspace{2cm}} = 48$$

$$L = \underline{\hspace{2cm}\text{e.}\hspace{2cm}}$$

Diet Delight: $\underline{\hspace{3cm}\text{f.}\hspace{3cm}} = \left(\begin{array}{c}\text{TOTAL} \\ \text{PRICE}\end{array}\right)$

Let D = PRICE PER OUNCE for Diet Delight.

$$\underline{\hspace{2cm}\text{g.}\hspace{2cm}} = \underline{\hspace{2cm}\text{h.}\hspace{2cm}}$$

$$D = \underline{\hspace{1.5cm}\text{i.}\hspace{1.5cm}}$$

$$\text{SAVINGS} = \left(\begin{array}{c}\text{COST PER OUNCE OF} \\ \text{MORE EXPENSIVE BEANS}\end{array}\right) - \underline{\hspace{3cm}\text{j.}\hspace{3cm}}$$

$$= \underline{\hspace{4cm}\text{k.}\hspace{4cm}}$$

$$= \underline{\hspace{2.5cm}\text{l.}\hspace{2.5cm}}$$

The Libby's green beans are about 1.1¢ per ounce less expensive than Diet Delight's.

5. A house and a lot are appraised at $112,200. If the house is worth five times the value of the lot, how much is the lot worth?

Solution:

(VALUE OF HOUSE) + (VALUE OF LOT) = (TOTAL VALUE)

(5 · VALUE OF LOT) + (_____**a.**_____) = 112,200

Let V = VALUE OF THE LOT.

$$\underline{\quad\quad\textbf{b.}\quad\quad} + \quad V \quad = \underline{\quad\textbf{c.}\quad}$$
$$6V = \underline{\quad\textbf{d.}\quad}$$
$$V = \underline{\quad\textbf{e.}\quad}$$

The value of the lot is $18,700.

6. Two persons are to run a race, but one can run 10 meters per second while the other can run 6 meters per second. If the slower runner has a 50 meter head start, how long will it be before the faster runner catches the slower runner if they begin at the same time?

Solution: $\underline{\quad\textbf{a.}\quad} = \underline{\qquad\textbf{b.}\qquad} + 50$

$\underline{\quad\textbf{c.}\quad} = \begin{pmatrix} \text{SLOWER} \\ \text{RUNNER'S} \\ \text{RATE} \end{pmatrix} \begin{pmatrix} \text{SLOWER} \\ \text{RUNNER'S} \\ \text{TIME} \end{pmatrix} + 50$

$\underline{\quad\textbf{d.}\quad} = \underline{\quad\textbf{e.}\quad} \,(\text{TIME FOR RACE}) + 50$

Let $T = $ TIME TO RUN THE RACE.

$$\underline{\quad\textbf{f.}\quad} = \underline{\quad\textbf{g.}\quad}$$
$$\underline{\quad\textbf{h.}\quad} = \underline{\quad\textbf{i.}\quad}$$
$$\underline{\quad\textbf{j.}\quad} = \underline{\quad\textbf{k.}\quad}$$

It takes 12.5 seconds for the race.

*Problems 7–12 are very similar to Problems 1–6. Use Problems 1–6 as models to illustrate your **technique** for solution. Notice that since you are not asked for **answers**, they are shown as a check on your procedure.*

7. The Datsun 510 sedan has an estimated MPG of 24, with a cruising range of 312 miles. What is the size of the tank?

   Answer: 13 gallons; you show the solution.

8. The Honda Accord has an estimated MPG of 23, with a fuel tank of 13.2 gallons. What is the cruising range?

   Answer: 304 miles; you show the solution.

9. Chicken of the Sea tuna sells for 75¢ for a 7 oz can while Star Kist diet pack costs 63¢ for a 5 oz can. Which is the better buy and by how much, rounded to the nearest $\frac{1}{10}$ cent?

   Answer: Chicken of the Sea is cheaper by 1.9¢ per ounce; you show the solution.

10. Which of the three sizes in the following ad is the least expensive?

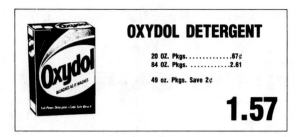

Answer: The 84 oz size; you show the solution.

11. A cabinet shop produced two types of custom-made cabinets for a customer. If one cabinet cost four times the cost of the other, and the total price for one of each type of cabinet is $2,075, how much does each cabinet cost?

    Answer: $415 and $1,660; you show the solution.

12. If the rangefinder on Starbuck's *Viper* shows the enemy Cylon 4,500 mi away, how long will it take to catch up to the Cylon if Starbuck's *Viper* travels at 15,000 mph and the Cylon travels at 12,000 mph?

    Answer: $1\frac{1}{2}$ hours; you show the solution.

CALCULATOR PROBLEMS

Currency exchange rates fluctuate daily, but we will use the accompanying table as the foreign currency exchange rate. In order to convert from a foreign currency into dollars, use the following equation:

$$\left(\begin{array}{c}\text{NUMBER OF}\\\text{DOLLARS}\end{array}\right)\left(\begin{array}{c}\text{EXCHANGE}\\\text{RATE}\end{array}\right) = \left(\begin{array}{c}\text{AMOUNT OF FOREIGN}\\\text{CURRENCY}\end{array}\right)$$

For example, if a ring costs 300 marks in West Germany, the equivalent amount in U.S. dollars is found as follows:

$$\left(\begin{array}{c}\text{NUMBER OF}\\\text{DOLLARS}\end{array}\right)(2.2475) = 300 \qquad \text{From the table}$$

$$\left(\begin{array}{c}\text{NUMBER OF}\\\text{DOLLARS}\end{array}\right) = 133.48 \qquad \begin{array}{l}\text{Divide both sides by 2.2475}\\\text{(rounded to the nearest cent)}\end{array}$$

Find the value of the items in Problems 13–20 expressed in U.S. dollars.

Foreign Currency Exchange Rate
Here is approximately what one U.S. dollar bought the week of Jan. 1, 1982.

Country	Currency	Rate	Country	Currency	Rate
Argentina	Peso	9,750	Italy	Lire	1,203.00
Australia	Dollar	.887	Japan	Yen	219.80
Austria	Schilling	15.86	Mexico	Peso	25.87
Belgium	Franc	38.47	New Zealand	Dollar	1.2162
Brazil	Cruzeiro	127.48	Norway	Kroner	5.8575
Canada	Dollar	1.185	Peru	Sole	505.42
Colombia	Peso	55.56	Philippine Islands	Peso	8.700
Denmark	Kroner	7.36	Portugal	Escudo	65.30
England	Pound	.523	Singapore	Dollar	2.0495
Finland	Markka	4.375	Spain	Peseta	97.30
France	Franc	5.73	Sweden	Kroner	5.5600
Greece	Drachma	57.00	Switzerland	Franc	1.7975
Holland	Guilder	2.4675	Venezuela	Bolivar	4.2940
Hong Kong	Dollar	5.680	W. Germany	Mark	2.2475
Israel	Pound	15.50			

13. A dinner in a Canadian restaurant is $12.00.

14. A sweater sells for $59 in Canada.

15. A hotel room in Mexico is 418 pesos.

16. A piece of lace in Belgium is marked 720 francs.

17. A prize tulip bulb in Holland is selling for 8 guilder.

18. A German Black Forest clock is selling for 435 marks.

19. A set of bone china in England is 346 pounds.

20. A tray in an Italian store is marked 24,120 lire.

*As gasoline prices rise, the gasoline mileage of our automobiles becomes more important. The gasoline consumption of a car is measured in **miles per gallon** (MPG). To find miles per gallon, divide the total miles driven by the number of gallons of gasoline used:*

$$\left(\begin{matrix} \text{MILES PER} \\ \text{GALLON} \end{matrix}\right) = \frac{\text{MILES DRIVEN}}{\text{GALLONS USED}}$$

Find the miles per gallon in Problems 21–30 to the nearest tenth.

	ODOMETER:		*Gallons*
	Start	*Finish*	*used*
21.	02316	02480	13.4
22.	47341	47576	17.3
23.	19715	19932	9.2
24.	13719	14067	11.1
25.	21812	22174	8.7
26.	16975	17152	18.6
27.	23485	24433	48.4
28.	08271.6	10373.0	53.2
29.	45678.9	48740.4	62.1
30.	13689.4	15274.9	46.2

Mind Bogglers

31. One day Perry White sent Lois Lane and Clark Kent out to Three-Mile Island to cover a big story. On the way, Clark, disguised as Superman, made a quick trip to K-Mart to pick up a present for Lois. It took Lois two hours to reach Three-Mile Island from the *Daily Planet*, but it only took Clark 10 minutes, even though he traveled twice as far as Lois. If Lois drove at 50 mph, how fast did Clark (Superman) travel?

● IF YOUR car gets 12 miles to a gallon at 70, and 10 percent better mileage at 55, you save less than $\frac{3}{4}$ of a gallon of gas every 100 miles. Assuming an average of $1\frac{1}{2}$ persons per car, that's a saving of $\frac{1}{2}$ gallon of gas for every 21 extra minutes you have wasted by driving at 55. At 75 cents a gallon, you are saving only about $1.12 per hour. Except that if your car gets twice as good mileage, you're only saving about 56 cents per hour. Any way you chop it, that's less than the minimum wage of $2.90. Considering that time is the only really valuable commodity you have, the extra time you consume at 55 pays you only $1.12 maximum per hour for your time. What kind of a schnook would work for $1.25 an hour these days? You can get a job raking leaves for the government for $2.90 an hour! If you sell your time for $5 an hour, you have lost $3.88 an hour by slowing down to 55. If you happen to be a professional person who charges from $25 to $150 an hour for your time, you could let that 55 mile gambit put you in the poorhouse.

Number of Miles per
Gallon Obtained by Traveling
at Different Speeds

Miles per hour	Miles per gallon
20	16.5
30	22.0
40	**22.5 Optimum**
50	21.5
60	19.5
70	17.3

The table shows the MPG *at various speeds. In Problems 32–34 suppose you commute 50 miles to work and 50 miles back each day.*

32. If you drive at the given speed, how many hours are spent driving in a regular 5 day week?
a. 40 mph　　**b.** 50 mph　　**c.** 60 mph　　**d.** 70 mph

33. If gasoline costs $1.75 per gallon, how much do you save per week by driving at 40 mph instead of at the given speed?
a. 50 mph　　**b.** 60 mph　　**c.** 70 mph

34. Using Problems 32 and 33, calculate your "salary" based on the extra time spent commuting by driving at 40 mph versus the "savings" in cost of gasoline.

4.5　Review Problems

Section 4.1　　**1.** Think of a number. Add ten. Multiply by three. Subtract twenty-four. Divide by three. Subtract two. The answer is your original number. Explain why this trick works.

Section 4.1　　**2.** Let x be the variable and let $D = \{-10, 0, 10, 100\}$ be the domain. Find the values of the variable expression $3x + 10$.

Section 4.1　　**3.** Let $x = -2$, $y = -3$, and $z = 4$. Find the value of $z(x - y)$.

Section 4.2　　**4.** Mentally solve each equation; then tell which property of equations you used.

　a.　$a - 10 = 40$　　**b.**　$\dfrac{b}{6} = 4$　　**c.**　$3c = 12$

　d.　$d + 5 = 2$

Section 4.3　　**5.** Solve each equation.
　a.　$3x + 2 = 8$　　**b.**　$3 - x = 12$

Section 4.3　　**6.** Solve each equation.
　a.　$\dfrac{y}{2} - 5 = 4$　　**b.**　$\dfrac{y - 5}{2} = 4$

Section 4.3　　**7.** Solve each equation.
　a.　$-z = 12$　　**b.**　$1 - 5z = 101$

Section 4.4　　**8.** The Mercedes-Benz 450SL has an estimated MPG of 12 and a fuel tank capacity of 23.8 gallons. What is the estimated cruising range? Show your solution as well as your answer.

Section 4.4　　**9.** If a 32 oz bottle of ketchup costs $0.99 and a 44 oz size costs $1.49, which size is the least expensive and by how much per ounce (rounded to the nearest $\frac{1}{10}$ cent)?

Section 4.4　　**10.** Suppose the odometer read 46312 at the start of your vacation and 48132 at the end. If your trip required 52 gallons, what was the car's MPG for this trip?

The Metric World

CHAPTER 5

5.1 Measuring Length

In the everyday world, numbers arise naturally in either of two ways—by counting or by measuring. In the first four chapters we concentrated on counting, but in this and the following chapter we'll focus on measuring.

Measure or Length

> To measure a length is to assign a number to its size. The number is called its **measure** or **length**.

Measuring is never exact, and you will need to decide how precise the measurement should be. You will also need to decide on some standard unit of measurement. The system of measurement used in the United States comes from the British system, but it originally dates back to the Babylonians and Egyptians. Measurements were made in terms of the human body (digit, palm, cubit, span, and foot). Eventually, measurements were standardized in terms of the physical measurement of certain monarchs. King Henry I, for example, decreed that one yard was the distance from the tip of his nose to the end of his thumb.

In 1790 the French Academy of Science was asked by the government to develop a logical system of measurement, and the original metric system came into being. By 1900 it was adopted by over 35 major countries. In 1906 there was a major effort to make the conversion to the metric system in the United States, but it was opposed by big business and the attempt failed. In 1960 the metric system was revised and simplified into what is now known as the **SI system** (an abbreviation of *Système International d'Unités*). The term *metric* in this book will refer to the SI system.

In 1972 and 1973, when the United States was the only major country not using the metric system (see Figure 5.1), further attempts to make it mandatory failed. However, in 1975 Congress declared conversion to the metric system to be "national policy." This time big business supported the drive toward metric conversion, and it appears inevitable that the metric system will eventually come into use in the United States.

ONE YARD

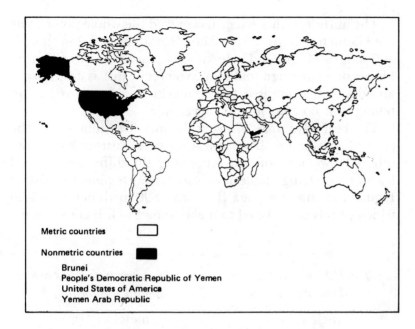

FIGURE 5.1 The metric world

The most difficult problem in changing from the British system to the metric system in the United States is not mathematical, but psychological. Many people fear that changing to the metric system will require complex multiplying and dividing and the use of confusing decimal points. For example, in a recent popular article, James Collier states:

> For instance, if someone tells me it's 250 miles up to Lake George, or 400 out to Cleveland, I can pretty well figure out how long it's going to take and plan accordingly. Translating all of this into kilometers is going to be an awful headache. A kilometer is about 0.62 miles, so to convert miles into kilometers you divide by six and multiply by ten, and even that isn't accurate. Who can do that kind of thing when somebody is asking me are we almost there, the dog is beginning to drool and somebody else is telling you you're driving too fast?
>
> Of course, that won't matter, because you won't know how fast you're going anyway. I remember once driving in a rented car on a superhighway in France, and everytime I looked down at the speedometer we were going 120. That kind of thing can give you the creeps. What's it going to be like when your wife keeps shouting, "Slow down, you're going almost 130"?
>
> But if you think kilometers will be hard to calculate. . . .

The author of this article has missed the whole point. Why are kilometers hard to calculate? How does he know that it's 400 miles to Cleveland? He knows because the odometer on his car or a road sign told him. Won't it be just as easy to read an odometer calibrated to kilometers or a metric road sign telling him how far it is to Cleveland?

The real advantage of using the metric system will be the ease of conversion from one unit of measurement to another. How many of you remember the difficulty you had learning to change tablespoons to cups? Or pints to gallons? Figure 5.2 shows a page from an 1890 arithmetic book in which pupils were asked to make some English conversions.

U.S. and Metric Standard Units of Length

The U.S. system has four standard units of length:	*The metric system has one standard unit of length:*
inch (in.) foot (ft) yard (yd) mile (mi)	meter (m)

In order to work with different lengths, you must memorize some conversion factors:

U.S. conversion factors for length	*Metric conversion factors for length*
12 in. = 1 ft 3 ft = 1 yd 1,760 yd = 1 mi 5,280 ft = 1 mi	*centi* means $\frac{1}{100}$; centimeter is abbreviated cm *kilo* means 1,000; kilometer is abbreviated km

In order to measure length, you need a measuring device appropriate to the length you are measuring. For example, you would not use the same ruler to measure the length of

FIGURE 5.2 A page from *First Book of Arithmetic for Pupils Uniting Oral and Written Exercises* by Emerson E. White (New York: American Book Co., 1890)

140   *FIRST BOOK OF ARITHMETIC.*

13. How many quarts in 10 bushels?
14. How many pints in 3 pecks?
15. How many bushels in 64 quarts?
16. What part of a quart is 1 pint? 3 pints?
17. What part of a peck is 1 quart? 3 quarts?
18. What part of a bushel is 1 peck? 2 pecks?
19. A man sold 1 bu. 3 pk. of clover-seed at 8 cents a quart: how much did he receive?
20. A fruit-dealer paid $7 for 3 bu. 3 pk. of peaches, and sold them at 60 cents a peck: what was his gain?
21. How many pecks of chestnuts can be bought for $15.60, at 40 cents a peck? How many bushels?

LIQUID MEASURE.

Liquid Measure is used in measuring liquids; as, oil, milk, alcohol, etc.

The denominations are *gills, pints, quarts,* and *gallons.*

TABLE.

4 gills (*gi.*) . . . . are 1 pint . . . . *pt.*
2 pints . . . . . . are 1 quart . . . . *qt.*
4 quarts . . . . . are 1 gallon . . . *gal.*
1 gal.   4 qt. = 8 pt. = 32 gi.

1 inch $\frac{1}{16}$ inch / $\frac{1}{8}$ inch / $\frac{1}{4}$ inch / $\frac{1}{2}$ inch

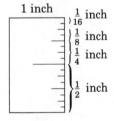

1 centimeter   .1 centimeter / .5 centimeter

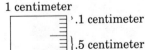

FIGURE 5.3 Actual size comparisons of inch and centimeter

this page, the length of a football field, and the distance between Los Angeles and San Francisco. You must also decide which measurement system to use—U.S. or metric. Finally, you need to decide on the precision of your measurement.

EXAMPLE 1

Measure the following line segments:

A ————————————————————

B ——————————————————

C ————————————————

a. To measure the segments to the nearest centimeter, place a ruler showing centimeters next to each segment.

A ————————————————————— End of A is nearest to 6

| 1 | 2 | 3 | 4 | 5 | 6 | 7 | 8 |

B ————————————————— End of B is nearer to 5 than to 6

| 1 | 2 | 3 | 4 | 5 | 6 | 7 | 8 |

C ——————————————— End of C is nearer to 5 than to 4

| 1 | 2 | 3 | 4 | 5 | 6 | 7 | 8 |

Centimeters is sometimes abbreviated cm. Segment A is 6 cm long, and B and C are 5 cm long (to the nearest centimeter).

b. To measure the segments to the nearest tenth of a centimeter, place a ruler showing tenths of a centimeter next to each segment. Notice that it looks like the length of segment A is right on 6 cm. When measuring to the nearest tenth of a centimeter, we write 6.0 cm to indicate that this measurement is correct to the nearest tenth.

B ————————————— B is 5.3 cm long

| 1 | 2 | 3 | 4 | 5 | 6 | 7 | 8 |

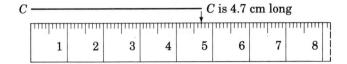

C —————————————————— C is 4.7 cm long

To measure distances that are too long to be conveniently measured in inches or in centimeters, we move to the next larger unit of measurement.

U.S. system	Metric system
12 in. = 1 ft 3 ft = 1 yd 36 in. = 1 yd	100 cm = 1 meter (m)

The important thing to remember about the conversion to the metric system is simply that you will use different measuring devices. Instead of a *yardstick* you will use a *meterstick*. **You should not convert from one system to the other.** Simply remeasure the length using a different standard unit, as we did in the previous examples. A comparison between the yardstick and the meterstick is shown in Figure 5.4. However, during the transition period you should remember that a meter is about 3 inches longer than a yard.

For measuring long distances, such as between cities, we move to the next larger unit of measurement.

U.S. system	Metric system
1,760 yd = 1 mile 5,280 ft = 1 mile 63,360 in. = 1 mile	1,000 m = 1 kilometer (km) 100,000 cm = 1 km

We don't usually measure miles or kilometers directly. We rely on road signs, maps, and the odometers on our cars to tell us the distance between cities. When the road signs, maps, and odometers give distances in kilometers, we will be able to estimate large metric distances as easily as we can

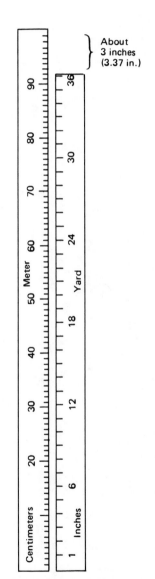

About
3 inches
(3.37 in.)

FIGURE 5.4 Comparison of a yard and a meter; these measuring sticks are about $\frac{1}{6}$ actual size

estimate miles. However, during the changeover period, there may be times when you need to convert from one system to the other.

To change from kilometers to miles:
  1 kilometer is about .6 mile, so multiply kilometers by .6

To change from miles to kilometers:
  1 mile is about 1.6 kilometers, so multiply miles by 1.6

The exact conversion is more complicated, but for all practical purposes you can use these estimates. Since these are just estimates and are not exact, we use the approximately equal symbol, $\approx$.

EXAMPLE 2

Change from kilometers to miles.

a. 600 km; multiply by .6: $600 \times .6 \approx 360$ mi
b. 200 km; $200 \times .6 \approx 120$ mi   ∎

EXAMPLE 3

Change from miles to kilometers.

a. 500 miles; multiply by 1.6: $500 \times 1.6 \approx 800$ km
b. 340 miles; $340 \times 1.6 \approx 544$ km   ∎

Problem Set 5.1

From memory, and without using any measuring devices, draw a line segment with the length indicated in Problems 1–6.

1. 1 in.   **2.** 2 in.   **3.** 3 in.   **4.** 1 cm
5. 5 cm   **6.** 10 cm

Measure the segments given in Problems 7–10 to the indicated accuracy.

7. **a.** To the nearest centimeter ⎫
    b. To the nearest $\frac{1}{10}$ centimeter ⎭ _____

8. **a.** To the nearest centimeter ⎫
    b. To the nearest $\frac{1}{10}$ centimeter ⎭ _____

9. **a.** To the nearest centimeter ⎫
    b. To the nearest $\frac{1}{10}$ centimeter ⎭ _____

10. **a.** To the nearest centimeter ⎫
    b. To the nearest $\frac{1}{10}$ centimeter ⎭ _____

Pick the best choices in Problems 11–22 by estimating. Do not measure or convert to the U.S. system, but try to visualize the length of a meter. The hardest part of the transition to the metric system is the transition to **thinking in metrics**.

11. The length of a VW bug is about:
    A. 1 m   B. 4 m   C. 10 m

12. The length of a 100 yard football field is:
    A. 100 m   B. More than 100 m   C. Less than 100 m

13. The height of a kitchen table is usually:
    A. 1 m   B. Less than 1 m   C. More than 1 m

14. The distance from floor to ceiling in a typical home is about:
    A. 2.5 m   B. .5 m   C. 4.5 m

15. An adult's height is most likely to be about:
    A. 6 m   B. 50 cm   C. 170 cm

16. The width of this book is about:
    A. 1.9 cm   B. 19 cm   C. 1.9 m

17. The width of a dollar bill is about:
    A. .65 cm   B. 6.5 cm   C. .65 m

18. The distance from your home to the nearest grocery store is most likely to be:
    A. 1 cm   B. 1 m   C. 1 km

19. The prefix *centi* means:
    A. One thousand   B. One thousandth   C. One hundredth

20. The prefix *kilo* means:
    A. One thousand   B. One thousandth   C. One hundredth

21. The distance from San Francisco to New York is about 3,000 miles. This distance in kilometers is:
    A. Less than 3,000 km   B. More than 3,000 km   C. About 3,000 km

22. In 1964 Bob Hayes ran the 100 meter dash in 10 seconds flat. At this same rate, he could run a 100 yard dash in:
    A. Less than 10 sec   B. More than 10 sec   C. 10 sec

*Change the distances given in Problems 23–28 from kilometers to miles.
Use the approximation given in this section.*

23. Boston to Chicago;   1,580 km
24. Dallas to Memphis;   750 km
25. Louisville to Birmingham;   600 km
26. Washington, D.C., to Moscow;   7,800 km
27. New York to Cape Town;   12,500 km
28. London to Sydney;   17,000 km

*Change the distances given in Problems 29–34 from miles to kilometers.
Use the approximation given in this section.*

29. San Francisco to Los Angeles;   400 miles
30. New York to Washington, D.C.;   230 miles
31. Miami to New Orleans;   890 miles
32. Honolulu to New York;   4,960 miles
33. San Francisco to Paris;   5,560 miles
34. Tokyo to Buenos Aires;   11,400 miles

Find your own metric measurements.

Women
35. Height
36. Bust
37. Waist
38. Hips
39. Distance from waist to hemline
40. Fist (measure around largest
part of hand over knuckles while
making a fist, excluding thumb)

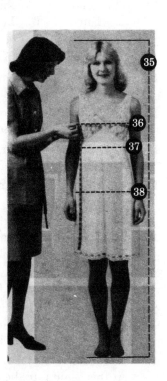

Men

35. Height
36. Chest
37. Waist
38. Seat
39. Neck
40. Length of shoe

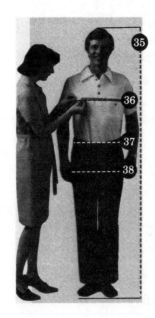

5.2 Measuring Capacity

In the last section we measured length. Another quantity we measure is **capacity**. The amount of liquid a container can hold measures the container's capacity. The capacities of a can of cola, a bottle of milk, an aquarium tank, the gas tank in your car, or the size of a swimming pool can all be measured by the amount of fluid they can hold.

The U.S. system has seven common units of capacity for liquid measure:	*The metric system has one standard unit of capacity:*	**U.S. and Metric Standard Units of Capacity**
teaspoon (tsp) tablespoon (tbsp) ounce (oz) cup (c) pint (pt) quart (qt) gallon (gal)	liter (ℓ)	

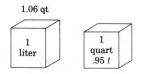

FIGURE 5.5   Comparison of a quart and a liter

In order to work with different capacities, you must memorize some conversion factors.

U.S. conversion factors for capacity	*Metric conversion factors for capacity*
3 tsp = 1 tbsp	*milli* means $\frac{1}{1,000}$; milliliter is abbreviated mℓ
1 oz = 2 tbsp	
16 tbsp = 1 c	*kilo* means 1,000; kiloliter is abbreviated kℓ
8 oz = 1 c	
16 oz = 1 pt	
2 c = 1 pt	
2 pt = 1 qt	
4 qt = 1 gal	

Most containers of liquid that you buy have capacities stated in both milliliters and ounces, or quarts and liters. Some of these size statements are shown in Table 5.1. The U.S. Bureau of Alcohol, Tobacco, and Firearms has made metric bottle sizes for liquor mandatory, so the half-pint, fifth, and quart will be replaced by 200 mℓ, 750 mℓ, and 1 ℓ sizes. A typical dosage of cough medicine would be 5 mℓ, and 1 kℓ is 1,000 ℓ or about the amount of water one person would use for all purposes in two or three days.

TABLE 5.1   Capacities of Common Grocery Items as Shown on Labels

Item	*U.S. capacity*	*Metric capacity*
Milk	$\frac{1}{2}$ gal	1.89 ℓ
Milk	1.06 qt	1 ℓ
Budweiser	12 oz	355 mℓ
Coke	67.6 oz	2 ℓ
Hawaiian Punch	1 qt	.95 ℓ
Del Monte Pickles	1 pt 6 oz	651 mℓ

Since it is common practice to label the capacities both in U.S. and in metric measuring units, it will generally not be necessary for you to make conversions from one system to another. But, if you do, it is easy to remember that a liter is just a little larger than a quart, just as a meter is a little larger than a yard.

For measuring capacity you use a measuring cup.

EXAMPLE 1

Measure the amount of liquid in the measuring cup both in the U.S. system and in the metric system.

5 marks = 100 mℓ, so each mark is 20 mℓ

mℓ	cups	
500	2	16 oz
400	$\frac{3}{4}$	14 oz
	$\frac{1}{2}$	12 oz
300	$\frac{1}{4}$	10 oz
	1	8 oz
200	$\frac{3}{4}$	6 oz
100	$\frac{1}{2}$	4 oz
	$\frac{1}{4}$	2 oz

Metric: 240 mℓ   U.S.:   About 1 c or 8 oz   ∎

The only places where you may have to make conversions from one system to another is at the gas pump or in the kitchen. The common conversions are shown in Table 5.2 on page 180.

EXAMPLE 2

Use Table 5.2 to make the indicated conversions.

a. 50 ℓ ≈ 13.2 gal
b. 15 gal ≈ 57 ℓ
c. 1 tbsp ≈ 15 mℓ
d. 1 oz ≈ 30 mℓ   ∎

My WIFE AND I stopped for lunch in a Nebraska town on our way to California, and I asked the waitress how much snow the area usually got. "About as deep as a meter," she replied. Impressed by her use of the metric system, I asked where she had learned it. She was momentarily baffled, then said, "That's the one I mean," and pointed out the window to the parking meter in front of the restaurant.
—N. A. Norris (Grand Marais, Mich.)

TABLE 5.2 Common Capacity Conversions

IN THE KITCHEN
1 tsp ≈ 5 mℓ
1 tbsp ≈ 15 mℓ
1 oz ≈ 30 mℓ
1 c ≈ 240 mℓ
1 pt ≈ 475 mℓ
1 qt ≈ 950 mℓ

AT THE GAS PUMP

Liters	Gallons	Liters	Gallons	Liters	Gallons	Liters	Gallons
1	.3	26	6.9	51	13.5	76	20.1
2	.5	27	7.1	52	13.7	77	20.3
3	.8	28	7.4	53	14.0	78	20.6
4	1.1	29	7.7	54	14.3	79	20.9
5	1.3	30	7.9	55	14.5	80	21.1
6	1.6	31	8.2	56	14.8	81	21.4
7	1.8	32	8.5	57	15.0	82	21.7
8	2.1	33	8.7	58	15.3	83	21.9
9	2.4	34	9.0	59	15.6	84	22.2
10	2.6	35	9.2	60	15.9	85	22.5
11	2.9	36	9.5	61	16.1	86	22.7
12	3.2	37	9.8	62	16.4	87	23.0
13	3.4	38	10.0	63	16.6	88	23.2
14	3.7	39	10.3	64	16.9	89	23.5
15	4.0	40	10.6	65	17.2	90	23.8
16	4.2	41	10.8	66	17.4	91	24.0
17	4.5	42	11.1	67	17.7	92	24.3
18	4.8	43	11.4	68	18.0	93	24.6
19	5.0	44	11.6	69	18.2	94	24.8
20	5.2	45	11.9	70	18.5	95	25.1
21	5.5	46	12.2	71	18.8	96	25.4
22	5.8	47	12.4	72	19.0	97	25.6
23	6.1	48	12.7	73	19.3	98	25.9
24	6.3	49	12.9	74	19.6	99	26.1
25	6.6	50	13.2	75	19.8	100	26.4

Problem Set 5.2

Measure the amounts given in Problems 1–12.

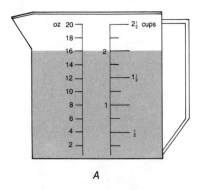

A

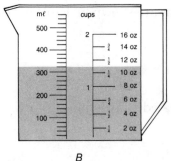

B

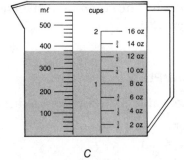

C

1. Container *A* in cups

2. Container *A* in ounces

3. Container *B* in ounces

4. Container *C* in ounces

5. Container *B* in milliliters

6. Container *C* in milliliters

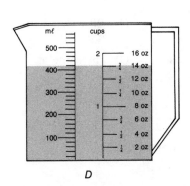

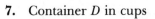

D   E   F   G

7. Container *D* in cups
8. Container *D* in ounces
9. Container *D* in milliliters
10. Container *E* in milliliters
11. Container *F* in milliliters
12. Container *G* in milliliters

Without measuring, pick the best answer in each of Problems 13–26 by estimating.

13. An average cup of coffee is about:
    A. 250 mℓ   B. 750 mℓ   C. 1 ℓ

14. If you want to paint some small bookshelves, how much paint would you probably need?
    A. 2 mℓ   B. 200 mℓ   C. 2 ℓ

15. A six-pack of beer would contain about:
    A. 2 mℓ   B. 200 mℓ   C. 2 ℓ

16. The dosage of a strong cough medicine might be:
    A. 2 mℓ   B. 200 mℓ   C. 2 ℓ

17. A glass of water served at a restaurant is about:
    A. 200 mℓ   B. 2 mℓ   C. 2 ℓ

18. Enough water for a bath would be about:
    A. 150 mℓ   B. 150 ℓ   C. 150 kℓ

19. Enough gas to fill your car's empty gas tank would be about:
    A. 15 ℓ   B. 200 mℓ   C. 70 ℓ

20. 50 kℓ of water would be about enough for:
    A. Taking a bath
    B. Taking a swim
    C. Supplying the drinking water for a large city

21. Which measurement would be appropriate for administering some medication?
    A. mℓ   B. ℓ   C. kℓ

22. You order some champagne for yourself and one companion. You would most likely order:
A. 2 mℓ B. 700 mℓ C. 20 ℓ

23. A liter is ___?___ a quart.
A. larger than   B. smaller than   C. the same as

24. The prefix *centi* means:
A. One thousand   B. One thousandth
C. One hundredth

25. The prefix *milli* means:
A. One thousand   B. One thousandth
C. One hundredth

26. The prefix *kilo* means:
A. One thousand   B. One thousandth
C. One hundredth

Use Table 5.2 to make the conversions in Problems 27–38.

27. 1 tsp ≈ _____ mℓ **28.** 1 qt ≈ _____ mℓ

29. 1 c ≈ _____ mℓ **30.** 1 oz ≈ _____ mℓ

31. 1 pt ≈ _____ mℓ **32.** 70 ℓ ≈ _____ gal

33. 10 gal ≈ _____ ℓ **34.** 40 ℓ ≈ _____ gal

35. 35 ℓ ≈ _____ gal   **36.** 5 gal ≈ _____ ℓ

37. 14 gal ≈ _____ ℓ **38.** 68 ℓ ≈ _____ gal

Mind Boggler

39. You have, no doubt, seen many examples of metric usage— for example, 100 mm cigarettes, 35 mm cameras, or liter steins of beer. Find some examples of the metric system in advertising or in other aspects of American life today.

5.3 Measuring Weight and Temperature

The third type of quantity that we measure is weight.

U.S. and Metric Standard Units of Weight

The U.S. system has three common weights:	*The metric system has one standard unit of weight:*
ounce (oz) pound (lb) ton (T)	gram (g)

In order to work with different weights, you must memorize some conversion factors.

U.S. conversion factors for weight	Metric conversion factors for weight
16 oz = 1 lb 2,000 lb = 1 T	*milli* means $\frac{1}{1,000}$; milligram is abbreviated mg *kilo* means 1,000; kilogram is abbreviated kg

About 2.2 lb

1 kilogram   1 pound

FIGURE 5.6   Comparison of a pound and a kilogram

A paperclip weighs about 1 g, a cube of sugar about 3 g, and a nickel about 5 g. This book weighs about 1 kg, and an average-sized person weighs from 50 to 100 kg. Some common weights of grocery items are shown in Table 5.3.

TABLE 5.3   Weights of Common Grocery Items as Shown on Labels

Item	U.S. weight	Metric weight
Kraft cheese spread	5 oz	142 g
Del Monte tomato sauce	8 oz	227 g
Campbell cream of chicken soup	$10\frac{3}{4}$ oz	305 g
Kraft marshmallow cream	11 oz	312 g
Bag of sugar	5 lb	2.3 kg
Bag of sugar	22 lb	10 kg

We use a scale to measure weight. In order to weigh items, or ourselves, in metric units, we need only to replace our U.S. weight scales with metric weight scales. As with other measures, we need to begin to think in terms of metric units. The multiple-choice questions in Problem Set 5.3 are designed to help you do this.

The final quantity of measure that we'll consider in this chapter is the measurement of temperature.

**U.S. and Metric
Standard Units of
Temperature**

The U.S. system has one common temperature unit:	*The metric system has one common temperature unit:*
Fahrenheit (°F)	Celsius (°C)

In order to work with temperatures, it is necessary to have some reference points.

U.S. temperature	*Metric temperature*
Water freezes: 32°F Water boils: 212°F	Water freezes: 0°C Water boils: 100°C

We are usually interested in measuring temperature in three areas: atmospheric temperature (usually given in weather reports), body temperature (used to determine illness), and oven temperature (used in cooking). The same

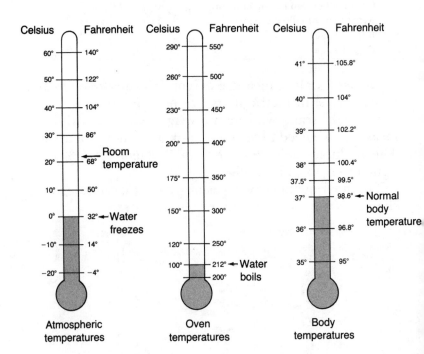

FIGURE 5.7 Temperature
comparisons between Celsius and
Fahrenheit

scales are used, of course, for measuring all these temperatures. But notice the difference in the range of temperatures we're considering. The comparisons for Fahrenheit and Celsius are shown in Figure 5.7.

As we've mentioned before, during the transition period from U.S. to metric measure, one of the few places where it will be necessary to calculate conversions will be in the kitchen. Even though there are exact conversion factors, most recipes can be rounded and changed from one system to another by use of a table. Table 5.4 can be used for this purpose; it doesn't show exact conversions, but it gives reasonable conversions for recipes. For example, 250°F is 121°C, but for practical purposes in the kitchen, a recipe would call for a 120°C oven and not a 121°C oven.

TABLE 5.4   U.S.–Metric Kitchen Conversions

Capacity		Weight		Temperature	
$\frac{1}{4}$ tsp	1.25 mℓ	1 oz	28 g	225°F	105°C
$\frac{1}{2}$ tsp	2.5 mℓ			250°F	120°C
1 tsp	5 mℓ	$\frac{1}{8}$ lb	57 g	275°F	135°C
1 tbsp	15 mℓ	$\frac{1}{4}$ lb	114 g	300°F	150°C
		$\frac{1}{3}$ lb	151 g	325°F	165°C
$\frac{1}{8}$ c	30 mℓ	$\frac{1}{2}$ lb	227 g	350°F	180°C
$\frac{1}{4}$ c	60 mℓ	$\frac{2}{3}$ lb	302 g	375°F	190°C
$\frac{1}{3}$ c	80 mℓ	$\frac{3}{4}$ lb	341 g	400°F	205°C
$\frac{1}{2}$ c	120 mℓ	1 lb	454 g	425°F	220°C
$\frac{2}{3}$ c	160 mℓ			450°F	230°C
$\frac{3}{4}$ c	180 mℓ	2.2 lb	1 kg	475°F	250°C
1 c	240 mℓ	or 1,000 g		500°F	260°C
				525°F	275°C
$\frac{1}{4}$ pt	120 mℓ			550°F	290°C
$\frac{1}{2}$ pt	240 mℓ			575°F	300°C
$\frac{3}{4}$ pt	360 mℓ			600°F	315°C
1 pt	475 mℓ				
2 pt	950 mℓ				
1 qt	950 mℓ				
1 gal	3,685 mℓ				

EXAMPLE 1

Use Table 5.4 to obtain the requested conversions.

a. $\frac{3}{4}$ lb $\approx$ ____?____ g
This is a weight conversion; from Table 5.4,
$\frac{3}{4}$ lb $\approx$ 341 g

b. 400°F $\approx$ ____?____ °C
This is a temperature conversion; from Table 5.4,
400°F $\approx$ 205°C

c. $\frac{1}{3}$ c $\approx$ ____?____ mℓ
This is a capacity conversion; from Table 5.4,
$\frac{1}{3}$ c $\approx$ 80 mℓ

d. 150 g $\approx$ ____?____ lb
This is a weight conversion. Notice that Table 5.4 shows
151 g and not 150 g, but, for recipe purposes, this is close
enough to $\frac{1}{3}$ lb.   ■

Problem Set 5.3

Without measuring, pick the best choice in each of Problems 1–20 by estimating.

1. A hamburger patty would weigh about:
   A. 170 g   B. 240 mg   C. 2 kg

2. A can of carrots at the grocery store most likely weighs about:
   A. 40 kg   B. 4 kg   C. .4 kg

3. A newborn baby would weigh about:
   A. 490 mg   B. 4 kg   C. 140 kg

4. John tells you he weighs 150 kg. If John is an adult, do you think he is:
   A. Underweight?   B. About average?
   C. Overweight?

5. You have invited 15 people over for Thanksgiving dinner. You should buy a turkey that weighs about:
   A. 795 mg   B. 4 kg   C. 12 kg

6. Water boils at:
   A. 0°C   B. 100°C   C. 212°C

7. If it is 32°C outside, you would most likely find people:
   A. Ice skating   B. Water skiing

8. If the doctor said that your child's temperature was 37°C, would you think your child's temperature was:
   A. Low?   B. Normal?   C. High?

9. You would most likely broil steaks at:
   A. 120°C   B. 500°C   C. 290°C

10. A kilogram is _____?_____ a pound.
   A. more than   B. less than   C. about the same as

11. The prefix used to mean 1,000 is:
   A. centi   B. milli   C. kilo

12. The prefix used to mean $\frac{1}{1,000}$ is:
   A. centi   B. milli   C. kilo

13. The prefix used to mean $\frac{1}{100}$ is:
   A. centi   B. milli   C. kilo

14. 15 kg is a measure of:
   A. Length   B. Capacity   C. Weight
   D. Temperature

15. 28.2 m is a measure of:
   A. Length   B. Capacity   C. Weight
   D. Temperature

16. 6 ℓ is a measure of:
   A. Length   B. Capacity   C. Weight
   D. Temperature

17. 38°C is a measure of:
   A. Length   B. Capacity   C. Weight
   D. Temperature

18. 7 mℓ is a measure of:
   A. Length   B. Capacity   C. Weight
   D. Temperature

19. 68 km is a measure of:
   A. Length   B. Capacity   C. Weight
   D. Temperature

20. 14.3 cm is a measure of:
   A. Length   B. Capacity   C. Weight
   D. Temperature

Name the metric unit you would use to measure each of the quantities in Problems 21–28.

21. The distance from New York to Chicago

22. The distance around your waist

23. The amount of gin in a martini

24. The amount of water in a swimming pool

25. The weight of a pencil

26. The weight of an automobile

27. The outside temperature

28. The temperature to bake a cake

Use Table 5.4 for Problems 29–46 to obtain the requested conversion.

29. $\frac{1}{8}$ c ≈ _____ mℓ   **30.** $\frac{3}{4}$ lb ≈ _____ g

31. $\frac{1}{3}$ lb ≈ _____ g   **32.** 350°F ≈ _____ °C

33. 550°F ≈ _____ °C   **34.** 275°F ≈ _____ °C

35. 1 lb ≈ _____ g   **36.** 1 tbsp ≈ _____ mℓ

37. 2 pt ≈ _____ mℓ   **38.** 5 mℓ ≈ _____ tsp

39. 65 mℓ ≈ _____ c   **40.** 950 mℓ ≈ _____ qt

41. 30 g ≈ _____ oz   **42.** 230 g ≈ _____ lb

43. 1 kg ≈ _____ lb   **44.** 190°C ≈ _____ °F

45. 220°C ≈ _____ °F   **46.** 300°C ≈ _____ °F

Mind Bogglers

47. Translate the following apple crisp recipe into a metric measurement recipe.

Apple Crisp Recipe

6 to 8 apples, sliced (about 1 qt)   1 tsp cinnamon
$\frac{1}{4}$ cup water   6 tbsp butter
$\frac{3}{4}$ cup sugar $\frac{1}{2}$ tsp salt
$\frac{1}{2}$ cup cake flour

Peel and slice the apples thinly. Put them in a baking dish. Add the water. Combine the sugar, flour, cinnamon and salt; blend in the butter until crumbly in consistency. Pour over the apples and press down. Bake uncovered in 375° oven for about 1 hour.

Mrs. Rigmor Sovndal

48. Translate the following sponge cake recipe into a U.S. measurement recipe.

Hot Milk Sponge Cake Recipe

6 or 7 eggs   10 mℓ baking powder
5 mℓ vanilla   pinch of salt
560 mℓ sifted cake flour   240 mℓ milk heated to boiling with
480 mℓ sugar   115 g butter

Beat eggs with electric mixer until light and fluffy. Gradually add sugar and beat until very light. Add vanilla. Set the mixer on slowest speed and beat in dry ingredients. Don't overbeat. Using same speed, add hot milk and butter. Pour batter into two 23 cm pans lined with wax paper and greased. Bake at 180° for 25 minutes.

Mrs. Linda Smith

49. Bake one of the recipes given in either Problem 47 or 48 using metric measurements and bring some samples to class.

5.4 Converting Units

When working with either the metric or the U.S. measurement systems, you sometimes need to change units of measurement within a particular system.

This section is divided into two parts so that you can work either with the U.S. system or with the metric system, or with both if you prefer. The conversion between these systems is optional and is given as an appendix to this section.

U.S. Measurement Conversions

To change from one unit to another, make substitutions of equal quantities using the U.S. conversion factors, which are repeated here for convenience.

Length	*Capacity*	*Weight*
12 in. = 1 ft	3 tsp = 1 tbsp	16 oz = 1 lb
3 ft = 1 yd	1 oz = 2 tbsp	2,000 lb = 1 T
1,760 yd = 1 mi	16 tbsp = 1 c	
5,280 ft = 1 mi	8 oz = 1 c	
	16 oz = 1 pt	
	2 c = 1 pt	
	2 pt = 1 qt	
	4 qt = 1 gal	

U.S. Measurement Conversions

EXAMPLE 1

Make the indicated conversions.

a. Change 8 ft to inches.

$$8 \text{ ft} = 8 \times 1 \text{ ft}$$
$$= 8 \times (\mathbf{12 \text{ in.}}) \qquad \text{Since 1 ft = 12 in., substitute 12 in.}$$
$$= 96 \text{ in.} \qquad \text{for 1 ft.}$$

b. Change 8 ft to yards.

$$8 \text{ ft} = 8 \times 1 \text{ ft}$$
$$= 8 \times \left(\frac{1}{3} \mathbf{yd}\right) \qquad \text{Since 1 ft = } \tfrac{1}{3} \text{ yd}$$
$$= \frac{8}{3} \text{ yd} = 2\tfrac{2}{3} \text{ yd}$$

c. Change 1,500 yd to feet.

$$1,500 \text{ yd} = 1,500 \times 1 \text{ yd}$$
$$= 1,500 \times (\mathbf{3 \text{ ft}}) \qquad \text{Since 1 yd = 3 ft}$$
$$= 4,500 \text{ ft}$$

d. Change 86 in. to feet.

$$86 \text{ in.} = 86 \times 1 \text{ in.}$$
$$= 86 \times \left(\frac{1}{12} \mathbf{ft}\right) \qquad \text{Since 1 in. = } \tfrac{1}{12} \text{ ft}$$
$$= \frac{86}{12} \text{ ft}$$
$$= 7\tfrac{2}{12} \text{ ft} = 7\tfrac{1}{6} \text{ ft}$$

e. Change 12 tsp to ounces.

$$12 \text{ tsp} = 12 \times 1 \text{ tsp}$$
$$= 12 \times \left(\frac{1}{3} \mathbf{tbsp}\right) \qquad \text{Since 1 tsp = } \tfrac{1}{3} \text{ tbsp}$$
$$= 12 \times \frac{1}{3} \times 1 \text{ tbsp}$$
$$= 12 \times \frac{1}{3} \times \left(\frac{1}{2} \mathbf{oz}\right) \qquad \text{Since 1 tbsp = } \tfrac{1}{2} \text{ oz}$$
$$= \frac{12}{6} \text{ oz} = 2 \text{ oz}$$

f. Change 5 c to pints.

$$5 \text{ c} = 5 \times 1 \text{ c}$$
$$= 5 \times \left(\frac{1}{2} \text{ pt}\right) \qquad \text{Since 1 c} = \frac{1}{2} \text{ pt}$$
$$= 2\frac{1}{2} \text{ pt}$$

g. Change 2 qt 1 pt to ounces.

$$2 \text{ qt } 1 \text{ pt} = \quad 2 \text{ qt} \quad + 1 \text{ pt}$$
$$= 2 \times (\textbf{2 pt}) + 1 \text{ pt} \qquad \text{Since 1 qt} = 2 \text{ pt}$$
$$= 4 \text{ pt} + 1 \text{ pt}$$
$$= 5 \text{ pt}$$
$$= 5 \times 1 \text{ pt}$$
$$= 5 \times (\textbf{16 oz}) \qquad \text{Since 1 pt} = 16 \text{ oz}$$
$$= 80 \text{ oz}$$

h. Change 48 oz to pints.

$$48 \text{ oz} = 48 \times 1 \text{ oz}$$
$$= 48 \times \left(\frac{1}{16} \text{ pt}\right) \qquad \text{Since 1 oz} = \frac{1}{16} \text{ pt}$$
$$= \frac{48}{16} \text{ pt}$$
$$= 3 \text{ pt}$$

i. Change 5 lb to ounces.

$$5 \text{ lb} = 5 \times 1 \text{ lb}$$
$$= 5 \times (\textbf{16 oz}) \qquad \text{Since 1 lb} = 16 \text{ oz}$$
$$= 80 \text{ oz}$$

j. Change 48 oz to pounds.

$$48 \text{ oz} = 48 \times 1 \text{ oz}$$
$$= 48 \times \left(\frac{1}{16} \text{ lb}\right) \qquad \text{Since 1 oz} = \frac{1}{16} \text{ lb}$$
$$= \frac{48}{16} \text{ lb}$$
$$= 3 \text{ lb}$$

k. Change $2\frac{1}{4}$ lb to ounces.

$$2\frac{1}{4} \text{ lb} = 2\frac{1}{4} \times 1 \text{ lb}$$
$$= 2\frac{1}{4} \times (\textbf{16 oz}) \qquad \text{Since 1 lb = 16 oz}$$
$$= \frac{9}{\overset{}{\underset{1}{\cancel{4}}}} \times \frac{\overset{4}{\cancel{16}}}{1} \text{ oz}$$
$$= 36 \text{ oz}$$

l. Change $8\frac{1}{2}$ T to pounds.

$$8\frac{1}{2} \text{ T} = 8\frac{1}{2} \times 1 \text{ T}$$
$$= 8\frac{1}{2} \times (\textbf{2,000 lb}) \qquad \text{Since 1 T = 2,000 lb}$$
$$= \frac{17}{2} \times \frac{2,000}{1} \text{ lb}$$
$$= 17,000 \text{ lb} \qquad\qquad\qquad \blacksquare$$

Metric Measurement Conversions

As you've seen, there are three basic metric units:

Length:	meter
Capacity:	liter
Weight:	gram

These units are combined with certain prefixes:

milli	means	$\frac{1}{1,000}$
centi	means	$\frac{1}{100}$
kilo	means	1,000

Other metric units are used less frequently but they should be mentioned:

deci	means	$\frac{1}{10}$
deka	means	10
hecto	means	100

A listing of all metric units (in order of size) is given in Table 5.5.

To convert from one metric unit to another, you **simply move the decimal point.**

TABLE 5.5   Metric Measurements

Length	Capacity	Weight	Meaning	Memory aid
kilometer (km)	**kilo**liter (kℓ)	**kilo**gram (kg)	1,000 units	**K**arl
hectometer (hm)	**hecto**liter (hℓ)	**hecto**gram (hg)	100 units	**H**as
dekameter (dkm)	**deka**liter (dkℓ)	**deka**gram (dkg)	10 units	**D**eveloped
meter (m)	**liter** (ℓ)	**gram** (g)	1 unit	**M**y
decimeter (dm)	**deci**liter (dℓ)	**deci**gram (dg)	0.1 unit	**D**ecimal
centimeter (cm)	**centi**liter (cℓ)	**centi**gram (cg)	0.01 unit	**C**raving for
millimeter (mm)	**milli**liter (mℓ)	**milli**gram (mg)	0.001 unit	**M**etrics

EXAMPLE 2

Write each given quantity using all the other prefixes.

a. 5 ℓ

Step 1. Place the decimal point on the given number if it is not shown: 5. ℓ

Step 2. Set up a chart as shown below and write the decimal point on the line corresponding to the given unit—the line just to the right of liter for this example:

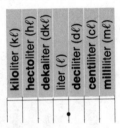

Step 3. Write the number in the chart so that the decimal point is correctly aligned—put one digit in each column:

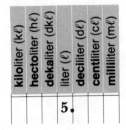

5 ℓ   Given

Step 4. Assume that there are 0s in all the other columns. To change to another unit, simply move the decimal point to the column corresponding to the unit to which you are changing.

To change to mℓ:

kiloliter (kℓ)	hectoliter (hℓ)	dekaliter (dkℓ)	liter (ℓ)	deciliter (dℓ)	centiliter (cℓ)	milliliter (mℓ)
0	0	0	5.	0	0	0

Move point to mℓ column

0	0	0	5	0	0	0.

Answer: 5,000 mℓ

To change to kℓ:

kiloliter (kℓ)	hectoliter (hℓ)	dekaliter (dkℓ)	liter (ℓ)	deciliter (dℓ)	centiliter (cℓ)	milliliter (mℓ)
0	0	0	5.	0	0	0

Move point to kℓ column

0.	0	0	5	0	0	0

Answer: .005 kℓ

These, as well as other possible conversions, are summarized in the single chart below.

kiloliter (kℓ)	hectoliter (hℓ)	dekaliter (dkℓ)	liter (ℓ)	deciliter (dℓ)	centiliter (cℓ)	milliliter (mℓ)		
0	0	0	5.	0	0	0	5ℓ	**Given**
0	0	0	5	0.	0	0	50 dℓ	
0	0	0	5	0	0.	0	500 cℓ	
0	0	0	5	0	0	0.	5,000 mℓ	
0	0	0.	5	0	0	0	.5 dkℓ	
0	0.	0	5	0	0	0	.05 hℓ	
0.	0	0	5	0	0	0	.005 kℓ	

Note: In practice, only the necessary zeros are filled in, as shown in the next example.

b. 34.71 kg

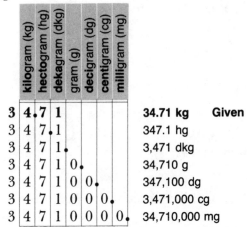

kilogram (kg)	hectogram (hg)	dekagram (dkg)	gram (g)	decigram (dg)	centigram (cg)	milligram (mg)		
3	4.7	1					34.71 kg	Given
3	4	7.1					347.1 hg	
3	4	7	1.				3,471 dkg	
3	4	7	1	0.			34,710 g	
3	4	7	1	0	0.		347,100 dg	
3	4	7	1	0	0	0.	3,471,000 cg	
3	4	7	1	0	0	0	0.	34,710,000 mg	

To help you remember this chart, try to memorize a short saying with the same first letters—like the one in the margin.

Karl	**K**ilometer
Has	**H**ectometer
Developed	**D**ekameter
My	**M**eter
Decimal	**D**ecimeter
Craving for	**C**entimeter
Metrics	**M**illimeter

EXAMPLE 3

Make the indicated conversions.

a. 46.4 cm = ____?____ m

Mentally picture a chart similar to the one shown in Example 2b. Then count the number of decimal places involved in this conversion.

Remember "**K**arl **H**as **D**eveloped **M**y **D**ecimal **C**raving for **M**etrics."

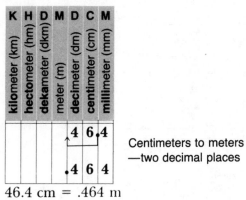

Centimeters to meters
—two decimal places

46.4 cm = .464 m

b. $6.32 \text{ km} = \underline{\quad ? \quad} \text{ m}$
Three decimal places on chart:

$$\text{K H D M D C M}$$
$$6.3 \ 2$$

$6.32 \text{ km} = 6,320 \text{ m}$

c. $503 \text{ m}\ell \underline{\quad ? \quad} \ell$

From memory aid: **K H D M D C M**
Change to liters: ℓ

$$5 \ 0 \ 3.$$

Three decimal places on chart: $503 \text{ m}\ell = .503 \ \ell$

d. $.031 \text{ k}\ell = \underline{\quad ? \quad} \ell$
Three decimal places on chart: $.031 \text{ k}\ell = 31 \ \ell$

e. $14 \text{ kg} = \underline{\quad ? \quad} \text{ dg}$
Four decimal places on chart: $14 \text{ kg} = 140,000 \text{ dg}$ ■

Problem Set 5.4

ODD-NUMBERED PROBLEMS are U.S. measurement conversions, and *EVEN-NUMBERED PROBLEMS* are metric conversions. Use reduced fractions for U.S. conversions, and use decimals for metric conversions.

In Problems 1–14 make the indicated conversions.

1. a. 1 in. = _____ ft
   b. 1 in. = _____ yd
   c. 1 in. = _____ mi

2. a. 1 cm = _____ mm
   b. 1 cm = _____ m
   c. 1 cm = _____ km

3. a. 1 ft = _____ in.
   b. 1 ft = _____ yd
   c. 1 ft = _____ mi

4. a. 1 mm = _____ cm
   b. 1 mm = _____ m
   c. 1 mm = _____ km

5. a. 1 yd = _____ in.
   b. 1 yd = _____ ft
   c. 1 yd = _____ mi

6. a. 1 m = _____ mm
   b. 1 m = _____ cm
   c. 1 m = _____ km

7. a. 1 tsp = _____ tbsp
   b. 1 tsp = _____ oz
   c. 1 tsp = _____ c

8. a. 1 mℓ = _____ dℓ
   b. 1 mℓ = _____ ℓ
   c. 1 mℓ = _____ kℓ

9. a. 1 oz = _____ c
   b. 1 oz = _____ pt
   c. 1 c = _____ pt

10. a. 1 ℓ = _____ mℓ
    b. 1 ℓ = _____ dℓ
    c. 1 ℓ = _____ kℓ

11. **a.** 1 pt = _____ qt
    b. 1 pt = _____ gal
    c. 1 qt = _____ gal
12. **a.** 1 kℓ = _____ mℓ
    b. 1 kℓ = _____ dℓ
    c. 1 kℓ = _____ ℓ
13. **a.** 1 oz = _____ lb
    b. 1 oz = _____ T
    c. 1 lb = _____ T
14. **a.** 1 g = _____ mg
    b. 1 g = _____ cg
    c. 1 g = _____ kg

In Problems 15–38 write each measurement in terms of all other units listed at the top of the problem. U.S. units will be changed to other U.S. units, and metric units will be changed to other metric units.

	mile (mi)	yard (yd)	feet (ft)	inches (in.)
15.		9		
17.				6
19.	4			
21.				150

	kilometer (km)	hectometer (hm)	dekameter (dkm)	meter (m)	decimeter (dm)	centimeter (cm)	millimeter (mm)
16.				9.			
18.						6.	
20.	4.						
22.			1	5	0.		

	quart (qt)	pint (pt)	cup (c)	ounce (oz)
23.				63
25.		3$\frac{1}{2}$		
27.			8	
29.	5			

	kiloliter (kℓ)	hectoliter (hℓ)	dekaliter (dkℓ)	liter (ℓ)	deciliter (dℓ)	centiliter (cℓ)	milliliter (mℓ)
24.						6	3.
26.			3.	5			
28.				8.			
30.		3.	1				

	ton (T)	pound (lb)	ounce (oz)
31.		4	
33.			96
35.		5$\frac{1}{4}$	
37.	4$\frac{1}{2}$		

	kilogram (kg)	hectogram (hg)	dekagram (dkg)	gram (g)	decigram (dg)	centigram (cg)	milligram (mg)
32.	6.	5					
34.			9	6.			
36.	5.	2	5				
38.	4.	5					

Appendix: U.S.–Metric Conversions

This section may be considered as an appendix for reference use. The emphasis in this chapter has been on everyday usage of the metric system. However, certain specialized applications require more precise conversions than we've considered. On the other hand, you don't want to become bogged down with arithmetic to the point where you say "nuts to the metric system."

The most difficult prospect concerning the change from the U.S. system to the metric system is not mathematical but psychological. If you have understood the chapter to this point, you realize that this is not the case. In fact, as you saw in the last section, *working within the metric system is much easier than working within the U.S. system.*

With these ideas firmly in mind, and realizing that your everyday work with the metric system is discussed in the other sections of this chapter, an appendix of conversion factors between these measurement systems is presented here. Many calculators will make these conversions for you— check your owner's manual.

LENGTH CONVERSIONS

U.S. to Metric

When you know	Multiply by	To find
in.	2.54	cm
ft	30.48	cm
ft	0.3048	m
yd	0.9144	m
mi	1.60934	km

Metric to U.S.

When you know	Multiply by	To find
cm	0.39370	in.
m	39.37	in.
m	3.28084	ft
m	1.09361	yd
km	0.62137	mi

CAPACITY CONVERSIONS

U.S. to Metric

When you know	Multiply by	To find
tsp	4.9289	mℓ
tbsp	14.7868	mℓ
oz	29.5735	mℓ
c	236.5882	mℓ
pt	473.1765	mℓ
qt	946.353	mℓ
qt	0.9464	ℓ
gal	3.7854	ℓ

Metric to U.S.

When you know	Multiply by	To find
mℓ	0.20288	tsp
mℓ	0.06763	tbsp
mℓ	0.03381	oz
mℓ	0.00423	c
mℓ	0.00211	pt
mℓ	0.00106	qt
ℓ	1.05672	qt
ℓ	0.26418	gal

WEIGHT CONVERSIONS

U.S. to Metric

When you know	Multiply by	To find
oz	28.3495	g
lb	453.59237	g
lb	0.453592	kg
T	907.18474	kg

Metric to U.S.

When you know	Multiply by	To find
g	0.0352739	oz
g	0.0022046	lb
kg	2.2046226	lb

TEMPERATURE CONVERSIONS

U.S. to Metric

1. Subtract 32 from degrees Fahrenheit
2. Multiply this result by $\frac{5}{9}$ to get Celsius

Metric to U.S.

1. Multiply Celsius degrees by $\frac{9}{5}$
2. Add 32 to get Fahrenheit

5.5 Review Problems

Section 5.1   **1.** Without using any measuring device, draw a segment about:
  a. 2 inches long   **b.** 3 centimeters long

Section 5.1   **2.** Measure the given segment to the: ——————————
  a. Nearest cm   **b.** Nearest $\frac{1}{10}$ cm

Sections 5.1–5.3   **3.** Fill in the following information about yourself using metric units:
  a. Height   **b.** Weight   **c.** Normal body temperature

Sections 5.1–5.3   **4.** Name a unit in the U.S. measurement system and a unit in the metric measurement system that you might use to measure each of the following:
  a. Size of notebook paper
  b. Weight of a dime
  c. Capacity of a can of Pepsi
  d. Distance from Boston to Chicago
  e. Outside temperature

Sections 5.1–5.3   **5.** State whether each measurement is measuring length, capacity, weight, or temperature.
  a. 4.2 mℓ   **b.** 24 km   **c.** 4.8 cm
  d. 9.3 kg   **e.** 23°C

Sections 5.1–5.3   **6.** Estimate the following quantities using U.S. and metric measurements:
  a. Distance between San Francisco and New York
  b. Capacity of a can of 7-UP
  c. Weight of a penny
  d. Comfortable room temperature
  e. Temperature on a hot summer day

Section 5.4   **7.** Arrange the following metric prefixes in order of largest to smallest and tell what each prefix means: milli, hecto, centi, kilo, deka, and deci.

Section 5.4   **8.** Write out in words what each abbreviation stands for:
  a. mg   **b.** c   **c.** in.   **d.** cℓ   **e.** mm
  f. dkg   **g.** cg   **h.** T   **i.** kℓ   **j.** hm

Sections 5.1–5.3   **9.** For each pair of measurements, name the larger one.
  a. Pound and kilogram   **b.** Inch and centimeter
  c. Quart and liter   **d.** Mile and kilometer
  e. Ounce and milliliter

Section 5.4   **10.** Make the indicated conversions.
  a. 4.8 km = _____ m   **b.** 35 mm = _____ cm
  c. 450 mℓ = _____ ℓ   **d.** 480 g = _____ kg
  e. 5.2 dm = _____ cm

Geometry and Problem Solving

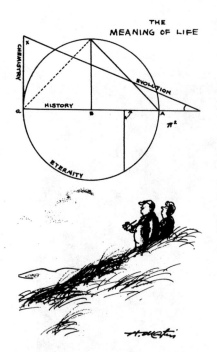

CHAPTER 6

THE
MEANING OF LIFE

"*I wish I could see the expression on the faces of my students who said there was no value in studying geometry.*"

6.1 Polygons and Angles

Geometry, or "earth measure," was one of the first branches of mathematics. Both the Egyptians and the Babylonians needed geometry for construction, for land measurement, and for commerce. Early civilizations observed from nature certain simple shapes, such as triangles, rectangles, and circles. The study of geometry began with the need to measure and understand the properties of these simple shapes.

In this book, we'll limit our investigation to polygons, circles, and certain simple three-dimensional objects. A **polygon** is a geometric figure having three or more straight sides that all lie on a flat surface, or **plane** so that the starting point and the ending point are the same. Polygons are classified according to the number of sides, as shown in Figure 6.1.

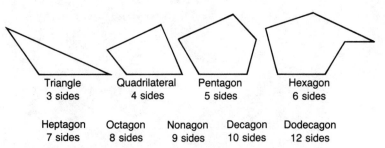

| Triangle | Quadrilateral | Pentagon | Hexagon |
| 3 sides | 4 sides | 5 sides | 6 sides |

| Heptagon | Octagon | Nonagon | Decagon | Dodecagon |
| 7 sides | 8 sides | 9 sides | 10 sides | 12 sides |

FIGURE 6.1 Definition and examples of polygons

A connecting point of two sides is called a **vertex** (plural **vertices**), and is usually designated by a capital letter. The **angles** between the sides are sometimes also denoted by a capital letter, but other ways of denoting angles are shown in Figure 6.2.

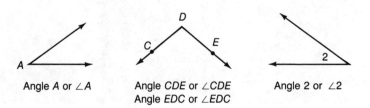

FIGURE 6.2 Ways of denoting angles

Angle A or $\angle A$

Angle CDE or $\angle CDE$
Angle EDC or $\angle EDC$

Angle 2 or $\angle 2$

EXAMPLE 1

Locate each of the following angles in the figure:

a. $\angle AOB$ **b.** $\angle COB$ **c.** $\angle BOA$

d. $\angle DOC$ **e.** $\angle 3$ **f.** $\angle 4$ ■

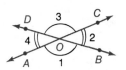

Two angles are said to be **equal** if they have the same measure. Notice in Example 1 that parts a and c name the same angle, so $\angle AOB = \angle BOA$. Also notice the single and double arcs used to mark the angles in Example 1; these are used to denote equal angles, so $\angle COB = \angle AOD$ and $\angle AOB = \angle COD$. Denoting an angle by a single letter is preferred except in the case of a situation as shown by Example 1 where several angles share the same vertex.

Angles are usually measured using a unit called a **degree**, which is defined to be $\frac{1}{360}$ of a full revolution. The symbol ° is used to designate degrees. To measure an angle, you can use a **protractor**, but in this book the angles whose measures we need will be labeled as shown in Figure 6.3.

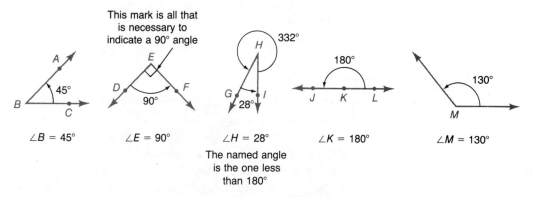

FIGURE 6.3 Examples of angles

Angles are sometimes classified according to their measures, as shown in Table 6.1.

TABLE 6.1 Classification of Angles

Type	Classification
Angles less than 90°	Acute
Angles equal to 90°	Right
Angles between 90° and 180°	Obtuse
Angles equal to 180°	Straight

EXAMPLE 2

Label the angles in Figure 6.3 by classification.

Solution

∠B is acute;   ∠E is right;   ∠H is acute;
∠K is straight;   ∠M is obtuse   ■

When working in geometry, whether you are dealing with angles or with other geometric figures, you often draw models or pictures to represent these geometric ideas. However, you must be careful not to try to prove assertions by looking at pictures. For example, ∠M in Figure 6.3 is 130° *not* because you can take a protractor and measure it, but because *it says* the angle is 130°. A figure may contain hidden assumptions or ambiguities. For example, consider Figure 6.4a. What do you see? A quadrilateral? Perhaps. But what if we take a slightly different view of the same object, as shown in Figure 6.4b. What do you see now? Still a quadrilateral?

  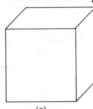

FIGURE 6.4   (a)   (b)   (c)

But what if we have in mind a **cube**, as shown clearly in Figure 6.4c. Even if you view this object as a cube, do you see the same cube as everyone else? For example, Figure 6.5 is also a cube. Is the fly inside or outside the cube? Or is it perhaps on the edge of the cube? Thus, although we may use a figure to help us understand a problem, we cannot prove results using this technique.

FIGURE 6.5   Fly on a cube

Problem Set 6.1

Name the polygons in Problems 1–12 according to the classifications shown in Figure 6.1.

1.   **2.**   **3.**

4. 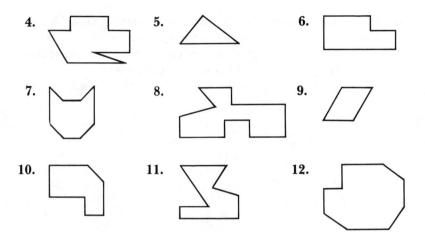 **5.** **6.**

7. **8.** **9.**

10. **11.** **12.**

13. The illustration in the margin is an example of an ambiguous figure. What do you see? A woman? Is she old or young?

Consider the following figure for Problems 14–17.

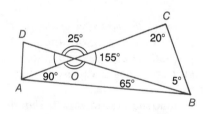

14. Classify the angles as acute, right, straight, or obtuse.
    a. ∠DOC   **b.** ∠AOB   **c.** ∠DBC   **d.** ∠OAB
    e. ∠DOB

15. Classify the angles as acute, right, straight, or obtuse.
    a. ∠C   **b.** ∠COB   **c.** ∠AOC   **d.** ∠DOA
    e. ∠OBA

16. Name a pair of equal angles.

17. Name an angle equal to ∠DOB.

18. The first illustration below shows a cube with the top cut off. Use solid lines and shading to depict seven other different views of a cube with one side cut off.

19. Show that *the sum of the measures of the interior angles of any triangle is 180°* by carrying out the following steps:
  a. Draw three triangles; one with all acute angles, one with a right angle, and a third with an obtuse angle.
  b. Cut or tear apart the angles of each triangle you've drawn.

  c. Place the pieces together to form a straight angle.

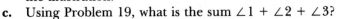

20. Show that *the sum of the measures of the interior angles of any quadrilateral is 360°* by carrying out the following steps:
  a. Draw any quadrilateral, as illustrated (but draw your quadrilateral so it has a different shape from the one shown here).

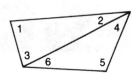

  b. Divide the quadrilateral into two triangles by drawing a **diagonal** (a line segment connecting two nonadjacent vertices). Label the angles of your triangles as shown in the illustration.
  c. Using Problem 19, what is the sum ∠1 + ∠2 + ∠3?
  d. Using Problem 19, what is the sum ∠4 + ∠5 + ∠6?
  e. The sum of the measures of the angles of the quadrilateral is

$$(\angle 1 + \angle 2 + \angle 3) + (\angle 4 + \angle 5 + \angle 6)$$

  What is this sum?
  f. Do you think this argument will apply for *any* quadrilateral?

Mind Bogglers

21. Look at Problems 19 and 20. What is the sum of the measures of the interior angles of any pentagon?

22. Look at Problems 19 and 20. What is the sum of the measures of the interior angles of any octagon?

23. Connect the nine dots shown in the illustration with four straight lines, but don't lift your pencil from the paper.

24. Two sets of three intersecting planes are shown below. Use solid lines and shading to illustrate different ways of viewing the planes. Can you find more than two ways?

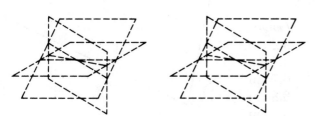

25. **a.** What is a regular polyhedron?
   b. How many regular polyhedra are possible? Name them.
   c. Construct models for the regular polyhedra.
       References: Coxeter, H. S. M., *Introduction to Geometry* (New York: Wiley, 1961).
           Trigg, Charles W., "Collapsible Models of the Regular Octahedron," *The Mathematics Teacher*, October 1972, pp. 530–533.
           Sobel and Maletsky, *Teaching Mathematics: A Sourcebook* (Englewood Cliffs, N.J.: Prentice-Hall, 1975), pp. 173–184.

6.2 Perimeter

One application of both measurement and geometry is finding the distance around a polygon. This distance is called the **perimeter** of the polygon.

The **perimeter** of a polygon is the sum of the lengths of the sides of that polygon.

Perimeter

EXAMPLE 1

Find the pe-
rimeter of the
given polygon
by measuring
each side to
the nearest $\frac{1}{10}$
centimeter.

Side	Length
A	3.2 cm
B	3.5 cm
C	6.7 cm
D	3.8 cm
E	3.0 cm
Total:	**20.2 cm** **Perimeter**

Here are some formulas for finding the perimeters of the
most common polygons.

Equilateral triangle

An **equilateral triangle** is a triangle with sides that are equal.

$$\text{Perimeter} = 3(\text{Side})$$
$$P = 3s$$

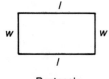

Rectangle

A **rectangle** is a quadrilateral with angles that are all
right angles.

$$\text{Perimeter} = 2(\text{Length}) + 2(\text{Width})$$
$$P = 2l + 2w$$

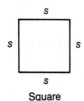

Square

A **square** is a rectangle with sides that are equal.

$$\text{Perimeter} = 4(\text{Side})$$
$$P = 4s$$

EXAMPLE 2

Find the perimeter of each polygon.

a.

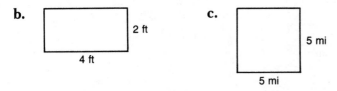

1 cm

1 cm

1 cm

b.

2 ft

4 ft

c.

5 mi

5 mi

d.

10 dm

10 dm   10 dm

e.

5 m

3 m

4 m

Solution

a. Rectangle is 2 cm by 9 cm, so

$$P = 2(9) + 2(2) = 18 + 4 = 22 \text{ cm}$$

b. Rectangle is 2 ft by 4 ft, so

$$P = 2(4) + 2(2) = 8 + 4 = 12 \text{ ft}$$

c. Square, so $P = 4(5) = 20$ mi.
d. Equilateral triangle, so $P = 3(10) = 30$ dm.
e. Triangle (add lengths of sides), so $P = 3 + 4 + 5 = 12$ m. ∎

While a **circle** is not a polygon, you sometimes need to find the distance around a circle. This distance is called the **circumference**. For *any circle*, if you divide the circumference by the diameter, you will get the *same number*. This number is given the name **pi** (pronounced "pie") and is symbolized by the Greek letter π. The number π is an irrational number and is about 3.14 or $\frac{22}{7}$. We need this number π to state a formula for the circumference, C.

$$C = d\pi \qquad \text{or} \qquad C = 2\pi r$$
$$d = \text{Diameter} \qquad\qquad r = \text{Radius}$$

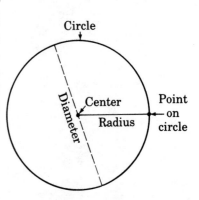

Circle

Diameter

Center

Radius

Point on circle

EXAMPLE 3

Find the circumference of each circle (or find the distance around each figure).

a.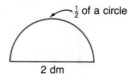
4 ft

b.
28 cm

c.
5 m

d.
$\frac{1}{2}$ of a circle

2 dm

e. 30 cm
$\frac{1}{8}$ of a circle

Solution

a. $C = 4\pi$

A decimal approximation (symbolized by $\approx$) is

$$C \approx 4(3.14) = 12.56$$

A fractional approximation is

$$C \approx 4\left(\frac{22}{7}\right) = \frac{88}{7} = 12\frac{4}{7}$$

The circumference is 4π ft, which is about $12\frac{1}{2}$ ft.

b. $C = 28\pi$

This is about 88 cm.

c. $C = 2\pi(5) = 10\pi$

This is about 31.4 m.

d. This is half of a circle (called a **semicircle**), thus the curved part is half of the circumference ($C = 2\pi \approx 6.28$), or about 3.14, and is added to the diameter:

$$3.14 + 2 = 5.14$$

The distance around the figure is about 5.14 dm.

e. This is one-eighth of a circle. The curved part is one-eighth of the circumference [$C = 2\pi(30) \approx 188.4$], or about 23.6. Thus, the distance around the figure is 83.6 cm (23.6 + 30 + 30). ∎

Many calculators have a single key marked π. If you press it, the display shows an approximation correct to several decimal places (the number of places depends on your calculator):

$\boxed{\pi}$ *Display:* 3.141592654

If your calculator doesn't have a key marked π, you may want to use the approximation above to obtain the accuracy you want. In this book, the answers are shown using an approximation for π (usually 3.14). If you use the π key on your calculator, you will sometimes obtain a slightly different answer.

These ideas involving perimeter and circumference are sometimes needed to solve certain types of problems.

EXAMPLE 4

Suppose you have enough material for 70 ft of fence and want to build a rectangular pen 14 ft wide. What is the length of this pen?

Solution

PERIMETER = 2(LENGTH) + 2(WIDTH)   This is the formula
  ↓   ↓   for perimeter.
  70   = 2(LENGTH) + 2(14)   Fill in the given
                                         information.

Let L = LENGTH OF PEN

$$70 = 2L + 28$$
$$42 = 2L$$
$$21 = L$$

The pen will be 21 ft long.   ■

Problem Set 6.2

Find the perimeter or circumference of each figure given in Problems 1–16 by using the appropriate formula. Use 3.14 as an approximation for π.

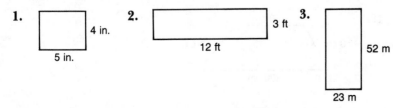

1.   4 in.
     5 in.

2.   12 ft

3 ft

3.   52 m
     23 m

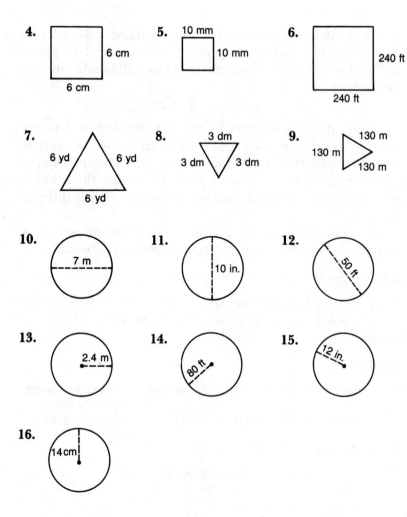

Find the distance around each figure given in Problems 17–28. Use 3.14 as an approximation for π.

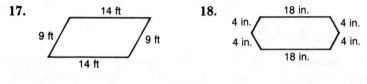

17. 14 ft, 9 ft, 9 ft, 14 ft

18. 18 in., 4 in., 4 in., 4 in., 4 in., 18 in.

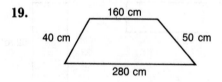

19. 160 cm, 40 cm, 50 cm, 280 cm

20.

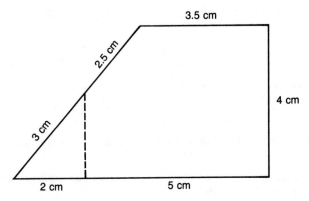

3.5 cm

2.5 cm

4 cm

3 cm

2 cm 5 cm

21.

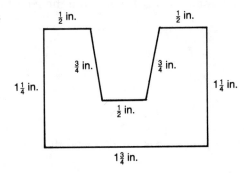

$\frac{1}{2}$ in. $\frac{1}{2}$ in.

$\frac{3}{4}$ in. $\frac{3}{4}$ in.

$1\frac{1}{4}$ in. $1\frac{1}{4}$ in.

$\frac{1}{2}$ in.

$1\frac{3}{4}$ in.

22.

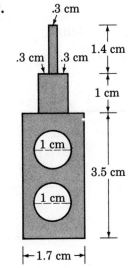

.3 cm

.3 cm .3 cm

1.4 cm

1 cm

1 cm

1 cm

3.5 cm

|← 1.7 cm →|

23.

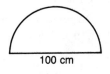

100 cm

24.

10 in.

25.

18 ft

26.

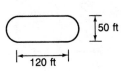

50 ft

120 ft

27. 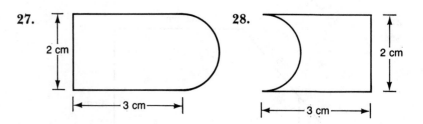　**28.**

29. What is the width of a rectangular lot that has a perimeter of 410 ft and a length of 140 ft?

30. What is the length of a rectangular lot that has a perimeter of 750 m and a width of 75 m?

31. The perimeter of triangle *ABC* is 117 in. Find the lengths of the sides.

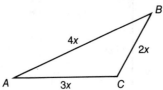

32. The perimeter of the pentagon is 280 cm. Find the lengths of the sides.

33. Find the dimensions of an equilateral triangle that has a perimeter measuring 198 dm.

34. Find the dimensions of a rectangle with a perimeter of 54 cm if the length is 5 less than three times the width.

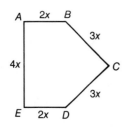

Find the perimeter of each polygon given in Problems 35–37 by measuring each side to the nearest $\frac{1}{10}$ centimeter.

35. 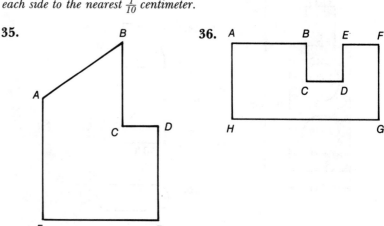　**36.**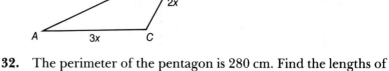

37.

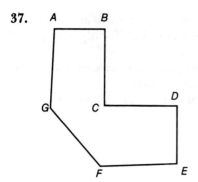

Mind Bogglers

38. Santa Rosa street problem. On Saturday evenings a favorite pastime of the high school students in Santa Rosa, California, is to cruise certain streets. The selected routes are shown on the map. Is it possible to choose a route so that all the permitted streets are traveled exactly once?

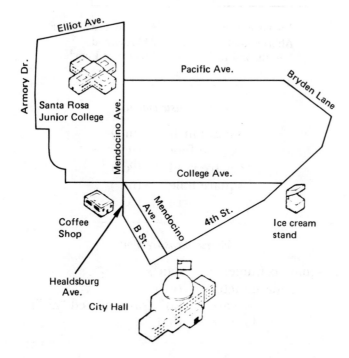

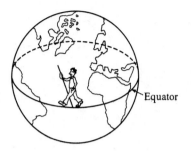

Equator

39. Suppose we fit a band tightly around the earth at the equator. We wish to raise the band so that it is uniformly supported 6 ft above the earth at the equator.

    a. Guess how much extra length would have to be added to the band (not the supports) to do this.

    b. Calculate the amount of extra material that would be needed.

6.3 Area

Suppose you want to carpet your living room. The price of carpet is quoted in terms of a price per square yard. A square yard is a measure of **area**. To measure the area of a plane figure, you fill it with **square units**.

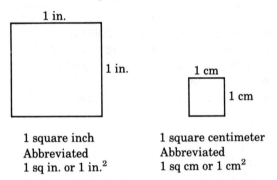

1 in.

1 in.

1 cm

1 cm

1 square inch
Abbreviated
1 sq in. or 1 in.^{2}

1 square centimeter
Abbreviated
1 sq cm or 1 cm^2

U.S. and Metric Standard Units of Area Measurement

U.S. measurement	
square inch	(in.^{2})
square foot	(ft^2)
square yard	(yd^2)
square mile	(mi^2)
acre	(A)

Metric measurement	
square centimeter	(cm^2)
square meter	(m^2)
are	(a) (pronounced "air")
hectare	(ha)

EXAMPLE 1

What is the area of the shaded region?

a.

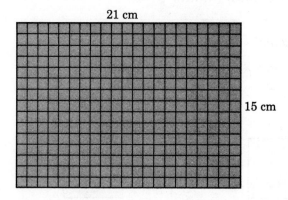

21 cm

15 cm

You can count the number of square centimeters contained in the shaded region; there are 315 squares. Also notice:

$$\begin{array}{cc} \textit{Across} & \textit{Down} \\ 21 \text{ cm} \times 15 \text{ cm} = 21 \times 15 \text{ cm} \times \text{cm} \\ = \textbf{315 cm}^2 \end{array}$$

b.

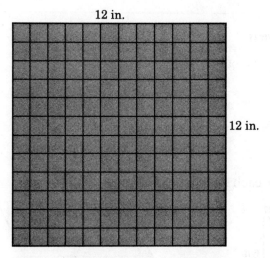

12 in.

12 in.

The shaded region is a **square foot**. You can count 144 square inches inside the region. Also notice:

$$\begin{array}{cc} \textit{Across} & \textit{Down} \\ 12 \text{ in.} \times 12 \text{ in.} = \textbf{144 in.}^2 \end{array}$$ ■

As you can see from Example 1, the area of a rectangular or square region is the product of the distance across (length) and the distance down (width).

Area of Rectangles and Squares

Rectangles	Squares
Area = Length × Width $A = l \times w$	Area = Side × Side = (Side)^2 $A = s^2$

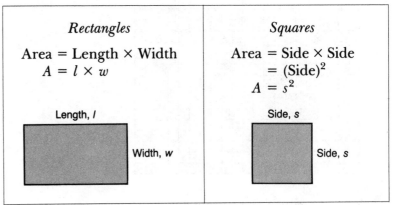

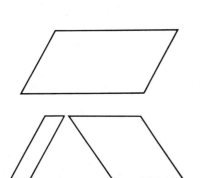

A **parallelogram** is a quadrilateral with two pairs of parallel sides. To find the area of a parallelogram, you can count the squares necessary to fill the region, rearrange part of the given parallelogram so that it forms a rectangle, or use the formula given below.

Area of Parallelograms

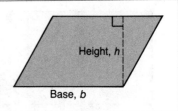

Parallelograms

Area = Base × Height
$A = b \times h$

EXAMPLE 2

Find the area of each shaded region.

a.
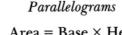

6 m

3 m

$A = 3 \text{ m} \times 6 \text{ m}$
$= 18 \text{ m}^2$

b.
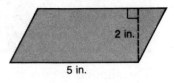

2 in.

5 in.

$A = 2 \text{ in.} \times 5 \text{ in.}$
$= 10 \text{ in.}^2$ ∎

You can find the area of a triangle by filling in square units or by rearranging the parts, but the easiest way to find the area of a triangle is to notice that *every* triangle has an area that is exactly half of a corresponding parallelogram. We can therefore state the following result:

Area of Triangles

Triangles 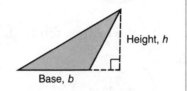
Area = $\frac{1}{2}$ × Base × Height
$A = \frac{1}{2} \times b \times h$

EXAMPLE 3

Find the area of each shaded region.

a.

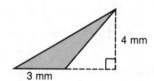

$A = \frac{1}{2} \times 4 \text{ mm} \times 3 \text{ mm}$
$\quad = \textbf{6 mm}^2$

b.

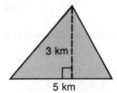

$A = \frac{1}{2} \times 3 \text{ km} \times 5 \text{ km}$
$\quad = \frac{15}{2} \textbf{ km}^2 \quad \text{or} \quad \textbf{7}\frac{1}{2} \textbf{ km}^2$

∎

Area of Circles

Circles 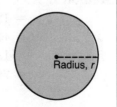
$A = r \times r \times \pi$
$\quad = r^2 \times \pi$
$\quad = \pi r^2 \quad \text{or} \quad A \approx 3.14(\text{Radius})^2$

EXAMPLE 4

Find the area of each shaded region to the nearest tenth unit.

a.

$r = 9$ yd

$$A = \pi \times (9 \text{ yd})^2$$
$$\approx 3.14 \times 81 \text{ yd}^2$$
$$= 254.34 \text{ yd}^2$$

To the nearest tenth, **the area is 254.3 yd².**

b.

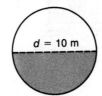

$d = 10$ m

$$d = 10 \text{ m}$$
$$r = 5 \text{ m}$$
$$A = \pi \times (5 \text{ m})^2$$
$$\approx 3.14 \times 25 \text{ m}^2$$
$$= 78.5 \text{ m}^2$$

However, the shaded portion is only half the area of the circle:

$$.5 \times 78.5 \text{ m}^2 = 39.25 \text{ m}^2$$

To the nearest tenth, **the area is 39.3 m².**   ∎

Sometimes we measure area using one unit of measurement and then want to convert the result to another. Consider Example 5.

EXAMPLE 5

Suppose your living room is 12 ft by 15 ft and you want to know how many square yards of carpet you need to cover this area.

Solution

Method I. $A = 12 \text{ ft} \times 15 \text{ ft}$
$= 180 \text{ ft}^2$
$= 180 \times 1 \text{ ft}^2$
$= 180 \times (\frac{1}{9} \text{ yd}^2)$
$= \textbf{20 yd}^2$

Since 1 yd = 3 ft,
$1 \text{ yd}^2 = (1 \text{ yd}) \times (1 \text{ yd})$
$1 \text{ yd}^2 = (3 \text{ ft}) \times (3 \text{ ft})$
$1 \text{ yd}^2 = 9 \text{ ft}^2$
$\frac{1}{9} \text{ yd}^2 = 1 \text{ ft}^2$

CARPET REMNANTS				
TYPE	COLOR	SIZE	AREA	PRICE
Commercial Print	Rust	12′×11′3″	15 yd²	$119
Plush	Camel	12′×11′8″	15.7 yd²	$72
Kitchen Print Rubberback	Earthtones	12′×9′6″	12.7 yd²	$68
Plush	Mardi Gras	12′×12′11″	17.2 yd²	$89
Cut & Loop	Lamplighter Orange	12′×7′9″	10.3 yd²	$49
Kitchen Print Rubberback	Earthtones	12′×12′	16 yd²	$82
Plush	Camel	12′×17′7″	23.4 yd²	$98
Kitchen Print Rubberback	Gold	12′×12′3″	16.3 yd²	$84
Plush	Camel	12′×9′6″	12.7 yd²	$132
Cut & Loop	Walnut	12′×9′9″	13 yd²	$69
JUST ARRIVED! LARGE SHIPMENT OF BEAUTIFUL CARPET REMNANTS				

Method II. Change feet to yards to begin the problem:

$$12 \text{ ft} = 4 \text{ yd} \qquad \text{and} \qquad 15 \text{ ft} = 5 \text{ yd}$$
$$A = 4 \text{ yd} \times 5 \text{ yd}$$
$$= \mathbf{20 \ yd^2} \qquad\qquad \blacksquare$$

Of course, Method II is easier for Example 5, but sometimes the conversion of units as a first step complicates the problem. Other times you don't have the individual measurements. Also, the biggest problem for most people is knowing whether to multiply or divide by 9. Table 6.2 (page 222) should help you with area conversions.

EXAMPLE 6

How many acres are there in a rectangular piece of property measuring 363 ft by 180 ft?

Solution

$$\text{Area} = 363 \text{ ft} \times 180 \text{ ft}$$

A calculator would be handy for this problem.

$$= 65,340 \text{ ft}^2$$

$$= 65,340 \times \left(\frac{1}{43,560} \right) \text{A}$$

This conversion factor is from Table 6.2.

$$= \mathbf{1.5 \ A} \qquad\qquad \blacksquare$$

TABLE 6.2   Conversion of Area Measurements

U.S. measurement		Metric measurement
$1\text{ ft}^2 = 144\text{ in.}^2$	$1\text{ in.}^2 = \dfrac{1}{144}\text{ ft}^2$	$1\text{ m}^2 = 10{,}000\text{ cm}^2$ or $(100\text{ cm})^2$
$1\text{ yd}^2 = 9\text{ ft}^2$	$1\text{ ft}^2 = \dfrac{1}{9}\text{ yd}^2$	$1\text{ km}^2 = 1{,}000{,}000\text{ m}^2$ or $(1000\text{ m})^2$
$1\text{ yd}^2 = 1{,}296\text{ in.}^2$	$1\text{ in.}^2 = \dfrac{1}{1{,}296}\text{ yd}^2$	$1\text{ a} = 100\text{ m}^2$ or $(10\text{ m})^2$
$1\text{ A} = 43{,}560\text{ ft}^2$	$1\text{ ft}^2 = \dfrac{1}{43{,}560}\text{ A}$	$1\text{ ha} = 100\text{ a}$ or $(100\text{ m})^2$ $= 10{,}000\text{ m}^2$
$1\text{ mi}^2 = 640\text{ A}$ $= 3{,}097{,}600\text{ yd}^2$ $= 27{,}878{,}400\text{ ft}^2$		

EXAMPLE 7

You want to paint 200 ft of a three-rail fence which is made up of 3 boards, each 6 inches wide. You want to know:

a. The number of square feet on one side of the fence
b. The number of square feet to be painted if the posts and edges of the boards comprise 100 ft^2
c. The number of gallons of paint to purchase if each gallon covers 325 ft^2

Solution

a. $A = (200\text{ ft}) \times (6\text{ in.}) \times 3$
$ = (200\text{ ft}) \times (\tfrac{1}{2}\text{ ft}) \times 3$
$ = \mathbf{300\text{ ft}^2}$

b. AMOUNT TO BE PAINTED
$ = 2(\text{AMOUNT ON ONE SIDE}) + \text{EDGES AND POSTS}$
$ = 2(300\text{ ft}^2) + 100\text{ ft}^2$
$ = \mathbf{700\text{ ft}^2}$

c. $\left(\begin{array}{c}\text{NUMBER OF SQUARE}\\ \text{FEET PAINTED}\end{array}\right) = \left(\begin{array}{c}\text{NUMBER OF SQUARE}\\ \text{FEET PER GALLON}\end{array}\right)\left(\begin{array}{c}\text{NUMBER OF}\\ \text{GALLONS}\end{array}\right)$

$700 = 325\left(\begin{array}{c}\text{NUMBER OF}\\ \text{GALLONS}\end{array}\right)$

$2.15 \approx \left(\begin{array}{c}\text{NUMBER OF}\\ \text{GALLONS}\end{array}\right)$

If paint must be purchased by the gallon (as implied by the question), the number of gallons to purchase is 3 gallons.   ■

Problem Set 6.3

Find the area of each shaded region in Problems 1–15 to the nearest $\frac{1}{10}$ unit. Use 3.14 as an approximation for π.

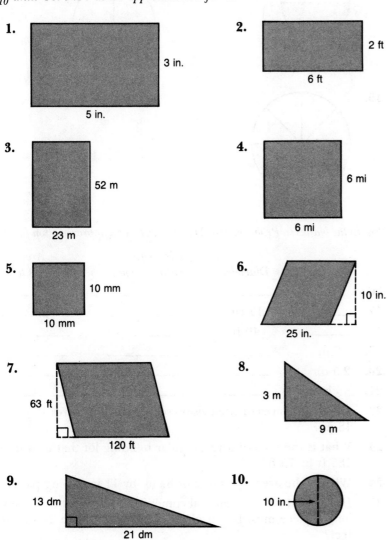

1.

3 in.

5 in.

2.

2 ft

6 ft

3.

52 m

23 m

4.

6 mi

6 mi

5.

10 mm

10 mm

6.

10 in.

25 in.

7.

63 ft

120 ft

8.

3 m

9 m

9.

13 dm

21 dm

10.

10 in.

11.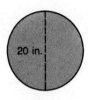

12.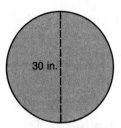

13.

14.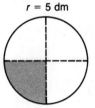

15.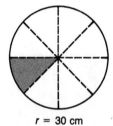

Fill in the blanks in Problems 16–21. Use 3.14 as an approximation for π.

	Radius	Diameter	Circumference (nearest tenth)	Area (nearest tenth)
16.	10 in.	_____	_____	_____
17.	_____	12 cm	_____	_____
18.	_____	40 ft	_____	_____
19.	3 m	_____	_____	_____
20.	2.5 dm	_____	_____	_____
21.	.5 km	_____	_____	_____

22. What is the area of a television screen that measures 12 in. by 18 in.?

23. What is the area of a rectangular building lot that measures 185 ft by 75 ft?

24. What is the area of a piece of $8\frac{1}{2}$ in. by 11 in. typing paper?

25. If a certain type of material comes in a bolt 3 feet wide, how long a piece must be purchased in order to have 24 square feet?

26. Find the cost of pouring a square concrete slab with sides of 25 ft if the cost is $1.75 per square foot.

Consider the house plan for Problems 27–33. Dimensions are in feet.

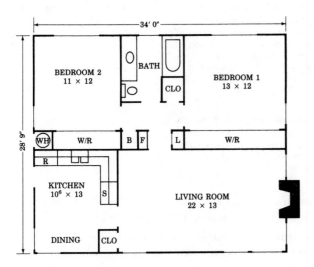

27. What is the area of the living room?

28. What is the area of Bedroom 1?

29. What is the area of Bedroom 2?

30. What is the area of the home?

31. The kitchen and dining area is labeled $10^6 \times 13$. This means 10 ft 6 in. by 13 ft. What is the area of the kitchen and dining area?

32. If the living room is carpeted with carpet that costs $21 per square yard (including labor and pad), what is the total cost (to the nearest dollar) for the carpeting of the living room in this home?

33. If Bedroom 1 is carpeted with carpet that costs $18 per square yard (including labor and pad), what is the total cost (to the nearest dollar) for carpeting this room?

34. How many square feet are in a $\frac{1}{4}$ acre piece of property?

35. What is the cost of seeding a rectangular lawn of 100 ft by 30 ft if one pound of seed costs $2.85 and covers 150 square feet?

Problems 36–45 refer to the Round Table® pizza menu.

The Pizzas of the Round Table

	Small Serves 1-2	Medium Serves 2-3	Large Serves 3-4
King Arthur's Supreme® The Round Table's finest: a glorious combination of cheeses, pepperoni, sausage, salami, beef, linguica, mushrooms, green peppers, onions and black olives. Plus if you'd like, shrimp and anchovy.	$3.95	$6.50	$7.90
Guinevere's Garden Delight The Vegetable Supreme: cheeses, mushrooms, olives, freshly sliced tomatoes, onions and green peppers.	$2.95	$4.95	$6.60

36. If a small pizza has a 10 in. diameter, what is the number of square inches (to the nearest square inch)?

37. If a medium pizza has a 12 in. diameter, what is the number of square inches (to the nearest square inch)?

38. If a large pizza has a 14 in. diameter, what is the number of square inches (to the nearest square inch)?

To find the price per square inch, divide the price by the number of square inches in the pizza.

39. What is the price per square inch for a small sausage pizza (King Arthur's Supreme)? (See Problem 36.)

40. What is the price per square inch for a medium sausage pizza? (See Problem 37.)

41. What is the price per square inch for a large sausage pizza? (See Problem 38.)

42. If you order a sausage pizza, what size should you order if you want the best price per square inch? (See Problems 39–41.)

43. If you order the vegetarian pizza (Guinevere's Garden Delight), what size should you order if you want the best price per square inch? (See Problems 39–41.)

44. Suppose you were the owner of a pizza restaurant and were going to offer a 16 inch pizza size. What would you charge for the sausage pizza if you wanted the price to be comparable with the prices of the other sizes?

45. Suppose you were the owner of a pizza restaurant and were going to offer an 8 inch pizza size. What would you charge for the sausage pizza if you wanted the price to be comparable with the prices of the other sizes?

CALCULATOR PROBLEMS

46. Which property has the least cost per square foot?

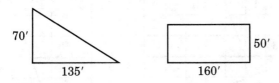

      Lot *A*: $4,500   Lot *B*: $8,500

47. Which property has the least cost per square foot?

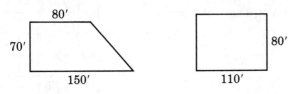

      Lot *C*: $15,000   Lot *D*: $18,000

48. If a rectangular piece of property is 750 ft by 1,290 ft, what is the acreage?

49. How many square feet are there in $4\frac{1}{2}$ acres?

50. If a rectangular piece of property is 150 m by 370 m, what is the size of the lot in hectares?

51. If a town has a rectangular boundary 4 miles by 3 miles, what is the acreage inside the city limits?

Mind Bogglers

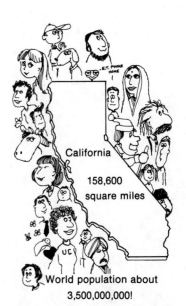

California
158,600
square miles

World population about
3,500,000,000!

52. What would happen if the *entire world population* moved to California?
  a. How much space would each person have (make a guess)?
    A. 12 sq in.   B. 12 sq ft   C. 125 sq ft
    D. 1,250 sq ft   E. 1 sq mi
  b. Now, using the information in the figure, calculate the answer that you guessed in part a; that is, how much space would be allocated to each person?

53. The accompanying figure illustrates a strange and interesting relationship. The square has an area of 64 cm² (8 cm by 8 cm). When *this same figure* is cut and rearranged as shown on the right, it appears to have an area of 65 cm². Where did this "extra" square centimeter come from?

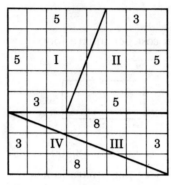

8 cm × 8 cm = 64 cm²

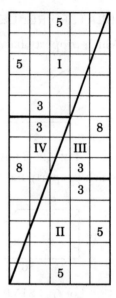

13 cm × 5 cm = 65 cm²

[*Hint:* Construct your own square 8 cm on a side, and then cut it into the four pieces as shown. Place the four pieces together as illustrated. Be sure to do your measuring and cutting very carefully. Satisfy yourself that this "extra" square centimeter has appeared. Can you explain this relationship?]

6.4 Volume

To measure area, we covered a region with square units and then found the area by using a mathematical formula. A similar procedure is used to find the **volume** of a solid object. We will fill a region with **cubes**. A **cubic inch** and a **cubic centimeter** are shown at the top of the next page.

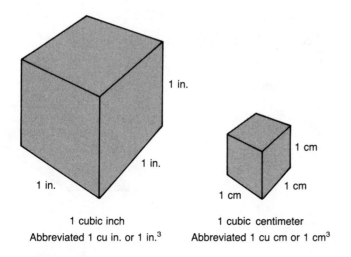

1 cubic inch
Abbreviated 1 cu in. or 1 in.^{3}

1 cubic centimeter
Abbreviated 1 cu cm or 1 cm^3

Volume of a Cube

The volume V of a cube with edge s is found by

$$\text{Volume} = \text{Edge} \times \text{Edge} \times \text{Edge}$$
$$= (\text{Edge})^3$$

or

$$V = s^3$$

If the solid is not a cube, but is a box with edges of different lengths (called a **rectangular parallelepiped**), the volume can be found similarly.

EXAMPLE 1

Find the volume of a box 4 ft by 6 ft by 4 ft.

Solution

There are 24 cubic feet on the bottom layer of cubes. Do you see how many layers of cubes will fill the solid?

Since there are four layers with 24 cubes in each, the total is

$$4 \times 24 = 96$$

The volume is 96 ft^3. ■

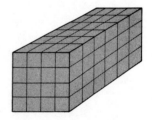

Volume of a Box (Parallelepiped)

The volume V of a box (parallelepiped) with edges l, w, and h is

$$\text{Volume} = \text{Length} \times \text{Width} \times \text{Height}$$

or

$$V = l \times w \times h$$

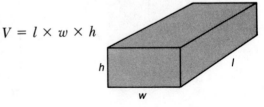

EXAMPLE 2

Find the volume of each solid.

a.

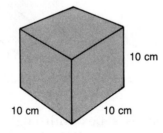

10 cm

10 cm   10 cm

$V = s^3$
$= (10 \text{ cm})^3$
$= (10 \times 10 \times 10) \text{ cm}^3$
$= \mathbf{1{,}000 \ cm^3}$

b.

4 cm

25 cm

10 cm

$V = l \times w \times h$
$= 25 \text{ cm} \times 10 \text{ cm} \times 4 \text{ cm}$
$= (25 \times 10 \times 4) \text{ cm}^3$
$= \mathbf{1{,}000 \ cm^3}$

c.

3 in.

7 in.   11 in.

$V = l \times w \times h$
$= 11 \text{ in.} \times 7 \text{ in.} \times 3 \text{ in.}$
$= (11 \times 7 \times 3) \text{ in.}^3$
$= \mathbf{231 \ in.^3}$

■

BROOM-HILDA © 1973 THE CHICAGO TRIBUNE

How much water is contained in Irwin's swimming pool? It is fairly easy to calculate the volume:

$$2 \times 2 \times 12 = 48 \text{ ft}^3$$

But this still doesn't tell us how much water the pool holds. That is, we need to find the relationship between volume measurement and the capacity measurements discussed in Section 5.2. The boxes in Examples 2a and 2b each have a volume of 1,000 cm^3, or 1 cubic decimeter. If you were to fill these boxes with water, they would hold exactly one liter. That is, a liter is sometimes *defined* as the volume contained in 1,000 cm^3 or 1 dm^3 (see Figure 6.6).

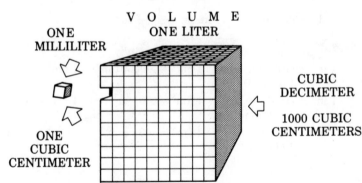

FIGURE 6.6 Relationship between metric measures of volume and capacity

One cubic centimeter is one-thousandth of a liter. Notice that this is the same as a milliliter. For this reason, you will sometimes see cc used to mean cm^3 or mℓ.

In the U.S. measurement system, the relationship between volume and capacity is a little more difficult to remember. The box in Example 2c has a volume of 231 in.^{3}. A box this size will hold exactly one gallon of water.

1 liter = 1,000 cm^3
1 gallon = 231 in.^{3}

Relationship between Volume and Capacity

In order to find the capacity of Irwin's swimming pool, we must change 48 ft³ to cubic inches:

$$48 \text{ ft}^3 = 48 \times (1 \text{ ft}) \times (1 \text{ ft}) \times (1 \text{ ft})$$
$$= 48 \times 12 \text{ in.} \times 12 \text{ in.} \times 12 \text{ in.}$$
$$= 82{,}944 \text{ in.}^3 \qquad \text{A calculator would help here.}$$

Since 1 gallon is 231 in.³, the final step (again with a calculator) is to divide 82,944 by 231 to obtain approximately 359 gallons. Table 6.3 shows some of the most common volume conversions. (Remember, you probably wouldn't bother to make these conversions unless you had a calculator available.)

TABLE 6.3   Conversions of Volume Measurement

U.S. measurement		
$1 \text{ ft}^3 = 1{,}728 \text{ in.}^3$	$1 \text{ in.}^3 = \dfrac{1}{1{,}728} \text{ ft}^3$	$1 \text{ ft}^3 \approx 7.48 \text{ gal}$
$1 \text{ yd}^3 = 27 \text{ ft}^3$	$1 \text{ ft}^3 = \dfrac{1}{27} \text{ yd}^3$	$231 \text{ in.}^3 = 1 \text{ gal}$
$1 \text{ yd}^3 = 46{,}656 \text{ in.}^3$	$1 \text{ in.}^3 = \dfrac{1}{46{,}656} \text{ yd}^3$	
Metric measurement		
$1 \text{ m}^3 = (100 \text{ cm})^3$ $= 1{,}000{,}000 \text{ cm}^3$	$1 \text{ cm}^3 = .000001 \text{ m}^3$	$1 \text{ dm}^3 = 1 \ \ell$
$1 \text{ m}^3 = (10 \text{ dm})^3$ $= 1{,}000 \text{ dm}^3$	$1 \text{ dm}^3 = .001 \text{ m}^3$	$1 \text{ cm}^3 = .001 \ \ell$
$1 \text{ dm}^3 = (10 \text{ cm})^3$ $= 1{,}000 \text{ cm}^3$	$1 \text{ cm}^3 = .001 \text{ dm}^3$	$1 \text{ m}^3 = 1{,}000 \ \ell$

EXAMPLE 3

How much water would each of the following containers hold?

a.
$$V = 90 \times 80 \times 40 \text{ cm}^3$$
$$= 288{,}000 \text{ cm}^3$$

Since each 1,000 cm³ is 1 liter,

$$\frac{288{,}000}{1{,}000} = 288$$

40 cm

90 cm   80 cm

The container will hold 288 liters.

b. $V = 7 \times 22 \times 6 \text{ in.}^3$
$= 924 \text{ in.}^3$

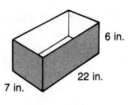

6 in.

22 in.

7 in.

Since each 231 in.^{3} is 1 gallon,

$$\frac{924}{231} = 4$$

The container will hold 4 gallons.   ■

EXAMPLE 4

An ecology swimming pool is advertised as being 20 ft × 25 ft × 5 ft. How many gallons will it hold?

Solution

$$V = 20 \text{ ft} \times 25 \text{ ft} \times 5 \text{ ft}$$
$$= 2{,}500 \text{ ft}^3$$

Since 1 ft$^3 \approx 7.48$ gal (see Table 6.3), the swimming pool contains about:

$$2{,}500 \times 7.48 \approx \textbf{18{,}700 gallons}$$ ■

EXAMPLE 5

Suppose you want to cover a rectangular walkway with gravel to a depth of 4 in. How many cubic yards of gravel do you need if the walkway is 4 ft by 36 ft?

Solution

The volume must be calculated by using similar units for the dimensions. **Working in feet** (which is probably the easiest unit for this problem):

$$4 \text{ in.} = \frac{4}{12} \text{ ft}$$

$$= \frac{1}{3} \text{ ft}$$

$$\text{Volume} = \frac{1}{3} \text{ ft} \times 4 \text{ ft} \times 36 \text{ ft} = \left(\frac{1}{3} \times 4 \times 36\right) \text{ ft}^3$$
$$= 48 \text{ ft}^3$$

Change to cubic yards:

$$48 \text{ ft}^3 = 48 \times \left(\frac{1}{27}\right) \text{yd}^3 \qquad \text{See Table 6.3}$$
$$= \frac{48}{27} \text{ yd}^3$$
$$= 1\frac{7}{9} \text{ yd}^3$$
$$\approx \mathbf{1.78 \text{ yd}^3}$$

You could also work in inches or yards to obtain the same answer. ■

Problem Set 6.4

In Problems 1–3 find the volume of each solid by counting the number of cubic centimeters in each box.

1.   **2.** 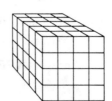 **3.**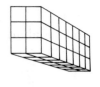

Find the volume of each solid in Problems 4–12.

4.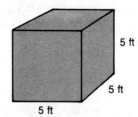

5 ft

5 ft

5 ft

5.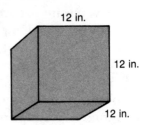

12 in.

12 in.

12 in.

6.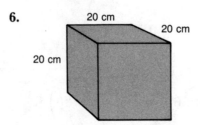

20 cm

20 cm

20 cm

7.

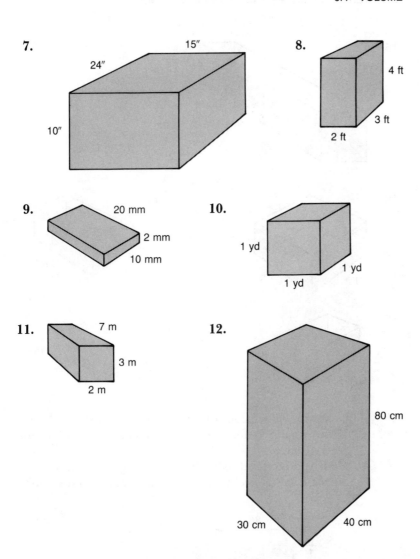

15"
24"
10"

8.

4 ft
3 ft
2 ft

9. 20 mm
2 mm
10 mm

10. 1 yd
1 yd
1 yd

11. 7 m
3 m
2 m

12. 80 cm
30 cm
40 cm

How much water would each of the containers in Problems 13–18 hold? (Give answers to the nearest $\frac{1}{10}$ gallon or $\frac{1}{10}$ liter.)

13.

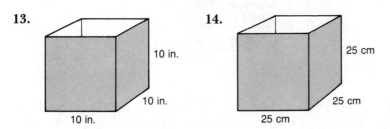

10 in.
10 in.
10 in.

14. 25 cm
25 cm
25 cm

15.

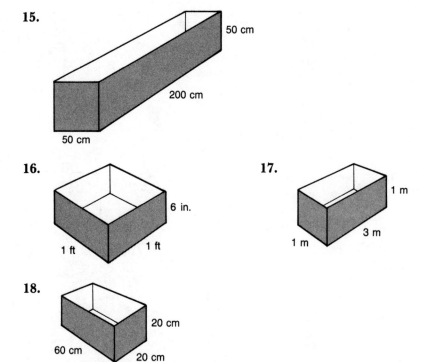

50 cm

200 cm

50 cm

16.

6 in.

1 ft   1 ft

17.

1 m

3 m

1 m

18.

20 cm

60 cm   20 cm

19. The exterior dimensions of a refrigerator are shown here.
  a. How many cubic feet are contained within the refrigerator?
  b. If it is advertised as a 19 cu ft refrigerator, how much space is taken up by the motor, insulation, and so on?

33 in.   36 in.

66 in.

20. The exterior dimensions of a freezer are 48 inches by 36 inches by 24 inches, and it is advertised as being 27.0 cu ft. Is the advertised volume correctly stated?

21. Suppose you must order concrete for a sidewalk 50′ by 4′ to a depth of 4″. How much concrete is required? (Answer to the nearest $\frac{1}{2}$ yd³.)

22. **a.** How much greater is the volume of container A than that of container B?

Inside dimension of B:
  $2′ \times 2′ \times 2′$

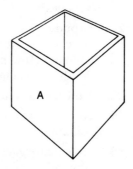

Inside dimension of A:
  $4′ \times 4′ \times 4′$

b. Consider the volume of any cube. If the length of each side of the cube is doubled, what is the increase in the volume of the cube?

CALCULATOR PROBLEMS

Use the plot plan shown in the figure to answer Problems 23–27. Answer to the nearest $\frac{1}{2}$ cubic yard.

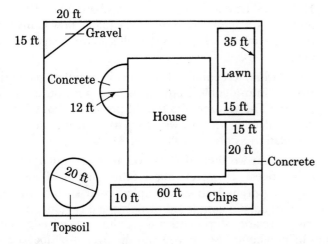

23. How many cubic yards of sawdust are needed for preparation of the lawn area if it is spread to a depth of 6 inches?

24. How many cubic yards of chips are necessary if they are placed to a depth of 3 inches?

25. How much topsoil needs to be hauled in if it is to be spread to a depth of 12 inches?

26. How much gravel is necessary if it is laid to a depth of 3 inches?

27. Suppose you wish to pave both of the areas labeled "concrete." The driveway is rectangular and the patio is semicircular. How much concrete is needed if it is poured to a depth of 4 inches?

28. How much water does a 1 cubic yard container hold?

29. How many kiloliters will a 7 m by 8 m by 2 m swimming pool contain?

30. How many gallons will a 21 ft by 24 ft by 4 ft swimming pool contain?

Mind Bogglers

31. The total population of the earth is about 3.5×10^9.
   a. If each person has the room of a prison cell (50 sq ft), and if there are about 2.5×10^7 sq ft in a sq mi, how many people could fit into a square mile?
   b. How many square miles would be required to accommodate the entire population of the earth?
   c. If the total land area of the earth is about 5.2×10^7 sq mi, and if all the land area is divided equally, how many acres of land would each person be allocated? (1 sq mi = 640 acres)

32. **a.** Guess what percentage of the world's population could be packed into a cubical box measuring $\frac{1}{2}$ mi on each side. [*Hint:* The volume of a typical person is about 2 cu ft.]
   b. Now calculate the answer to part a, using the earth's population as given in Problem 31.

33. In this section we limited ourselves to rectangular parallelepipeds (boxes). However, many other shapes of solids are possible. A **polyhedron** is a simple closed surface in space whose boundary is composed of polygonal regions. A rather surprising relationship exists among the vertices, edges, and sides of polyhedra. See if you can discover it by looking for patterns in the figures shown at the top of the next page and filling in the blanks in the following table.

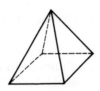

a. Triangular pyramid

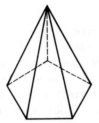

b. Quadrilateral pyramid

c. Pentagonal pyramid

d. Regular tetrahedron

e. Regular hexahedron (cube)

f. Regular octahedron

g. Regular dodecahedron

h. Regular icosahedron

Figure	Sides	Vertices	Edges
a.	4	4	6
b.	5	___	8
c.	___	6	10
d.	4	4	___
e.	___	___	12
f.	___	6	___
g.	___	___	30
h.	___	___	30

34. Mathematical forms in nature. What forms in nature represent mathematical patterns? What shapes in nature are

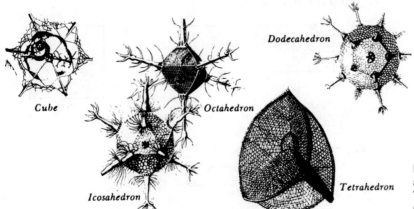

Examples of the five regular solids as they can be found in nature. These are skeletons of minute marine animals called *radiolaria*.

geometric and/or symmetric? What activities in nature illustrate mathematical functions?

Exhibit suggestions: samples of spiral shells, crystals, ellipsoidal stones or eggs, spiraling sunflower seed pods, branch distributions illustrating Fibonacci series, snowflake patterns, body bones acting as levers, ratio of food consumed to body size

References: Bergamini, David, *Mathematics*, Chap. 4 (New York: Time, Inc., Life Science Library, 1963).

Carson, Judithlynne, "Fibonacci Numbers and Pineapple Phyllotaxy," *Two Year College Mathematics Journal*, June 1978, pp. 132–136.

Newman, James, *The World of Mathematics* (New York: Simon and Schuster, 1956).

"Crystals and the Future of Physics," pp. 871–881.

"On Being the Right Size," pp. 952–957.

"On Magnitude," pp. 1001–1046.

"The Soap Bubble," pp. 891–900.

See also the *Fibonacci Quarterly*.

6.5 Review Problems

Section 6.1

1. a. Name the polygon.
   b. Classify $\angle A$.
   c. Classify $\angle B$.
   d. Classify $\angle C$.
   e. What is the measure of $\angle D$?

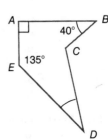

Section 6.2

2. Find the perimeter.

a.

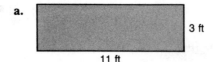

3 ft

11 ft

b.

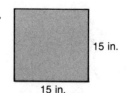

15 in.

15 in.

c.

6 m

4 m

8 m

3. Find the circumference or the distance around each figure.   Section 6.2
(Use 3.14 for π.)

a.   **b.**   **c.**

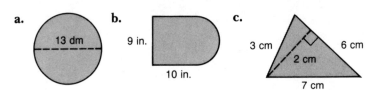

4. Find the perimeter of the given figure. Measure each side to   Section 6.2
the nearest $\frac{1}{10}$ centimeter.

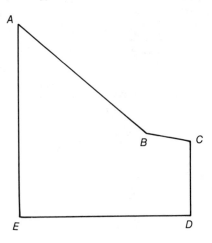

5. Find the area of each figure in Problem 2.   Section 6.3

6. Find the area of each figure in Problem 3 correct to the   Section 6.3
nearest tenth unit. Use 3.14 as an approximation for π.

7. How many square yards of carpet are necessary to carpet a   Section 6.3
13′ by 14′ room? (Assume that you cannot purchase part of a
square yard.)

8. Find the volume of each figure.   Section 6.4

a.

b.

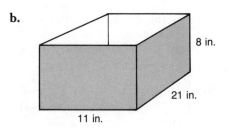

8 in.

21 in.

11 in.

c.

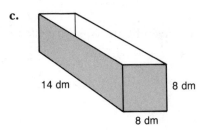

14 dm   8 dm

8 dm

Section 6.4   **9.** How many gallons or liters does each box in Problem 8 hold (answer to the nearest tenth unit)?

Section 6.4   **10.** How many cubic yards of concrete are necessary for a rectangular driveway 20′ by 25′ by 4″? Assume that you can order in units of $\frac{1}{2}$ cubic yard.

Percents and Problem Solving

7.1 Ratio and Proportion

One of the most common uses of mathematics in our every-day lives is in reference to **percents**. If you want to develop your problem-solving ability to apply it to everyday problems, you will first need to understand the ideas of **ratio** and **proportion**.

A **ratio** expresses a size relationship between two sets. It is written using the word "to," a colon, or as a fraction. That is, if the ratio of men to women is 5 **to** 4, this could also be written as 5:4 or $\frac{5}{4}$. We will emphasize the idea that a ratio is a fraction (or a quotient of two numbers). Since a fraction can be reduced, a ratio also can be reduced.

EXAMPLE 1

Reduce the given ratios to lowest terms.

a. A ratio of 4 to 52   **b.** A ratio of 15 to 3

$$\frac{4}{52} = \frac{1}{13}$$ $$\frac{15}{3} = 5$$

A ratio of 1 to 13   Write this as $\frac{5}{1}$ because a ratio compares two numbers.
A ratio of 5 to 1

c. A ratio of $1\frac{1}{2}$ to 2   **d.** A ratio of $1\frac{2}{3}$ to $3\frac{3}{4}$

$$\frac{1\frac{1}{2}}{2} = 1\frac{1}{2} \div 2$$ $$\frac{1\frac{2}{3}}{3\frac{3}{4}} = 1\frac{2}{3} \div 3\frac{3}{4}$$

$$= \frac{3}{2} \times \frac{1}{2}$$ $$= \frac{5}{3} \div \frac{15}{4}$$

$$= \frac{3}{4}$$ $$= \frac{5}{3} \times \frac{4}{15}$$

A ratio of 3 to 4 $$= \frac{4}{9}$$

A ratio of 4 to 9 ■

A **proportion** is a statement of equality between ratios. In symbols,

$$\underset{\uparrow}{\frac{a}{b}} \quad \overset{=}{\underset{\uparrow}{}} \quad \underset{\uparrow}{\frac{c}{d}}$$

which is read   *"a is to b"* *"as"* *"c is to d"*

The notation used in some books is

$$a : b :: c : d$$

Even though we won't use this notation, we will use words associated with this notation to name the terms:

$$
\begin{array}{c}
\text{Extremes} \\
\text{Means} \\
a : b :: c : d
\end{array}
$$

In the more common fractional notation:

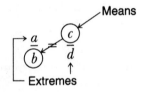

EXAMPLE 2

Read each proportion and name the means and the extremes.

a. $\dfrac{2}{3} = \dfrac{10}{15}$

    Read: Two is to three as ten is to fifteen.
    Means: 3 and 10
    Extremes: 2 and 15

b. $\dfrac{m}{5} = \dfrac{3}{8}$

    Read: *m* is to five as three is to eight.
    Means: 5 and 3
    Extremes: *m* and 8   ■

The following property is fundamental to our study of proportions and percents:

Property of Proportions

> If the product of the means equals the product of the extremes, then the ratios form a proportion.
>
> ALSO
>
> If the ratios form a proportion, then the product of the means equals the product of the extremes.
>   In symbols,
>
> $$\frac{a}{b} = \frac{c}{d}$$
>
> $$\underbrace{b \times c} = \underbrace{a \times d}$$
>
> Product of means = Product of extremes

EXAMPLE 3

Tell whether each pair of ratios forms a proportion.

a. $\dfrac{3}{4}$; $\dfrac{36}{48}$

Means	*Extremes*
4×36	3×48
144	=	144

Thus,

$$\frac{3}{4} = \frac{36}{48}$$

They form
a proportion.

b. $\dfrac{5}{16}$; $\dfrac{7}{22}$

Means	*Extremes*
16×7	5×22
112	$\neq$	110

Thus,

$$\frac{5}{16} \neq \frac{7}{22}$$

They do not form
a proportion. ■

CALCULATOR COMMENT

The remainder of this section is optional for those who have calculators. Since ratios are fractions, they can be written in decimal form. This is a particularly useful representation of a ratio when using a calculator.

EXAMPLE 4

Write the decimal forms for the given ratios.

a. 3 to 5

$\boxed{3}\ \boxed{\div}\ \boxed{5}\ \boxed{=}\ \text{.6}$

b. 2 to 3

$\boxed{2}\ \boxed{\div}\ \boxed{3}\ \boxed{=}\ \text{.666666667}$ ■

   There are two methods for using a calculator to determine whether two ratios form a proportion. The first method is to use the product of the means and the extremes, as shown in Example 3. The second method is to write each ratio in decimal form and then compare the decimals. This second method will be applied to Example 3b (as well as a new example) so that you can compare these methods.

EXAMPLE 5

Tell whether each pair of ratios forms a proportion.

a. $\dfrac{5}{16}$; $\dfrac{7}{22}$

$\boxed{5}\ \boxed{\div}\ \boxed{16}\ \boxed{=}\ \text{.3125}$

$\boxed{7}\ \boxed{\div}\ \boxed{22}\ \boxed{=}\ \text{.318181818}$

They do not form a proportion.

b. $\dfrac{4}{5}$; $\dfrac{5}{6\frac{1}{4}}$

$\boxed{4}\ \boxed{\div}\ \boxed{5}\ \boxed{=}\ \text{.8}$

$\boxed{5}\ \boxed{\div}\ \boxed{(}\ \boxed{25}\ \boxed{\div}\ \boxed{4}\ \boxed{)}\ \boxed{=}\ \text{.8}$

If your calculator doesn't have parentheses, write

$$\frac{5}{6\frac{1}{4}} = 5 \div \frac{25}{4} = 5 \times \frac{4}{25}$$

$\boxed{5}\ \boxed{\times}\ \boxed{4}\ \boxed{\div}\ \boxed{25}\ \boxed{=}\ \text{.8}$

They do form a proportion.   ■

Problem Set 7.1

Write the ratios given in Problems 1–14 as reduced fractions.

1. The ratio of "yes" to "no" answers is 3 to 2.
2. The ratio of cars to people is 1 to 3.
3. The ratio of cats to dogs is 4 to 7.
4. The ratio of dogs to cats is 7 to 4.
5. The ratio of gallons to miles is 60 to 4.
6. The ratio of dollars to people is 92 to 4.
7. The ratio of wins to losses is 12 to 4.
8. The ratio of apples to oranges is 6 to 2.
9. The ratio of sand to gravel is 2 to 4.
10. The ratio of children to adults is $2\frac{1}{5}$ to 2.
11. The ratio of cars to people is 1 to $2\frac{1}{2}$.
12. Find the ratio of $2\frac{3}{4}$ to $6\frac{1}{4}$.
13. Find the ratio of $6\frac{1}{4}$ to $2\frac{3}{4}$.
14. Find the ratio of $5\frac{2}{3}$ to $2\frac{1}{3}$.

In Problems 15–18 read each proportion and name the means and extremes.

15. $\dfrac{5}{8} = \dfrac{35}{56}$ 16. $\dfrac{94}{47} = \dfrac{2}{1}$ 17. $\dfrac{5}{3} = \dfrac{2}{x}$ 18. $\dfrac{5}{x} = \dfrac{1}{2}$

Tell whether each pair of ratios in Problems 19–30 forms a proportion.

19. $\dfrac{7}{1}$; $\dfrac{21}{3}$ 20. $\dfrac{6}{8}$; $\dfrac{9}{12}$ 21. $\dfrac{3}{6}$; $\dfrac{5}{10}$

22. $\dfrac{7}{8}$; $\dfrac{6}{7}$ 23. $\dfrac{9}{2}$; $\dfrac{10}{3}$ 24. $\dfrac{6}{5}$; $\dfrac{42}{35}$

25. $\dfrac{20}{70}$; $\dfrac{4}{14}$ 26. $\dfrac{3}{4}$; $\dfrac{75}{100}$ 27. $\dfrac{2}{3}$; $\dfrac{67}{100}$

28. $\dfrac{5}{3}$; $\dfrac{7\frac{1}{3}}{4}$ 29. $\dfrac{3}{2}$; $\dfrac{5}{3\frac{1}{2}}$ 30. $\dfrac{5\frac{1}{5}}{7}$; $\dfrac{4}{5}$

31. A cement mixture calls for 60 pounds of cement for 3 gallons of water. What is the ratio of cement to water?
32. What is the ratio of water to cement in Problem 31?
33. If you drive 119 miles on $8\frac{1}{2}$ gallons of gas, what is the simplified ratio of miles to gallons?
34. If you drive 180 miles on $7\frac{1}{2}$ gallons of gas, what is the simplified ratio of miles to gallons?

35. If you drive 279 miles on $15\frac{1}{2}$ gallons of gas, what is the simplified ratio of miles to gallons?

36. About 106 baby boys are born for every 100 baby girls. Write this as a simplified ratio of males to females.

37. Use Problem 36 to write a simplified ratio of females to males.

A baseball player's **batting average** *is the ratio of* **hits** *to* **times at bat.** *For example, how many hits per times at bat does a player with a batting average of .275 have?*

$$.275 = \frac{275}{1,000} = \frac{11}{40}$$

This is a ratio of 11 to 40, which means the player has 11 hits for every 40 times at bat. Find the number of hits per times at bat for the batting champions named in Problems 38–41.

38. Jake Daubert (Brooklyn, 1913);   .350

39. Ty Cobb (Detroit, 1912);   .390

40. Ernie Lombardi (Boston, 1942);   .330

41. Pete Runnels (Boston, 1960);   .320

Mind Boggler

42. A proportion machine. Take two sheets of lined paper. Label one *numerator* and one *denominator*, and number the lines consecutively from 0, as shown in Figure 7.1. Fasten the pages together at 0 so that the top page pivots at this point (or else always be sure that the 0 lines coincide).

Suppose you wish to find a fraction equal to $\frac{5}{6}$ using a denominator of 18. Slide the top page down until the lines 5 (numerator) and 6 (denominator) coincide. Now you can read other equivalent fractions from the *proportion machine* by noting the places where the lines coincide. From Figure 7.1 we see that

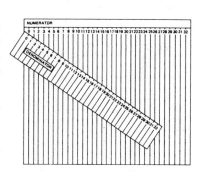

FIGURE 7.1 A proportion machine

$$\frac{5}{6} = \frac{15}{18}$$

Also notice from the figure that $\frac{5}{6} = \frac{10}{12}$ and $\frac{5}{6} = \frac{20}{24}$. Use the proportion machine you have constructed to fill in the following blanks:

a. $\frac{5}{7} = \frac{?}{21}$ **b.** $\frac{4}{9} = \frac{?}{18}$ **c.** $\frac{5}{8} = \frac{?}{4}$ **d.** $\frac{3}{4} = \frac{1}{?}$

Demonstrate a proportion machine to the class. Can you think of any other uses for a proportion machine?*

Reference: "Reducing Fractions Can Be Easy, Maybe Even Fun," by M. Wassmansdorf, *Arithmetic Teacher*, February 1974, pp. 99–102.

7.2 Problem Solving with Proportions

In the last section, we tested pairs of ratios to see if they formed a proportion. However, the usual setting for a proportion is where three of the terms of a proportion are known and one of the terms is unknown. It is always possible to find the missing term.

EXAMPLE 1

Find the missing term of each proportion.

a. $\dfrac{3}{4} = \dfrac{w}{20}$

PRODUCT OF MEANS $=$ PRODUCT OF EXTREMES

$$4w = 3(20)$$

$$w = \frac{3(\cancel{20}^{5})}{\cancel{4}_{1}}$$

Solve the equation by dividing both sides by 4. Notice that 4 is the number opposite the unknown:

$$w = \mathbf{15}$$

$$\frac{3}{4} = \frac{w}{20} \leftarrow$$

b. $\dfrac{3}{4} = \dfrac{27}{y}$

PRODUCT OF MEANS $=$ PRODUCT OF EXTREMES

$$4(27) = 3y$$

$$\frac{4(\cancel{27}^{9})}{\cancel{3}_{1}} = y$$

Divide both sides by 3. Notice that 3 is the number opposite the unknown:

$$\mathbf{36} = y$$

$$\frac{3}{4} = \frac{27}{y}$$

*My thanks to Mark Wassmansdorf of the Philadelphia School District for this problem.

c. $\dfrac{2}{x} = \dfrac{8}{9}$

PRODUCT OF MEANS = PRODUCT OF EXTREMES

$$8x = 2(9)$$

$$x = \dfrac{\overset{1}{\cancel{2}}(9)}{\underset{4}{\cancel{8}}}$$

Divide both sides by 8. Notice that 8 is the number opposite the unknown:

$$x = \dfrac{9}{4}$$

$$\dfrac{2}{x} = \dfrac{8}{9} \leftarrow$$

d. $\dfrac{t}{15} = \dfrac{3}{5}$

PRODUCT OF MEANS = PRODUCT OF EXTREMES

$$3(15) = 5t$$

$$\dfrac{3(\cancel{15})}{\underset{1}{\cancel{5}}} = t$$

Divide both sides by 5. Notice that 5 is the number opposite the unknown:

$$\mathbf{9} = t$$

$$\dfrac{t}{15} = \dfrac{3}{5}$$

Notice that the unknown terms can be in any one of four positions, as illustrated by the four parts of Example 1. Even though you can find the missing term of a proportion (called **solving the proportion**) by the technique used in Example 1, it is easier to think in terms of **the cross-product divided by the number opposite the unknown.**

---

1. Find the product of the means or the product of the extremes, whichever does not contain the unknown term.

2. Divide this product by the number that is opposite the unknown term.

Procedure for Solving Proportions

EXAMPLE 2

Solve the proportion for the unknown term.

a. $\dfrac{5}{6} = \dfrac{55}{y}$

$$y = \frac{6 \times 55}{5} \quad \begin{array}{l} \longleftarrow \text{Product of the means} \\ \longleftarrow \text{Number opposite the unknown} \end{array}$$

$$= \frac{6 \times \overset{11}{\cancel{55}}}{\underset{1}{\cancel{5}}} \qquad \begin{array}{l} \text{You can cancel to simplify} \\ \text{many of these problems.} \end{array}$$

$$= \mathbf{66}$$

b. $\dfrac{5}{b} = \dfrac{3}{4}$

$$b = \frac{5 \times 4}{3} \quad \begin{array}{l} \longleftarrow \text{Product of the extremes} \\ \longleftarrow \text{Number opposite the unknown} \end{array}$$

$$= \mathbf{\frac{20}{3}} \quad \text{or} \quad \mathbf{6\frac{2}{3}}$$

Notice that your answers don't have to be whole numbers. This means that the correct proportion is

$$\frac{5}{6\frac{2}{3}} = \frac{3}{4}$$

c. $\dfrac{2\frac{1}{2}}{5} = \dfrac{a}{8}$

$$a = \frac{2\frac{1}{2} \times 8}{5} \quad \begin{array}{l} \longleftarrow \text{Product of the extremes} \\ \longleftarrow \text{Number opposite the unknown} \end{array}$$

$$= \frac{\frac{5}{2} \times \frac{8}{1}}{5}$$

$$= \frac{20}{5}$$

$$= \mathbf{4} \qquad \blacksquare$$

Many applied problems can be solved using a proportion. When setting up a proportion with units, be sure that like units occupy corresponding positions, as illustrated by Examples 3–5.

EXAMPLE 3

If 4 cans of cola sell for 89¢, how much will 12 cans cost?

Solution

$$\text{Cans} \longrightarrow \frac{4}{89} \underset{\longleftarrow \text{Cents}}{\overset{\longleftarrow \text{Cans}}{=}} \frac{12}{x}$$

$$x = \frac{89 \times \overset{3}{\cancel{12}}}{\underset{1}{\cancel{4}}}$$

$$= 267$$

12 cans cost \$2.67.   ■

EXAMPLE 4

If a 120 mile trip took $8\frac{1}{2}$ gallons of gas, then how much gas is
needed for a 240 mile trip?

Solution

$$\text{Miles} \longrightarrow \frac{120}{8\frac{1}{2}} = \frac{240}{x} \longleftarrow \text{Miles}$$

$$\text{Gallons} \qquad \qquad \qquad \text{Gallons}$$

$$x = \frac{8\frac{1}{2} \times \overset{2}{\cancel{240}}}{\underset{1}{\cancel{120}}}$$

$$x = 17$$

The trip will require 17 gallons.   ■

EXAMPLE 5

If the property tax on a \$65,000 home is \$416, what is the tax
on an \$85,000 home?

Solution

$$\text{Value} \longrightarrow \frac{65,000}{416} = \frac{85,000}{x} \longleftarrow \text{Value}$$

$$\text{Tax} \qquad \qquad \qquad \text{Tax}$$

$$x = \frac{416 \times 85,000}{65,000}$$

You can do the arithmetic on a calculator or by canceling:

$$x = \frac{416 \times \overset{85}{\cancel{85,000}}}{\underset{65}{\cancel{65,000}}}$$

$$= \frac{416 \times \overset{17}{\cancel{85}}}{\underset{13}{\cancel{65}}}$$

$$= \frac{\overset{32}{\cancel{416}} \times 17}{\underset{1}{\cancel{13}}}$$

$$= 544$$

The tax is $544.   ■

CALCULATOR COMMENT

Proportions are quite easy to solve if you have a calculator. You can multiply the cross-terms and divide by the number opposite the variable all in one calculator sequence. We will show this process in Example 6 by repeating the calculations from Examples 2–5.

EXAMPLE 6

From Example 2a: $\dfrac{5}{6} = \dfrac{55}{y}$

Press: $\boxed{6}$ $\boxed{\times}$ $\boxed{55}$ $\boxed{\div}$ $\boxed{5}$ $\boxed{=}$ *Display:* $66.$

From Example 2b: $\dfrac{5}{b} = \dfrac{3}{4}$

Press: $\boxed{5}$ $\boxed{\times}$ $\boxed{4}$ $\boxed{\div}$ $\boxed{3}$ $\boxed{=}$ *Display:* 6.6666667

From Example 2c: $\dfrac{2\frac{1}{2}}{5} = \dfrac{a}{8}$

Press: $\boxed{2.5}$ $\boxed{\times}$ $\boxed{8}$ $\boxed{\div}$ $\boxed{5}$ $\boxed{=}$ *Display:* $4.$

From Example 3: $\dfrac{4}{89} = \dfrac{12}{x}$

Press: `89` `×` `12` `÷` `4` `=`   *Display:* $267.$

From Example 4: $\dfrac{120}{8\frac{1}{2}} = \dfrac{240}{x}$

Press: `8.5` `×` `240` `÷` `120` `=`   *Display:* $17.$

From Example 5: $\dfrac{65,000}{416} = \dfrac{85,000}{x}$

Press: `85000` `×` `416` `÷` `65000` `=` *Display:* $544.$ ∎

Problem Set 7.2

Solve each proportion in Problems 1–24 for the item represented by a letter.

1. $\dfrac{5}{1} = \dfrac{A}{3}$ **2.** $\dfrac{1}{9} = \dfrac{4}{B}$ **3.** $\dfrac{C}{2} = \dfrac{5}{1}$ **4.** $\dfrac{7}{D} = \dfrac{1}{8}$

5. $\dfrac{12}{18} = \dfrac{E}{12}$ **6.** $\dfrac{12}{15} = \dfrac{20}{F}$ **7.** $\dfrac{G}{24} = \dfrac{14}{16}$ **8.** $\dfrac{4}{H} = \dfrac{3}{15}$

9. $\dfrac{2}{3} = \dfrac{I}{7}$ **10.** $\dfrac{4}{5} = \dfrac{3}{J}$ **11.** $\dfrac{3}{K} = \dfrac{2}{5}$ **12.** $\dfrac{L}{2} = \dfrac{5}{6}$

13. $\dfrac{7\frac{1}{5}}{9} = \dfrac{M}{5}$ **14.** $\dfrac{4}{2\frac{2}{3}} = \dfrac{3}{N}$ **15.** $\dfrac{P}{4} = \dfrac{4\frac{1}{2}}{6}$ **16.** $\dfrac{5}{2} = \dfrac{Q}{12\frac{3}{5}}$

17. $\dfrac{5}{R} = \dfrac{7}{12\frac{3}{5}}$ **18.** $\dfrac{1\frac{1}{3}}{\frac{1}{9}} = \dfrac{S}{2\frac{2}{3}}$ **19.** $\dfrac{33}{2\frac{1}{5}} = \dfrac{3\frac{3}{4}}{T}$ **20.** $\dfrac{U}{1\frac{1}{2}} = \dfrac{\frac{1}{2}}{\frac{3}{4}}$

21. $\dfrac{\frac{1}{5}}{\frac{2}{3}} = \dfrac{\frac{3}{4}}{V}$ **22.** $\dfrac{\frac{3}{5}}{\frac{1}{2}} = \dfrac{X}{\frac{2}{3}}$ **23.** $\dfrac{9}{Y} = \dfrac{1\frac{1}{2}}{3\frac{2}{3}}$ **24.** $\dfrac{Z}{2\frac{1}{3}} = \dfrac{1\frac{1}{2}}{4\frac{1}{5}}$

25. This problem will help you check your work in Problems 1–24. Fill in the capital letters from Problems 1–24 to correspond with their numerical values in the boxes on the next page. For example, if

$$\frac{W}{7} = \frac{10}{14}$$

then

$$W = \frac{\overset{1}{\cancel{7}} \times \overset{5}{\cancel{10}}}{\underset{\underset{1}{2}}{\cancel{14}}}$$

$$W = 5$$

Now find the box or boxes with the number 5 in the corner and fill in the letter W. This has already been done for you. (The letter O has also been filled in for you.) Some letters may not appear in the boxes. When you are finished filling in the letters, darken in all the blank spaces to separate the words in the secret message. Notice that one of the blank spaces has also been filled in to help you.

$\frac{1}{4}$	$\frac{1}{4}$	20	$\frac{14}{3}$	32	7	3	9	O	36	$\frac{5}{3}$	8	4	7	■	13
5 **W**	$\frac{14}{3}$	$\frac{5}{3}$	$\frac{5}{3}$	$\frac{12}{5}$	1	2	56	O	1	36	$\frac{1}{4}$	8	56	$\frac{5}{3}$	22
12	15	32	32	$\frac{14}{3}$	32	$\frac{1}{4}$	6	22 O	1	$\frac{1}{2}$	7	$\frac{14}{3}$	2	18
25	$\frac{14}{3}$	2	56	$\frac{14}{3}$	2	21	11	8	9	9 O	9	32	!	!

26. If a 184 mile trip took $11\frac{1}{2}$ gallons of gas, then how much gas is needed for a 160 mile trip?

27. If a 121 mile trip took $5\frac{1}{2}$ gallons of gas, how many miles can be driven with a full tank of 13 gallons?

28. If a family uses $3\frac{1}{2}$ gallons of milk per week, how much milk will this family need for four days?

29. If 2 gallons of paint are needed for 75 ft of fence, how many gallons are needed for 900 ft of fence?

30. If Jack jogs 3 miles in 40 minutes, how long will it take him (to the nearest minute) to jog 2 miles at the same rate?

31. If Jill jogs 2 miles in 15 minutes, how long will it take her (to the nearest minute) to jog 5 miles at the same rate?

32. A moderately active 140 pound person will use 2,100 calories per day to maintain that body weight. How many calories per day are necessary to maintain a moderately active 165 pound person?

33. You've probably seen advertisements for posters that can be made from any photograph. If the finished poster will be 2 ft by 3 ft, it's likely that part of your original snapshot will be cut off. Suppose you send in a photo that measures 3″ by 5″. If the shorter side of the enlargement will be 2 ft, what size should the longer side of the enlargement be so that the entire snapshot is shown in the poster?

34. Suppose you wish to make a scale drawing of your living room, which measures 18 ft by 25 ft. If the shorter side of the drawing is 6 inches, how long is the longer side of the scale drawing?

Mind Boggler

35. At a certain hamburger stand the owner sold soft drinks out of two 16 gallon barrels. At the end of the first day she wished to increase her profit, so she filled the soft-drink barrels with water, thus diluting the drink served. She repeated the procedure at the end of the second and third days. At the end of the fourth day she had 10 gallons remaining in the barrels, but they contained only 1 pint of pure soft drink. If the same amount was served each day, how much pure soft drink was served in the four days?

7.3 Percent

Percent is a commonly used word, which you can find daily in any newspaper.

Meaning of Percent

Percent is a ratio of a given number to 100.

The symbol % is used to indicate percent. Consider some examples from one issue of a local newspaper.

EXAMPLE 1

Illustrate the meaning of percent in these quotes.

a. "The President recommended an 8 percent cost of living raise in Social Security payments."

"8 percent" means the "ratio of 8 to 100."

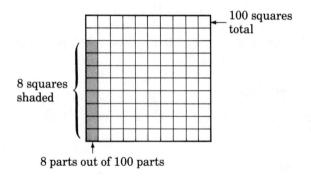

b. "$33\frac{1}{3}$% OFF"

"$33\frac{1}{3}\%$" means the "ratio of $33\frac{1}{3}$ to 100."

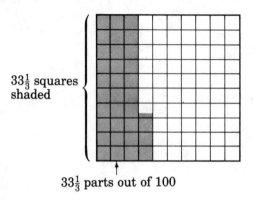

Since a percent is a ratio, percents can be easily written in fractional form.

EXAMPLE 2

Write the percents as simplified fractions.

a. **"SALE 75% OFF"**

"75%" means a "ratio of 75 to 100":

$$\frac{75}{100} = \frac{3}{4}$$

b. **"SALARIES UP 6.8%"**

"6.8%" means a "ratio of 6.8 to 100":

$$6.8 \div 100 = 6\tfrac{8}{10} \div 100$$
$$= 6\tfrac{4}{5} \div 100$$
$$= \frac{\overset{17}{\cancel{34}}}{5} \times \frac{1}{\underset{50}{\cancel{100}}}$$
$$= \frac{17}{250} \qquad \blacksquare$$

Percents can also be written as decimals. Since percent is a ratio of a number to 100, we can apply the short method of dividing by 100 that was introduced and discussed in Chapter 2.

To express a percent as a decimal, shift the decimal point two places to the *left*, and delete the % symbol.

Procedure for Changing a Percent to a Decimal

EXAMPLE 3

Write each percent in decimal form.

a. 8 percent

8%

— If a decimal point is not shown, it is always understood to be at the right of the whole number.

.08.%

Shift two places to the left; add zeros as place-holders, if necessary. Delete percent symbol.

Answer: 8% = .08

b. 6.8%

Think: **6.8%**

Shift two places; add place-holders as necessary; and delete percent symbol.

Answer: 6.8% = **.068**

c. $33\frac{1}{3}\%$

Think: **$33.\frac{1}{3}\%$**

Decimal point is understood

Answer: $33\frac{1}{3}\% = .33\frac{1}{3}$ or   **.333 . . .**

d. $\frac{1}{2}\%$

$\frac{1}{2} = .5,$ so $\frac{1}{2}\% = .5\%$ *Think:*   **00.5%**

Answer: .5% = **.005**   ■

As you can see from the examples above, every number can be written in each of three forms: fractional, decimal, and percent. Even though we discussed changing from fractional to decimal form earlier in the text, we'll review the three forms in this section.

Fraction	Decimal	Percent
1. To change to a decimal: Divide numerator by denominator (review Section 2.2, if necessary).	**1. To change to a percent:** Shift decimal point two places to the *right*, and add a percent symbol.	**1. To change to a decimal:** Shift decimal point two places to the *left*, and delete the percent symbol.
2. To change to a percent: First change to a decimal; then look under the Decimal heading in this chart.	**2. To change a terminating decimal to a fraction:** **a.** Name the decimal position of the last digit. **b.** Multiply the digits by the decimal name of the last digit (review Section 2.4).	**2. To change to a fraction:** Write as a ratio of 100 and reduce the fraction. If the percent involves a decimal, first write the decimal in fractional form; then write as a ratio of 100.

EXAMPLE 4

Change each fraction to percent form.

a. $\dfrac{5}{8}$

Look under the Fraction heading in the chart, and follow the directions for changing to a percent.

Step 1. $\dfrac{5}{8} = .625$

Step 2. $.625 = 62.5\%$

Two places; add zeros as place-holders as necessary; also add percent symbol.

$$
\begin{array}{r}
.6\,2\,5 \\
8\overline{)5.0\,0\,0} \\
4\,8 \\
\overline{2\,0} \\
1\,6 \\
\overline{4\,0} \\
4\,0 \\
\hline
\end{array}
$$

Answer: **62.5%**

b. $\dfrac{5}{6}$

Step 1. $\dfrac{5}{6} = .8333\ldots$

Sometimes this is written as $.83\frac{1}{3}$.

Step 2. $.8333\ldots = 83.333\ldots\%$
or $83\frac{1}{3}\%$
The decimal point is understood.

$$
\begin{array}{r}
.8\,3\,3 \\
6\overline{)5.0\,0\,0} \\
4\,8 \\
\overline{2\,0} \\
1\,8 \\
\overline{2\,0} \\
1\,8 \\
\overline{2}\;\; \text{Repeats} \\
\end{array}
$$

Answer: **$83\frac{1}{3}\%$**   ■

EXAMPLE 5

Change the decimal forms to percents and fractions.

a. $.85$

Step 1. $.85 = 85\%$

Two places   Decimal understood

Step 2. 85% means $85 \times \dfrac{1}{100} = \dfrac{85}{100}$

$$= \dfrac{17}{20}$$

b. 2.485

Step 1. $2.485 = 248.5\%$

Step 2. 248.5% means $248.5 \times \dfrac{1}{100} = 248\tfrac{1}{2} \times \dfrac{1}{100}$

$$= \dfrac{497}{2} \times \dfrac{1}{100}$$

$$= \dfrac{497}{200} \text{ or } 2\tfrac{97}{200}$$

■

Because you will often need to make the kinds of changes illustrated in the previous examples, it is convenient to have a table to refer to. You probably know some of the entries in Table 7.1 already, and perhaps you will want to memorize others for your future work.

CALCULATOR COMMENT

Many calculators have a percent key

$$\boxed{\%}$$

This key will convert a number from a percent to a decimal.

EXAMPLE 6

Convert 43.5% to decimal form.

Press	Display
$\boxed{43.5}$	43.5
$\boxed{\%}$	.435

■

TABLE 7.1   Fraction, Decimal, and Percent Equivalents

Denominator	Numerator									
	1	2	3	4	5	6	7	8	9	10
2	.50 50%	1.00 100%								
3	$.33\frac{1}{3}$ $33\frac{1}{3}\%$	$.66\frac{2}{3}$ $66\frac{2}{3}\%$	1.00 100%							
4	.25 25%	.50 50%	.75 75%	1.00 100%						
5	.20 20%	.40 40%	.60 60%	.80 80%	1.00 100%					
6	$.16\frac{2}{3}$ $16\frac{2}{3}\%$	$.33\frac{1}{3}$ $33\frac{1}{3}\%$	.50 50%	$.66\frac{2}{3}$ $66\frac{2}{3}\%$	$.83\frac{1}{3}$ $83\frac{1}{3}\%$	1.00 100%				
7	$.14\frac{2}{7}$ $14\frac{2}{7}\%$	$.28\frac{4}{7}$ $28\frac{4}{7}\%$	$.42\frac{6}{7}$ $42\frac{6}{7}\%$	$.57\frac{1}{7}$ $57\frac{1}{7}\%$	$.71\frac{3}{7}$ $71\frac{3}{7}\%$	$.85\frac{5}{7}$ $85\frac{5}{7}\%$	1.00 100%			
8	$.12\frac{1}{2}$ $12\frac{1}{2}\%$	.25 25%	$.37\frac{1}{2}$ $37\frac{1}{2}\%$	.50 50%	$.62\frac{1}{2}$ $62\frac{1}{2}\%$	.75 75%	$.87\frac{1}{2}$ $87\frac{1}{2}\%$	1.00 100%		
9	$.11\frac{1}{9}$ $11\frac{1}{9}\%$	$.22\frac{2}{9}$ $22\frac{2}{9}\%$	$.33\frac{1}{3}$ $33\frac{1}{3}\%$	$.44\frac{4}{9}$ $44\frac{4}{9}\%$	$.55\frac{5}{9}$ $55\frac{5}{9}\%$	$.66\frac{2}{3}$ $66\frac{2}{3}\%$	$.77\frac{7}{9}$ $77\frac{7}{9}\%$	$.88\frac{8}{9}$ $88\frac{8}{9}\%$	1.00 100%	
10	.10 10%	.20 20%	.30 30%	.40 40%	.50 50%	.60 60%	.70 70%	.80 80%	.90 90%	1.00 100%

Problem Set 7.3

Write the percents in Problems 1–6 as simplified fractions.

1. **SALE   50% OFF**

2. In 1974 there were two persons out of work in Luxembourg, and by the end of the year unemployment had spiraled to 11, making the unemployment figure jump by 550%.

3. There was a 13.4 percent increase in the use of electricity.

4. The weight of a certain model of automobile was decreased by 3.7 percent.

5. Ivory Snow is $99\frac{44}{100}\%$ pure.

6. In 1978 the sales tax in Missouri was $3\frac{1}{8}\%$.

In Problems 7–30 change the given form into the two missing forms.

	Fraction	Decimal	Percent
7.	$\frac{1}{2}$	_____	_____
8.	$\frac{1}{4}$	_____	_____
9.	_____	.75	_____
10.	_____	.85	_____
11.	_____	_____	65%
12.	_____	_____	45%
13.	_____	_____	120%
14.	_____	_____	250%
15.	_____	.7	_____
16.	_____	.9	_____
17.	$\frac{1}{3}$	_____	_____
18.	$\frac{2}{3}$	_____	_____
19.	$\frac{1}{5}$	_____	_____
20.	$\frac{2}{5}$	_____	_____
21.	_____	.05	_____
22.	_____	.025	_____
23.	_____	_____	4%
24.	_____	_____	8%
25.	_____	_____	$6\frac{1}{2}\%$
26.	_____	_____	$12\frac{1}{2}\%$
27.	_____	.375	_____
28.	_____	.875	_____
29.	$\frac{3}{20}$	_____	_____
30.	$\frac{1}{12}$	_____	_____

Grades in a classroom are often given according to the percentage score obtained by the student. Suppose a teacher grades according to the following scheme, where all scores are rounded to the nearest percent:

A	90%–100%
B	80%–89%
C	65%–79%
D	50%–64%
F	0%–49%

To determine a percent grade, form a ratio of score received to possible score and write this ratio as a percent. For example, a student gets seven answers right out of a possible ten. This score is

$$\frac{7}{10} = 70\%$$

Calculate the percent (nearest percent) and determine the letter grade for each score given in Problems 31–33.

31. Possible 10 points:
    a. 7   **b.** 8   **c.** $7\frac{1}{2}$ **d.** $8\frac{1}{4}$ **e.** $9\frac{1}{2}$

32. Possible 100 points:
    a. 75   **b.** 92   **c.** 38   **d.** $66\frac{1}{2}$ **e.** $95\frac{1}{2}$

33. Possible 200 points:
    a. 160   **b.** 150   **c.** 155   **d.** 120   **e.** 195

Mind Bogglers

34. A man goes into a store and says to the salesperson "Give me as much money as I have with me and I will spend $10 here." It is done. The operation is repeated in a second and a third store, after which he has no money left. How much did he have originally?

35. This is an old problem, but it is still fascinating. One day three men went to a hotel and were charged $30 for their room. The desk clerk then realized that he had overcharged them $5, and he sent the refund up with the bellboy. Now, the bellboy, being an amateur mathematician, realized that it would be difficult to split the $5 three ways. Therefore, he kept a $2 "tip" and gave the men only $3. Each man had originally paid $10 and was returned $1. Thus it cost each man $9 for the room. This means that they spent $27 for the room plus the $2 "tip." What happened to the other dollar?

The Percent Problem

7.4 Problem Solving with Percents

The following quotation was found in a recent publication: "An elected official is one who gets 51 percent of the vote cast by 40 percent of the 60 percent of voters who registered."

Certainly, most of us will have trouble understanding the percentage given in this quotation, but nevertheless you can't pick up a newspaper without seeing dozens of examples of ideas that require some understanding of working with percents. The most difficult job for most of us is knowing whether to multiply or divide by the given numbers. In this section, I will provide you with a sure-fire method for knowing what to do. The first step is to understand what is meant by the percent problem.

$$A \quad \text{is} \quad P\% \quad \text{of} \quad W$$

This is the given *amount*. | The *percent* is written $\frac{P}{100}$ | This is the *whole quantity*. It always follows the word *of*.

The percent problem won't always be stated in this form. But notice that there are three quantities associated with it:

1. The *amount*—sometimes called the **percentage**.
2. The *percent*—sometimes called the **rate**.
3. The *whole quantity*—sometimes called the **base**.

Now, regardless of the form in which you are given the percent problem, follow these steps to write a proportion:

1. Identify the *percent* first—it will be followed by the symbol % or the word "percent." Write it as a fraction:

$$\frac{P}{100}$$

2. Identify the *whole quantity* next—it is preceded by the word "of." It is the denominator of the second fraction in the proportion:

$$\frac{P}{100} = \frac{}{W}$$

3. The remaining number is the partial amount—it is the numerator of the second fraction in the proportion:

$$\frac{P}{100} = \frac{A}{W}$$

EXAMPLE 1

Write a proportion for each question.

a. What number is 12% of 94?

$$\frac{12}{100} = \frac{A}{94}$$

b. 12% of 94 is what number?

$$\frac{12}{100} = \frac{A}{94}$$

c. 94 is 12% of what number?

$$\frac{12}{100} = \frac{94}{W}$$

Whole number follows the word "of." Compare with Example 1a.

d. 14 is what percent of 420?

$$\frac{P}{100} = \frac{14}{420}$$

e. 18% of what number is 72?

$$\frac{18}{100} = \frac{72}{W}$$

f. 120 is what percent of 60?

$$\frac{P}{100} = \frac{120}{60}$$ ■

Since there are only three letters in the proportion

$$\frac{P}{100} = \frac{A}{W}$$

there are three types of percent problems. These possible types were illustrated in Example 1. To answer a question involving a percent, write a proportion and then solve the proportion.

EXAMPLE 2

In a certain class there are 500 points possible. The lowest C grade is 65% of the possible points. How many points are equal to the lowest C grade?

Solution

What is 65% of 500 points?

$$\frac{65}{100} = \frac{A}{500} \longleftarrow \text{The whole amount follows the word "of."}$$

$$A = \frac{65 \times \overset{5}{\cancel{500}}}{\underset{1}{\cancel{100}}}$$

$$= 325$$

The lowest C grade is 325 points. ∎

EXAMPLE 3

If your monthly salary is $1,500 and 21% is withheld for taxes and Social Security, how much money will be withheld from your check on pay day?

Solution

How much is 21% of $1,500?

$$\frac{21}{100} = \frac{A}{1,500}$$

$$A = \frac{21 \times \overset{15}{\cancel{1,500}}}{\underset{1}{\cancel{100}}}$$

$$= 315$$

The withholding is $315. ∎

EXAMPLE 4

You make a $25 purchase, and the clerk adds $2.25 sales tax. This doesn't seem right to you, so you want to know what percent tax has been charged.

Solution

What percent of $25 is $2.25?

$$\frac{P}{100} = \frac{2.25}{25}$$

$$\frac{\overset{4}{\cancel{100}} \times 2.25}{\underset{1}{\cancel{25}}} = P$$

$$9 = P$$

The tax charged was 9%. ∎

EXAMPLE 5

Your neighbors tell you that they paid $4,500 in taxes last year, and this amounted to 30% of their total income. What was their total income?

Solution

30% of total income is $4,500.

$$\frac{30}{100} = \frac{4,500}{W}$$

$$\frac{100 \times \overset{150}{\cancel{4,500}}}{\underset{1}{\cancel{30}}} = W$$

$$15,000 = W$$

Their total income was $15,000. ∎

CALCULATOR COMMENT

The rest of this section is optional for those who have a calculator. You can use a calculator to find the percent of a number as shown in Example 6.

EXAMPLE 6

a. What is 8% of 250?

Press: | 250 | | × | | 8 | | % | | = |
Display: 250,   250,   8,   .08   20,

b. Find $6\frac{1}{2}$% of $150.

When percents are in fractional form, they must first be converted to decimal form.

Press: | 150 | | × | | 6.5 | | % | | = |
Display: 150,   150,   6.5   .065   9.75

The answer is $9.75.   ∎

Sometimes you want to find the percent of a number and then add it to the total, such as when you are adding the sales tax to the cost of an item.

EXAMPLE 7

Suppose you want $6\frac{1}{2}$% of $150 (tax) added to $150 to give the total amount.

Press: | 150 | | + | | 6.5 | | % | | = |
Display: 150,   150,   6.5 9.75 159.75

The total amount is $159.75.   ∎

Problem Set 7.4

Write each sentence in Problems 1–18 as a proportion and then solve to answer the question.

1. 4 is what percent of 5?   **2.** 2 is what percent of 5?

3. What percent of 12 is 9?   **4.** What percent is 25 of 5?

5. 14% of what number is 21?   **6.** 40% of what number is 60?

7. 49 is 35% of what number?   **8.** 3 is 12% of what number?

9. What number is 15% of 64?   **10.** What number is 120% of 10

11. 10 is what percent of 5?   **12.** What percent of $20 is $1.2

13. 120% of what number is 16?   **14.** 21 is $66\frac{2}{3}$% of what number

15. 12 is $33\frac{1}{3}$% of what number?   **16.** What is 8% of $2,425?

17. What is 6% of $8,150?   **18.** 400% of what number is 15

19. If you were charged $151 tax on a $3,020 purchase, what percent tax were you charged?

20. If a government worker will receive a pension of 80% of her present salary, what will the pension be if her monthly salary is $1,250?

21. Government regulations require that, for certain companies to receive federal grant money, 15% of the employees need to meet minority requirements. If a company employs 390 people, how many minority people should be employed to meet the minimum requirements?

22. If $14,300 has been contributed to the United Way fund drive and this amount represents 22% of the goal, what is the United Way goal?

23. In October of 1978 the Department of Transportation announced that 39,100 miles of interstate highways are now open to traffic. This amounts to 92% of the interstate system. How many miles are there in the interstate system?

24. One recipe for a "perfect" martini calls for 15% vermouth and 85% gin. How much gin should be mixed with 3 oz of vermouth to make a "perfect" martini? How many 4 oz servings would you have?

25. Shannon Smith received an 8% raise, which amounted to $100 per month. What was his old wage, and what will his new wage be?

26. In a certain class there are 500 points possible. The lowest B grade is 80%. How many points are needed to obtain the lowest B grade?

27. A certain test is worth 125 points. How many points are needed to obtain a score of 75%?

28. If you correctly answer 8 out of 12 questions on a quiz, what is your percent right?

29. If Carlos answered 18 out of 20 questions on a test correctly, what was his percentage right?

30. If Wendy answered 15 questions correctly and obtained 75%, how many questions were on the test?

Mind Bogglers

31. If a drop from 50 to 10 is a loss of 80%, what is the percent gain from 10 to 50?

32. A saleswoman complained to her friend that she had a bad day. She had only two sales for $1,500 each. On the first sale she made a profit of 30% on the cost price, but on the second one she took a 30% loss on the cost price. "That doesn't seem to be any loss at all," said the friend. "Your profit and loss balance each other."

"On the contrary," said the saleswoman, "I lost almost $300." Who was right, the saleswoman or the friend? Justify your answer.

7.5 Review Problems

Section 7.1   **1.** Reduce each ratio to lowest terms.
   a. The ratio of cement to water is 120 to 6.
   b. The gear ratio is 34 teeth to 17 teeth.
   c. The ratio of miles to gallons is 117 to $6\frac{1}{2}$.

Section 7.1   **2. a.** Read $\frac{5}{8} = \frac{A}{2}$ as a proportion.
   b. List the extremes of the proportion in part a.
   c. Do $\frac{3}{2}$ and $\frac{9}{6}$ form a proportion?

Section 7.2   **3.** Solve the proportions.
   a. $\frac{5}{8} = \frac{A}{2}$ **b.** $\frac{4}{5} = \frac{3}{B}$ **c.** $\frac{C}{100} = \frac{2}{3}$

Section 7.2   **4.** If a car went 351 miles on $13\frac{1}{2}$ gallons of gas, then how many miles will it go on 5 gallons of gas?

Section 7.3   **5. a.** Write .005 as a fraction and as a percent.
   b. Write $\frac{1}{6}$ as a decimal and as a percent.

Section 7.4   **6. a.** 45% of 120 is what number?
   b. 60 is what percent of 80?

Section 7.4   **7. a.** What percent of 64 is 16?
   b. 90 is 120% of what number?

Section 7.4   **8. a.** What number is 25% of 300?
   b. 25 is what percent of 20?

Section 7.4   **9.** If the sales tax on a $59 item is 6%, what is the total price, including tax?

Section 7.4   **10.** Suppose you get 16 correct out of 20 on a test. What is your score, expressed as a percent?

Applications: The Utility of Mathematics

PART II

How are you doing so far? The first part of the book was concerned with problem solving and the mechanics of mathematics. This second part of the book is quite different from the first part. Each section of this part involves a real-life situation. It is my hope that you will be able to put yourself into the setting described and think about how you could change the facts to fit your own situation. If you have not yet encountered the situation in your own experience, you will, no doubt, encounter it sometime in the future. Each section discusses how to solve the problem raised in the given situation. I've tried to keep each section as practical as possible without using theory, but many of the ideas from Part I, "The Power of Mathematics," come into play. The chapters in this part are mostly independent, but you will need to study Chapter 9 before you study Chapter 10.

Remember the things we talked about in the first section of this book? Keep them in mind as you are confronted with everyday situations requiring mathematics. Also keep in mind the following quotation (William F. White, *A Scrap-book of Elementary Mathematics*, Chicago, 1908, p. 215):

> Mathematics, the science of the ideal, becomes the means of investigating, understanding and making known the world of the real. The complex is expressed in terms of the simple. From one point of view mathematics may be defined as the science of successive substitutions of simpler concepts for more complex . . . .

Chapter 8
Personal Money
Management

Chapter 9
Interest

Chapter 10
Consumer Applications

Chapter 11
Probability

Chapter 12
Statistics

Chapter 13
Graphs

Chapter 14
Computers

Personal Money Management

8.1 Checking Accounts

> SITUATION   Don and Linda were just married and were discussing who should handle the money for the family. They agreed to do it together, but first they would have to open a joint checking account. Don asked Linda to show him what to do when they received their bank statement because he had been rather haphazard about keeping records in the past. In this section, Don and Linda will learn how to write a check, keep a record of checks written, fill out a deposit slip, and reconcile a bank statement.

There are two main types of checking accounts, **personal** and **business**. The major difference between these two types is the cost and amount of service the bank gives to each account. The major choice you have when opening a checking account is the way you'll pay for the checks:

1. **Fixed monthly charge:** A flat monthly fee (usually $2 or $3) with no limit on the number of checks written—sometimes this comes with a "package deal" giving you a variety of bank services for the payment of this single fee.
2. **Fixed charge per check:** A fixed charge per check (usually 10¢ or 15¢ per check)—this is best for those with small balances or who write few checks.
3. **Free checking:** Usually a minimum balance is required ($200–$300). Sometimes depositing a minimum amount (often $1,000) in a savings account will qualify the depositor for a free checking account.
4. **Variable-fee checking:** The amount of the service charge depends on the number of checks written and the amount of money held in the account. It combines the features of free checking and fixed monthly charge accounts.
5. **Interest-bearing checking:** These accounts pay interest on the balance in the checking account. They are really a combination checking/savings account. A minimum balance is often required, and if the balance falls below this amount, the bank may impose one or more charges.

The proper format for a check is shown in Figure 8.1.

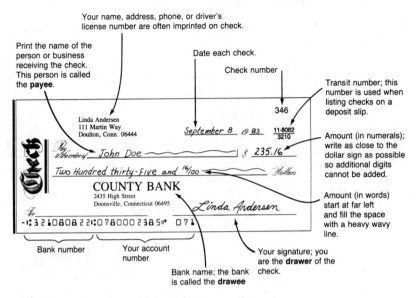

FIGURE 8.1 Proper format for a check

When you write a check, some record of it must be made. Several types of forms are used and you should be able to use a variety of these forms, some of which are shown in Figure 8.2.

(a) Check stub (c) Column check register

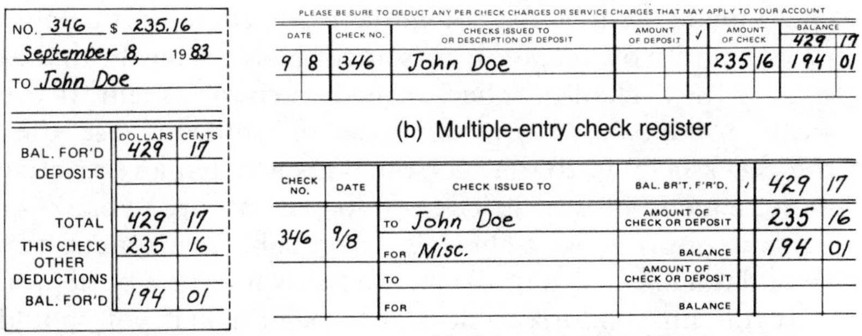

FIGURE 8.2 Samples of check registers

EXAMPLE 1

Write a check number 487 for $16.42 dated January 28, 1983 to Montgomery Ward, and complete the bank stub.

NO. 487 $ 16.42		
January 28 19 83		
TO Montgomery Ward		

	DOLLARS	CENTS
BAL. FOR'D	604	38
DEPOSITS		
TOTAL	604	38
THIS CHECK	16	42
OTHER DEDUCTIONS		
BAL. FOR'D	587	96

January 28 19 83   NO. **487**   90-468 / 1211

PAY TO THE ORDER OF *Montgomery Ward* $ *16.42*

Sixteen and 42/100 ———————————— DOLLARS

COUNTY BANK
Tulsa, Oklahoma

MEMO *Acct # 7823-5701-4952*   *K J Smith*

⑆0141003101⑆   903⑈391⑈

◼

In order to write checks, you will need to make deposits. A typical deposit slip is shown in Figure 8.3.

CURRENCY	
COIN	
CHECKS	
TOTAL FROM OTHER SIDE	
TOTAL	90-468 2 / 1211 DEPOSIT TICKET
LESS CASH RECEIVED	PLEASE ITEMIZE ADDITIONAL CHECKS ON REVERSE SIDE
Total Deposit	

DATE _____ 19____
CHECKS AND OTHER ITEMS ARE RECEIVED FOR DEPOSIT SUBJECT TO THE TERMS AND CONDITIONS OF THIS INSTITUTION'S COLLECTION AGREEMENT.

SIGN HERE ONLY IF CASH RECEIVED FROM DEPOSIT

SOUTHWEST STATE
Pine Bluff, Louisiana

⑆0141003911⑆   903⑈310⑈

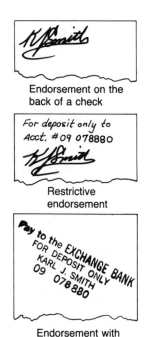

Endorsement on the
back of a check

For deposit only to
Acct. #09 078880

Restrictive
endorsement

Pay to the EXCHANGE BANK
FOR DEPOSIT ONLY
KARL J. SMITH
09 078880

Endorsement with
a rubber stamp

FIGURE 8.4   Check endorsements

You will fill in the date, the amount of currency and the coins deposited, and list the checks by transit number. A check made out to you is a **negotiable instrument**. It belongs to you until you release it to someone else. You do this by signing the back of the check, called an **endorsement**. If you write on the back "Pay to the order of" someone else, then the check belongs to that person. This is called a **restrictive endorsement**. Most restrictive endorsements are to banks or bank accounts. Since a check is a negotiable instrument, you should not endorse it until you are ready to cash it or deposit it. If you must endorse a check at home or mail it, you should use a restrictive endorsement, as shown in Figure 8.4. If you are depositing a large number of checks for a business, you may want to use an endorsement stamp.

EXAMPLE 2

Fill out a deposit slip for the following checks on February 15, 1983:

Check	Amount
$\dfrac{90\text{-}555/2}{1211}$	$83.55
$\dfrac{53\text{-}556}{113}$	$125.89
$\dfrac{35\text{-}6686}{3130}$	$18.35
$\dfrac{79\text{-}7905}{711}$	$582.79

CURRENCY		
COIN		
90-555/2	83	55
53-556	125	89
35-6686	18	35
TOTAL FROM OTHER SIDE	582	79
TOTAL	810	58
LESS CASH RECEIVED		
Total Deposit	810	58

DATE _February 15,_____ 19 _83_

CHECKS AND OTHER ITEMS ARE RECEIVED FOR DEPOSIT SUBJECT TO THE TERMS AND CONDITIONS OF THIS INSTITUTION'S COLLECTION AGREEMENT.

11-808 / 3210

DEPOSIT TICKET

PLEASE ITEMIZE ADDITIONAL CHECKS ON REVERSE SIDE

SIGN HERE ONLY IF CASH RECEIVED FROM DEPOSIT

MOUNTAIN BANK
Albuquerque, N.M.

⑂ 3 2 10808 2 2⑃ 0 76 500 2385 ⑆ 0 6 2385

Once a month (or some other agreed upon time interval) the bank will send you a bank statement, along with some checks you've written. These are returned to you by your bank after the amounts have been deducted from your account. They are called **canceled checks**. The next step is to **reconcile** your bank statement with your checkbook. Most banks provide a form for this purpose on the back of the bank statement, as shown in Example 3.

EXAMPLE 3

Don and Linda's bank statement on February 28, 1983 shows $65.75 in their account. Their checkbook shows $12.92. The bank service charge is $3.00 and the outstanding checks are:

No. 131	Thompson Cleaners,	$17.38
No. 135	Acy Garage,	$20
No. 138	Rick's Restaurant,	$12.45
No. 140	Cash,	$6.00

The following bank reconciliation statement has been completed according to the five steps outlined on the form. Study it until you can see that it **balances** (the amounts shown by * on the form are the same).

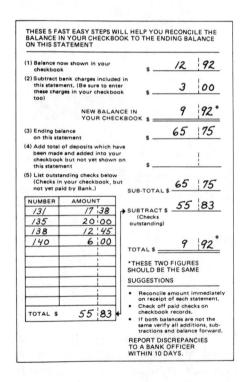

At times the amount shown by * on your reconciliation statement will not be the same. This means there is an error someplace. The following checklist will help you find your mistake (or the bank's).

1. Check for an arithmetic mistake on the reconciliation statement.
2. Check for an arithmetic mistake in subtraction on your check stubs or check register.
3. Check for deposits, checks, or service charges not accounted for. This includes:
   a. Not entering a deposit or check.
   b. Forgetting to subtract bank charges.
   c. Forgetting to list a check written several months ago and still not returned.
4. Check to see that the amount you recorded in your check register was the same as the amount of the check.
5. After you have made certain that the error is not any of the above, and your bank statement balanced last month, check with your bank for the possibility of a bank error.

Steps to Check When Reconciliation Statement Does Not Balance

EXAMPLE 4

Niels Andersen received the following bank statement:

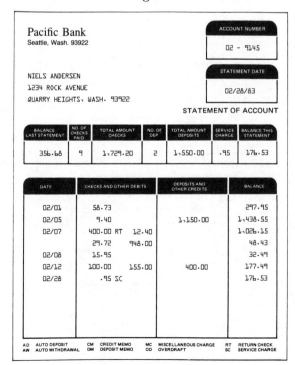

Pacific Bank
Seattle, Wash. 93922

ACCOUNT NUMBER
02 - 9145

NIELS ANDERSEN
1234 ROCK AVENUE
QUARRY HEIGHTS, WASH. 93922

STATEMENT DATE
02/28/83

STATEMENT OF ACCOUNT

BALANCE LAST STATEMENT	NO. OF CHECKS PAID	TOTAL AMOUNT CHECKS	NO. OF DEP	TOTAL AMOUNT DEPOSITS	SERVICE CHARGE	BALANCE THIS STATEMENT
356.68	9	1,729.20	2	1,550.00	.95	176.53

DATE	CHECKS AND OTHER DEBITS		DEPOSITS AND OTHER CREDITS	BALANCE
02/01	58.73			297.95
02/05	9.40		1,150.00	1,438.55
02/07	400.00 RT	12.40		1,026.15
	29.72	948.00		48.43
02/08	15.95			32.49
02/12	100.00	155.00	400.00	177.49
02/28	.95 SC			176.53

AD	AUTO DEPOSIT	CM	CREDIT MEMO	MC	MISCELLANEOUS CHARGE	RT	RETURN CHECK
AW	AUTO WITHDRAWAL	DM	DEPOSIT MEMO	OD	OVERDRAFT	SC	SERVICE CHARGE

Niels' check register is shown below:

DATE		CHECK NUMBER	CHECKS ISSUED TO OR DEPOSIT RECEIVED FROM	AMOUNT OF DEPOSIT	✓	AMOUNT OF CHECK	BALANCE 356 99	
			BE SURE TO DEDUCT ANY PER CHECK CHARGES OR MAINTENANCE CHARGES THAT MAY APPLY					
2	1	138	Penny's			58 73	298	26
2	1	139	St. Francis Cleaners			9 40	288	86
2	1	140	Board of Utilities	1,150 00		29 72	1,409	14
2	1	141	World Savings			948 00	461	14
2	1	142	United Gas & Electric			12 90	448	24
2	1	143	Empire Disposal			15 95	432	39
2	1	144	Eveline Grocott			100 00	332	39
1	31	—	January Service Charge			31	332	08
2	12	146	Littles Cleaning Service	400 00		155 00	577	08
2	26	147	Asians Appliance	750 00		18 00	1,309	08
2	7	145	Cash			12 40	1,296	68

When Niels carried out his bank reconciliation, the amounts didn't balance, as shown below. The $400RT on the bank statement means that a check was returned. Find the error(s) by making the amounts balance.

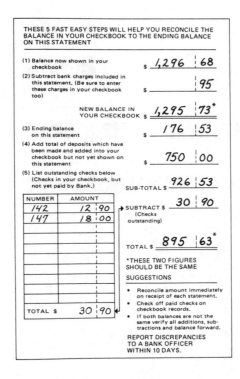

THESE 5 FAST EASY STEPS WILL HELP YOU RECONCILE THE BALANCE IN YOUR CHECKBOOK TO THE ENDING BALANCE ON THIS STATEMENT

(1) Balance now shown in your checkbook $ 1,296 68

(2) Subtract bank charges included in this statement. (Be sure to enter these charges in your checkbook too) $ 95

NEW BALANCE IN YOUR CHECKBOOK $ 1,295 73*

(3) Ending balance on this statement $ 176 53

(4) Add total of deposits which have been made and added into your checkbook but not yet shown on this statement $ 750 00

(5) List outstanding checks below (Checks in your checkbook, but not yet paid by Bank.)   SUB-TOTAL $ 926 53

NUMBER	AMOUNT
142	12 90
147	18 00
TOTAL $	30 90

SUBTRACT $ 30 90 (Checks outstanding)

TOTAL $ 895 63*

*THESE TWO FIGURES SHOULD BE THE SAME

SUGGESTIONS

- Reconcile amount immediately on receipt of each statement.
- Check off paid checks on checkbook records.
- If both balances are not the same verify all additions, subtractions and balance forward.

REPORT DISCREPANCIES TO A BANK OFFICER WITHIN 10 DAYS.

Solution

There are two errors. The returned check is not subtracted from the checkbook. Subtract $400.00.

$$
\begin{array}{r}
\$1,\!295.73 \\
-\quad 400.00 \\
\hline
\$\;\;895.73
\end{array}
$$

The total still doesn't balance. There is an error in subtraction on the line for check 143 in the check register. The total should be $432.29. To correct this error, subtract $.10 from the balance:

$$
\begin{array}{r}
\$895.73 \\
-\quad .10 \\
\hline
\$895.63
\end{array}
$$

The statement now balances.   ■

Problem Set 8.1

Fill in the appropriate word or words in Problems 1–5.

1. There are two main types of checking accounts, _____ and _____.

2. The person writing a check is called the _____ and the person receiving the check is the _____. The bank is called the _____.

3. When you sign a check on the back before depositing it, your signature is called the _____.

4. A check that has been cashed and returned to you with your bank statement is called a _____ check.

5. To _____ your bank statement means to bring the bank statement into agreement with your checkbook.

6. Record the following personal checks in the check register shown at the top of the next page.

   a. #214 9/1 Bank of America, $364.25
   b. #215 9/1 Edison, $65.24
   c. #216 9/3 Macy's, $100.06
   d. #217 9/5 Bethlehem Church, $85.00
   e. #218 9/10 Jack Foley, $200.00

		PLEASE BE SURE TO DEDUCT ANY PER CHECK CHARGES OR SERVICE CHARGES THAT MAY APPLY TO YOUR ACCOUNT			
CHECK NO.	DATE	CHECK ISSUED TO	BAL. BR'T. F'R'D.	✓	*4,728* *43*
		TO	AMOUNT OF CHECK OR DEPOSIT		
		FOR	BALANCE		
		TO	AMOUNT OF CHECK OR DEPOSIT		
		FOR	BALANCE		
		TO	AMOUNT OF CHECK OR DEPOSIT		
		FOR	BALANCE		
		TO	AMOUNT OF CHECK OR DEPOSIT		
		FOR	BALANCE		
		TO	AMOUNT OF CHECK OR DEPOSIT		
		FOR	BALANCE		

Write the business checks given in Problems 7–9 and record them in the accompanying check stubs. Use the current date and your own signature.

7. #3496   10/1   Great Western Savings, $1,069.65

NO. ____ $ ____	
____ 19 ____	____ 19 ___ NO. ____ 90-468 / 1211
TO ____	PAY TO THE ORDER OF ____ $ ____
BAL. FOR'D	DOLLARS 38,410 CENTS 58	____ DOLLARS
DEPOSITS	
TOTAL	FIRST NATIONAL BANK
THIS CHECK	Pine Bluff, Louisiana
OTHER DEDUCTIONS	MEMO ____
BAL. FOR'D	⑊01410031⑊ 903⑊391⑊

8. #3497   10/6   First American Title, $10,400.35

NO. ____ $ ____	
____ 19 ____	____ 19 ___ NO. ____ 90-468 / 1211
TO ____	PAY TO THE ORDER OF ____ $ ____
BAL. FOR'D	DOLLARS 37,340 CENTS 93	____ DOLLARS
DEPOSITS	
TOTAL	FIRST NATIONAL BANK
THIS CHECK	Pine Bluff, Louisiana
OTHER DEDUCTIONS	MEMO ____
BAL. FOR'D	⑊01410031⑊ 903⑊391⑊

9. #3498   10/8   U.S. Postal Service, $100.38

NO. ____ $ ____	
____ 19 ____	____ 19 ___ NO. ____ 90-468 / 1211
TO ____	PAY TO THE ORDER OF ____ $ ____
BAL. FOR'D	DOLLARS 26,940 CENTS 58	____ DOLLARS
DEPOSITS	
TOTAL	FIRST NATIONAL BANK
THIS CHECK	Pine Bluff, Louisiana
OTHER DEDUCTIONS	MEMO ____
BAL. FOR'D	⑊01410031⑊ 903⑊391⑊

Use the following check register for Problems 10–13. You will need to do all these problems in sequence in order to do Problem 13.

		BE SURE TO DEDUCT ANY PER CHECK CHARGES OR MAINTENANCE CHARGES THAT MAY APPLY					
DATE	CHECK NUMBER	CHECKS ISSUED TO OR DEPOSIT RECEIVED FROM	AMOUNT OF DEPOSIT	✓	AMOUNT OF CHECK	BALANCE 1,843	64

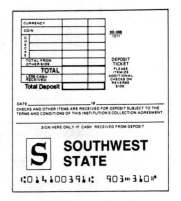

10. On January 3, prepare a deposit slip for the following checks:

Transit number	Amount
$\dfrac{112\text{-}01}{131}$	$250.00
$\dfrac{6113\text{-}76}{484}$	$250.00
$\dfrac{8\text{-}31}{1001}$	$265.00
$\dfrac{438\text{-}7}{28}$	$265.00
$\dfrac{681\text{-}40}{1031}$	$485.00

Record this deposit in the check register.

11. Show a restrictive endorsement for one of the checks in Problem 10. Use your own signature and deposit it to account 09-310.

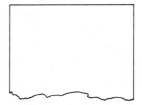

12. Record the following checks and charges into the check register. Keep a running balance.
   a. #412   1/3   Summit Savings, $965.00
   b. #413   1/3   City Sanitation, $27.95

c.		2/29	December service charge, $.37
d.	#414	1/5	Southern Gas & Electric, $48.27
e.	#415	1/12	Ken's Maytag, $75.00
f.	#416	1/18	Sears, $27.35
g.	#417	1/23	Jerpback Realty, $400.00
h.	#418	1/29	State Farm Insurance, $339.00
i.	#419	2/2	Summit Savings, $965.00

13. Reconcile the following bank statement with the check register:

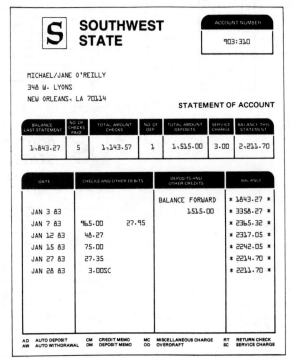

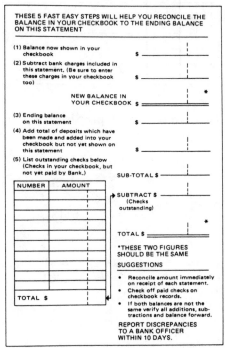

For Problems 14–16 use the check register shown on page 287. The beginning balance is $816.28 and the following checks are outstanding:

    #82   Russ Martin, $30.48
    #92   United Way, $45.00
    #97   Curry Company, $47.35

Don't show these checks on the check register, since they have already been deducted. However, you will need to use these checks when reconciling the bank statement in Problem 16.

DATE	CHECK NUMBER	CHECKS ISSUED TO OR DEPOSIT RECEIVED FROM	AMOUNT OF DEPOSIT	✓	AMOUNT OF CHECK	BALANCE
		BE SURE TO DEDUCT ANY PER CHECK CHARGES OR MAINTENANCE CHARGES THAT MAY APPLY				

14. On March 1, 1983, prepare a deposit slip for the following checks:

Transit number	Amount
$\dfrac{106\text{-}8}{112}$	$2,239.46
$\dfrac{681\text{-}3}{1007}$	$243.91
$\dfrac{6\text{-}0411}{8}$	$100.30

Enter this deposit into the check register.

15. Record in the check register the following checks. Keep a running total.
   a. #108   3/1   Mills Bank, $450.38
   b. #109   3/1   National Bank, $135.14
   c. #110   3/1   Marlow County Tax Collector, $1,400.53
   d. #111   3/1   Edison, $48.30
   e. #112   3/1   Central Bell, $39.45
   f. #113   3/10   Morning Times, $9.00
   g. #114   3/12   RLI Insurance, $32.00
   h. #115   3/15   National Title Company, $1,000.00
   i. #116   3/18   Cash, $200.00

16. Reconcile the bank statement shown at the top of the next page with the check register.

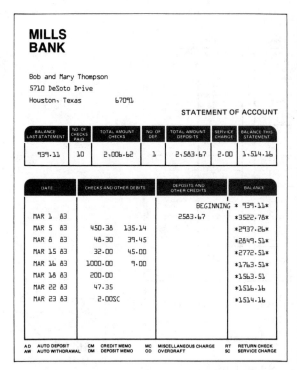

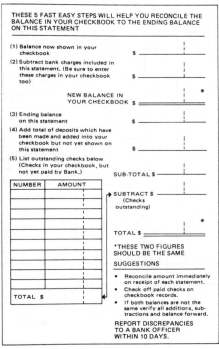

17. Andy Gomez received the following bank statement:

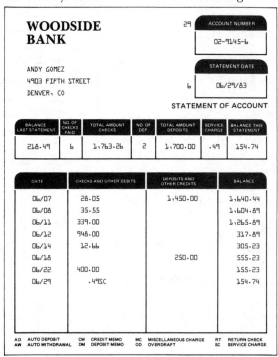

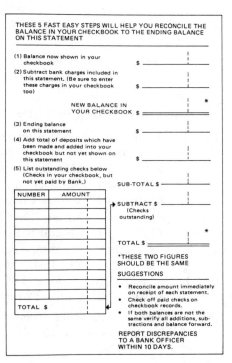

Andy's check register is shown below:

			BE SURE TO DEDUCT ANY PER CHECK CHARGES OR MAINTENANCE CHARGES THAT MAY APPLY				
DATE	CHECK NUMBER	CHECKS ISSUED TO OR DEPOSIT RECEIVED FROM	AMOUNT OF DEPOSIT	✓	AMOUNT OF CHECK	BALANCE 218 97	
6 1	483	National City Bank	1,450 00		948 00	720 97	
6 1	484	Board of Public Utilities			35 55	685 42	
6 1	485	Empire Disposal			28 05	657 37	
6 5	486	Allstate Insurance			339 00	318 30	
6 8	487	Mountain Gas			12 66	305 64	
6 18	489	New Value Realty	250 00		400 00	155 64	
5 31	—	Service Charge (May)			48	155 16	
6 25	490	New Valley Brake			89 50	65 66	

Use the form shown above to reconcile the bank statement and check register. Find and correct any errors.

8.2 Discount, Sale Price, and Sales Tax

SITUATION Dan is building a storage shed and needs to buy the hardware to complete the shed. He figures that it will cost about $175.00. Mary tells Dan to wait until the first Saturday of the month because Central Hardware Store sells its entire stock at 20% off on the first Saturday of each month. How much will Dan save by waiting for the sale? In this section, Dan will learn how to determine the sale price of an item if it is given as a percent markdown. He will also see how to determine the regular price if he is given the sale price. Dan will find out how to calculate the sales tax on the items he purchases.

SALE 20% OFF

As consumers, we are all familiar with sales. In order to encourage us to buy items at one time rather than another, to introduce us to a particular store, to entice us into a store to buy nonsale items, to sell old or damaged merchandise, or to meet or beat the competition, a retailer will often sell merchandise ON SALE at a *reduced price*. The amount an item is reduced from the regular, or original, price is called the **discount**.

DISCOUNT = (ORIGINAL PRICE) × (PERCENT MARKDOWN)

Finding the Discount

The **percent markdown** is written as a decimal in order to calculate the discount.

EXAMPLE 1

Find the amount of discount.

a. Hardware, $175;   20% off

$$\text{DISCOUNT} = 175 \times .20$$
$$= 35$$

The discount is $35.

b. Shoes, $25;   15% discount

$$\text{DISCOUNT} = 25 \times .15$$
$$= 3.75$$

The discount is $3.75.   ∎

Sometimes the markdown is given as a fraction, as shown in Example 2.

EXAMPLE 2

Blouses with a regular price of $35 are on sale marked $\frac{1}{3}$ off. What is the discount?

Solution

$$\text{DISCOUNT} = 35 \times \frac{1}{3}$$
$$\approx 11.67 \qquad \text{Round the discount to the nearest cent.}$$

The discount is $11.67.   ∎

As consumers, we are usually concerned with the sale price rather than the amount of discount. The sale price can be found by subtracting the discount from the original price. For example, the hardware in Example 1a had an original price of $175 and a discount of $35, so the sale price is

$$\$175 - \$35 = \$140$$

However, there is an easier way to find the sale price, which involves something called the **complement**. Two numbers less than 1 are called **complements** if their sum is 1.

EXAMPLE 3

Find the complements.

a. .7;   complement is .3
b. .15;   complement is .85
c. 40%;   complement is 60% or .6
d. $\frac{1}{4}$;   complement is $\frac{3}{4}$ or .75   ■

$$\text{SALE PRICE} = \left(\begin{matrix} \text{ORIGINAL} \\ \text{PRICE} \end{matrix} \right) \times \left(\begin{matrix} \text{COMPLEMENT} \\ \text{OF MARKDOWN} \end{matrix} \right)$$

Finding the Sale Price

EXAMPLE 4

Find the sale price.

a. $175;   20% discount

$$\text{SALE PRICE} = 175 \times .8$$
$$= 140$$

The sale price is $140.

b. $45;   sale 40% off

$$\text{SALE PRICE} = 45 \times .6$$
$$= 27$$

The sale price is $27.   ■

   This same formula for finding the sale price of an item can be used for finding the original price or the percent of markdown. In each case you will be able to fill in two of the three quantities of the formula while being asked to find the other.

EXAMPLE 5

During a 20% off sale an item is marked $52. What is the regular price?

Solution

$$\text{SALE PRICE} = \begin{pmatrix} \text{ORIGINAL} \\ \text{PRICE} \end{pmatrix} \times \begin{pmatrix} \text{COMPLEMENT} \\ \text{OF MARKDOWN} \end{pmatrix}$$

$$52 = (\text{ORIGINAL PRICE}) \times .80$$

$$\frac{52}{.8} = (\text{ORIGINAL PRICE}) \qquad \text{Divide both sides by .8.}$$
Press: 52 ÷ .8 =

$$65 = (\text{ORIGINAL PRICE})$$

The regular price is $65.   ■

EXAMPLE 6

If the sale price of an item is $2,288 and the regular price is $3,520, what is the percent markdown?

Solution

$$\text{SALE PRICE} = \begin{pmatrix} \text{ORIGINAL} \\ \text{PRICE} \end{pmatrix} \times \begin{pmatrix} \text{COMPLEMENT} \\ \text{OF MARKDOWN} \end{pmatrix}$$

$$2,288 = 3,250 \times \begin{pmatrix} \text{COMPLEMENT} \\ \text{OF MARKDOWN} \end{pmatrix}$$

$$\frac{2,288}{3,520} = \begin{pmatrix} \text{COMPLEMENT} \\ \text{OF MARKDOWN} \end{pmatrix}$$

$$.65 = \begin{pmatrix} \text{COMPLEMENT} \\ \text{OF MARKDOWN} \end{pmatrix} \qquad \begin{array}{l} \text{Divide both sides by 3,250} \\ \textit{Press:} \ \boxed{2288} \ \boxed{÷} \ \boxed{3520} \ \boxed{=} \end{array}$$

Thus, if the complement is .65, then the percent markdown is 35%.   ■

Sales tax is levied by almost every state, as shown in Table 8.1. The procedure for finding the amount of sales tax is identical to the procedure for finding the discount. Convert the percent sales tax to a decimal and multiply.

TABLE 8.1   State Retail Sales Taxes: 1981*

Alabama	4%	Indiana	4%	Nebraska	3%	South Carolina	4%
Alaska	0%	Iowa	3%	Nevada	3%	South Dakota	5%
Arizona	4%	Kansas	3%	New Hampshire	0%	Tennessee	3%
Arkansas	3%	Kentucky	5%	New Jersey	5%	Texas	4%
California	6%	Louisiana	3%	New Mexico	$3\frac{3}{4}$%	Utah	4%
Colorado	3%	Maine	5%	New York	4%	Vermont	3%
Connecticut	$7\frac{1}{2}$%	Maryland	5%	North Carolina	3%	Virginia	3%
Delaware	0%	Massachusetts	5%	North Dakota	3%	Washington	$4\frac{1}{2}$%
Florida	4%	Michigan	4%	Ohio	4%	West Virginia	3%
Georgia	3%	Minnesota	4%	Oklahoma	2%	Wisconsin	4%
Hawaii	4%	Mississippi	5%	Oregon	0%	Wyoming	3%
Idaho	3%	Missouri	$4\frac{1}{8}$%	Pennsylvania	6%	District of	
Illinois	4%	Montana	0%	Rhode Island	6%	Columbia	5%

*Does not include local sales taxes.

> $$\begin{pmatrix} \text{AMOUNT OF} \\ \text{SALES TAX} \end{pmatrix} = (\text{ORIGINAL PRICE}) \times (\text{TAX RATE})$$
>
> $$\text{TOTAL PRICE} = (\text{ORIGINAL PRICE}) \times (\text{TAX RATE} + 1)$$
> (including tax)

Sales Tax

EXAMPLE 7

Find the Pennsylvania sales tax for building materials costing $84.65.

Solution

TAX $= 84.65 \times .06$	Find the sales tax rate from Table 8.1.
$= 5.079$	Round money answers to the nearest cent.

The sales tax is $5.08.   ∎

   Notice that when you want to find the total price (including tax) the procedure is to multiply the original price times 1 added to the tax rate, as shown in Example 8.

EXAMPLE 8

In New Jersey, a set of tires at Sears costs $168.00. What is the total amount, including tax?

Solution

In Table 8.1, we find that the New Jersey sales tax is 5%. Write this as a decimal and add 1:

$$1 + .05 = 1.05$$

Multiply this by the original price.

$$\text{TOTAL PRICE} = 168 \times 1.05$$
$$= 176.40$$

The total price, including tax, is \$176.40.   ■

Problem Set 8.2

Find the discount of the items in Problems 1–10, given the original price and the percent markdown.

1. \$450; 5%	**2.** \$65; 20%
3. \$55; 25%	**4.** \$250; 35%
5. \$25; 15%	**6.** \$14.95; 30%
7. \$12.50; 40%	**8.** \$45.50; 50%
9. \$9.95; 10%	**10.** \$16.95; 15%

Find the sale price of the items in Problems 11–20, given the original price and the percent markdown. Items 11–16 were all found in one issue of a local newspaper.

11.

5% DISCOUNT ON ITEMS ALREADY ON SALE SAVE!

ITEM:
Golf clubs
\$230

12.

SALE 20% OFF

ITEM:
Dress
\$85

13.

22 %

ITEM:
Coat
\$95

14.

STEREO COMPONENTS
36% OFF

ITEM:
Stereo
Amplifier
$205

15.

50% OFF
on CHEMIN de FER
JEANS
AND CORDS

ITEM:
Jeans
$15

16.

CALCULATORS
45% OFF

ITEM:
Calculator
$65

17. $350 skis; $\frac{1}{2}$ off   **18.** $875 go-cart; $\frac{1}{2}$ off

19. $12.95 sweater; $\frac{1}{4}$ off   **20.** $9.50 belt; $\frac{1}{3}$ off

21. A dealer selling an automobile for $5,830 offers a $500 rebate. What is the percent markdown (to the nearest tenth of a percent)?

22. If a swimming pool is marked down to $695 from $1,150, what is the percent markdown (to the nearest tenth of a percent)?

23. If an item is marked $420 during a 20% off sale, what is the regular price?

24. If an item is marked $165 during a $\frac{1}{3}$ off sale, what is the regular price?

25. The advertisement shown here claims $\frac{1}{2}$ price. However, the sale price is $5.25 and the regular price is $10.00. What is the actual percent markdown (to the nearest tenth of a percent)?

26. What is the percent markdown (to the nearest tenth of a percent) for a cable reconnect as shown on the advertisement?

Find the sales tax for the items in Problems 27–29. Refer to Table 8.1.

27. Golf clubs; $230 in California

28. Dress; $95 in Texas

29. Coat; $85 in Washington state

Total Television of Santa Rosa

JUNE
INSTALLATION
SPECIAL

1/2 **PRICE**
for Cable TV
Installed

NOW! **$5²⁵***
Reg. $10.00

... New Customers Only ...
—Reconnect Existing Cable—

First TV Only   NOW! **$2⁶³***
Reg. $5.25

*Offer Expires June 15

Find the sales tax for the items in Problems 30–32 using the rate in your own state. If your state has no sales tax, use the California rate.

30. Shoes; $17.99

31. Set of pots and pans; $109.99

32. Stereo; $559.95

Find the total amount, including tax, for the items in Problems 33–38.

33. $12.50; 6%   **34.** $1,925; 5%

35. $4,312; $3\frac{3}{4}$%   **36.** $91.60; 3%

37. $25.36; $4\frac{1}{2}$%   **38.** $365; $4\frac{1}{8}$%

8.3 Household Budgeting

SITUATION   Clint and Jean are a recently married couple who are both working. Their combined income is $2,100 per month, but they are having trouble meeting their expenses. They don't seem to have too much trouble paying their bills month-to-month, but Jean has trouble with the bills that don't come due each month—for example, the semiannual insurance payments, the car registration, birthday and anniversary gifts, Christmas, and vacations. Clint and Jean decide that they need to do a better job of planning for these expenses. In this section, Clint and Jean will see how to prepare a household budget, how to average variable expenses, and finally, how to compare their own budget with those made from national averages.

A **budget** is an organized plan for spending the available income. Expenses are divided into two categories:

1. Fixed expenses
2. Variable expenses

Some of these expenses are monthly expenses, some of these are paid on some other regular time interval, while still others are one-time expenses.

The first step in preparing a budget is to analyze current spending. This is necessary in order to create a realistic budget. Next, if the income exceeds the expenses (which rarely seems to be the case), then you can decide where to spend the excess money. On the other hand, if the expenses

exceed the income, then certain of the expenses must be cut
in order to bring them within the limitations of the income.

You can use Figure 8.5 to help you carry through this
process.

BUDGET

INCOME
Monthly income before taxes $_____   Other income $_____
                            $_____ $_____
Total other ÷ 12 $_____   Total other $_____
TOTAL MONTHLY INCOME $_____

EXPENSES

MONTHLY   NONMONTHLY

Fixed   *Variable*

Rent or   Food $_____   Car repairs $_____
mortgage $_____ Car gasoline $_____   Home repairs $_____
Car payments $_____   Car license $_____
Income tax   UTILITIES   Medical $_____
withholding $_____ a. Electricity $_____ Dental $_____
Social Security   b. Gas $_____ Clothing $_____
withholding   c. Water &   Gifts $_____
(or other   Sewer $_____ Contributions $_____
retirement) $_____ d. Telephone $_____ Vacation $_____
Contributions $_____ e. Cable TV $_____ Professional
Insurance $_____ f. Other $_____ tools or
                        Entertainment $_____ organizations $_____
INSTALLMENTS
a. _____ $_____   OTHER   INSURANCE
b. _____ $_____   a. _____ $_____   a. _____ $_____
c. _____ $_____   b. _____ $_____   b. _____ $_____
                        c. _____ $_____   c. _____ $_____
SAVINGS
a. _____ $_____   TOTAL   OTHER
b. _____ $_____   VARIABLE   a. _____ $_____
                        EXPENSES $_____   b. _____ $_____
OTHER   c. _____ $_____
a. _____ $_____   RESERVED
b. _____ $_____   MONTHLY   TOTAL
TOTAL   (Total annual   NONMONTHLY
FIXED EXPENSES $_____ expenses   EXPENSES $_____
                        divided by 12) $_____

                        TOTAL
                        FIXED EXPENSES $_____

                        TOTAL
                        MONTHLY
                        EXPENSES $_____

FIGURE 8.5
Sample budget form

EXAMPLE 1

The following is a list of variable expenses for Clint and Jean
for one month. Total up the entries in each of the categories

shown in the sample budget form under monthly variable expenses. If an entry is not listed as a category, total it under the miscellaneous category. Restaurant expenses should be considered entertainment and not food.

Groceries, $65.12	Gas, $12.58
Paper boy, $3.75	Movies, $5.00
Clint's allowance, $100	Dry cleaners, $5.70
Boat rental, $6.50	Groceries, $46.35
Water & Sewer, $35.18	Gas, $15.30
Groceries, $96.30	Jean's allowance, $100
Gas, $14.60	Parking, $9.50
Restaurant, $5.30	Phone, $23.16
Restaurant, $8.40	Gas & Electric, $83.41
Gas, $10.30	Restaurant, $7.75
Groceries, $73.75	Bowling, $6.75
Cable TV, $15.00	Gas, $13.20

Solution

The totals have been entered in the budget categories:

Food
$65.12
96.30
73.75
46.35
$281.52

Gas
$14.60
10.30
12.58
15.30
13.20
$65.98

Entertainment
$6.50
5.30
8.40
5.00
7.75
6.75
$39.70

Other
Clint: $100.00
Jean: $100.00

Miscellaneous: $ 3.75
5.70
9.50
$18.95

Variable

Food	$ 281.52
Car gasoline	$ 65.98
UTILITIES	
a. Electricity	$ 83.41
b. Gas	$
c. Water & Sewer	$ 35.18
d. Telephone	$ 23.16
e. Cable TV	$ 15.00
f. Other	$
Entertainment	$ 39.70
OTHER	
a. *Clint's allowance*	$ 100.00
b. *Jean's allowance*	$ 100.00
c. *Miscellaneous*	$ 18.95
TOTAL VARIABLE EXPENSES	$ 762.90

When budgeting variable expenses you should consider the average expenses for each item. The average we'll use for this purpose is called the *mean* and will be discussed in

Chapter 12. But for our purposes here, you'll need to add the variable expenses for a particular item for several months and then divide by the number of months you've added, as illustrated by Example 2.

EXAMPLE 2

a. Suppose the amounts spent for food for the past three months have been $240, $295, and $270. The average amount spent is

$$\frac{\$240 + \$295 + \$270}{3} = \frac{\$805}{3}$$
$$\approx \$268$$

Rounded to the nearest dollar.

b. Suppose the phone bills for the last six months (rounded to the nearest dollar) were $23, $28, $53, $24, $31, and $35. Then, the average spent for the phone is

$$\frac{\$23 + \$28 + \$53 + \$24 + \$31 + \$35}{6} = \frac{\$194}{6}$$
$$\approx \$32$$

Rounded ∎

When certain expenses on a budget are paid out other than monthly, a certain amount out of each month's pay check should be reserved to pay those expenses.

EXAMPLE 3

a. If a family spends $1,000 on Christmas presents, how much should be reserved each month?

$$\frac{\$1,000}{12} \approx \$83 \qquad \text{Rounded to the nearest dollar.}$$

b. If you pay car insurance twice a year and the premiums are $236 and $285, how much should be reserved each month?

$$\frac{\$236 + \$285}{12} = \frac{\$521}{12}$$
$$\approx \$43 \qquad \text{Rounded to the nearest dollar.} \quad ∎$$

Instead of reserving a monthly amount for each of these nonmonthly expenses, usually they are all totaled together and one amount set aside each month to cover them, as illustrated by Example 4.

EXAMPLE 4

Suppose Clint and Jean had the following nonmonthly expenses last year:

Doctor, $10.00
Tax preparation, $40
Car insurance, $195.00
Car repair, $61.50
Jean's birthday, $20
Car license, $45.00
Car insurance, $195.00
Doctor, $10.00
Home insurance, $125
Prescription, $3.80
Clothes, $38.50
Union dues, $360.00
Dentist, $55.00
Vacation, $835.00

Doctor, $10.00
Clothes, $83.20
Clint's birthday, $25.00
Contribution, $20.00
State taxes, $260.00
Wedding gift, $38.25
Clothes, $106.40
Contribution, $25.00
Life insurance, $107.00
Prescription, $12.90
Dishwasher repair, $35
Car repair, $128.00
Christmas presents, $450

What is the amount they should reserve monthly to meet these expenses?

Solution

Car repairs	$ *189.50*	INSURANCE	
Home repairs	$ *35.00*	a. *Car* $ *390.00*	
Car license	$ *45.00*	b. *Life* $ *107.00*	
Medical	$ *46.70*	c. *Home* $ *125.00*	
Dental	$ *55.00*		
Clothing	$ *228.10*	OTHER	
Gifts	$ *533.25*	a. *State taxes* $ *260.00*	
Contributions	$ *45.00*	b. *Tax prep* $ *40.00*	
Vacation	$ *835.00*	c. _____ $____	
Professional		TOTAL	
tools or		NONMONTHLY	
organizations	$ *360.00*	EXPENSES $*3,294.55*	

RESERVED MONTHLY   = $275
(Total Annual Expenses divided by 12) ∎

Suppose you complete a budget similar to the example shown in Figure 8.6 and find that your expenses exceed your income. Notice in this example that Clint and Jean thought their income ($2,100 per month) was sufficient for their expenses, but they hadn't considered the necessary $275 reserve.

MONTHLY			
Fixed		*Variable*	
Rent or mortgage	$ 341	Food	$ 280
Car payments	$ 125	Car gasoline	$ 70
Income tax withholding	$ 170	UTILITIES	
		a. Electricity	$ 80
		b. Gas	$
Social Security withholding (or other retirement)	$ 140	c. Water & Sewer	$ 36
		d. Telephone	$ 25
Contributions	$ 100	e. Cable TV	$ 15
Insurance	$ 16	f. Other	$
INSTALLMENTS		Entertainment	$ 40
a. *Wells Fargo*	$ 60		
b. *Sears*	$ 45	OTHER	
c. *Mastercard*	$ 55	a. *Clint's allowance*	$ 100
		b. *Jean's allowance*	$ 100
SAVINGS		c. *Miscellaneous*	$ 54
a. *Credit Union*	$ 100		
b.	$	TOTAL VARIABLE EXPENSES	$ 800
OTHER			
a.	$	RESERVED MONTHLY (Total annual expenses divided by 12)	$ 275
b.	$		
TOTAL FIXED EXPENSES	$ 1,152		
		TOTAL FIXED EXPENSES	$ 1,152
		TOTAL MONTHLY EXPENSES	$ 2,227

FIGURE 8.6 Expense portion of Clint and Jean's budget

How can you decide where to cut expenses? Of course, it is a very personal matter, and each person must set his or her own priorities, but sometimes comparing your budget to what others are spending can help you decide which categories are too high and which are too low. Table 8.2 shows some typical amounts spent for essential items. Clint and Jean would probably compare their expenses with the middle-income category.

TABLE 8.2   Monthly Spending for Urban Families of Four

Item	Low annual income, $12,000	Middle annual income, $21,000	High annual income, $29,000
Food	$280	$360	$450
Housing	$180	$490	$600
Transportation	$150	$170	$190
Clothing	$100	$150	$220
Taxes	$100	$230	$460
All other	$190	$350	$500

Problem Set 8.3

1. The following is a list of variable expenses for Ron and Lorraine. Total up the entries in each of the categories shown in the sample budget form (Figure 8.5) under monthly variable expenses. If an entry is not listed as a category, total it under the miscellaneous category.

   Groceries, $38.40   Gas, $18.30
   Magazines, $4.30   Water & Sewer, $14.30
   Cosmetics, $6.20   Groceries, $16.50
   Groceries, $36.12   Gas, $15.40
   Gas, $12.50   Groceries, $29.35
   Electricity, $48.30   Gas, $14.00
   Groceries, $42.90   Newspapers, $4.65
   Laundromat, $12.00   Ron's allowance, $50
   Groceries, $88.40   Lorraine's allowance, $50

2. The following is a list of variable expenses for Bob and Marsha. Total up the entries in each of the categories shown in the sample budget form (Figure 8.5) under monthly variable expenses. If an entry is not listed as a category, total it under the miscellaneous category.

   Groceries, $75.38   Gas, $15.40
   Allowances: Groceries, $82.60
     Bob, $45   Allowances:
     Marsha, $45   Bob, $45
     Children, $10   Marsha, $40
   Amusement park, $75   Children, $10
   Electricity, $120.40   Tolls & Parking, $17.00
   Natural gas, $45.50   Restaurant, $15.30

Water & Sewer, $29.40
Gas, $15.80
Gas, $18.90
Groceries, $123.40
Restaurant, $17.60
Dry cleaners, $14.30
Cable TV, $18.00

Gas, $14.50
Groceries, $65.30
Movies, $15.00
Phone, $85.35
Allowances:
  Bob, $45
  Marsha, $45
  Children, $10

3. The following is a list of variable expenses for Jack and Jackie. Total up the entries in each of the categories shown in the sample budget form (Figure 8.5) under monthly variable expenses. If an entry is not listed as a category, total it under the miscellaneous category.

Groceries, $35.18
Gas & Electricity, $61.30
Water & Sewer, $23.50
Cable TV, $9.00
Phone, $23.50
Groceries, $25.10
Restaurant, $16.30
Parking and Tolls, $75.00
Groceries, $146.30

School lunches, $40.00
Beach party, $28.40
Gas, $145.38
Allowances:
  Jack, $100
  Jackie, $200
  Children, $25
Groceries, $38.90
Groceries, $8.50

Find the average monthly amounts (rounded to the nearest dollar) spent for each budget item in Problems 4–12 for the number of months listed.

4. Food: $260, $340, $320

5. Food: $460, $420, $395

6. Food: $390, $320, $350

7. Clothes: $35, $140, $65, $20

8. Clothes: $110, $350, $120, $12

9. Clothes: $320, $0, $32, $29

10. Electricity: $230, $210, $150, $120, $100, $50

11. Electricity: $40, $50, $52, $100, $150

12. Electricity: $180, $150, $110, $80, $60, $50

If the total nonmonthly expenses are those given in Problems 13–18, indicate the necessary monthly reserve (rounded to the nearest dollar).

13. $600   14. $400   15. $1,200

16. $1,500   17. $2,500   18. $900

Determine the amount of monthly reserve (rounded to the nearest dollar) necessary for the annual nonmonthly budget expenses in Problems 19–24.

19. Gifts of $35, $20, $5, $85, and $10

20. Christmas presents of $800

21. Insurance payments of $150 and $150

22. Tax payments of $1,530, $750, $625, and $750

23. Credit card purchases of $35, $65, $235, and $130

24. Propane gas payments of $51, $49, $45, and $62

25. Assume your income is $1,000 per month. Prepare a sample budget using the sample budget form shown in Figure 8.5.

26. Assume your income is $1,750 per month. Prepare a sample budget using the sample budget form shown in Figure 8.5.

27. Assume your income is $2,500 per month. Prepare a sample budget using the sample budget form shown in Figure 8.5.

28. Prepare a sample budget using the sample budget form shown in Figure 8.5. Use your own income or make up some income and use that amount in preparing your budget.

8.4 Income Taxes

SITUATION   Jim and Betty are talking about marriage and Jim says that since they are both working, they will pay more income tax as a married couple than they do as single people. Betty agrees that a few years ago that was true, but the new tax regulations have done away with a "marriage penalty." In this section, Jim and Betty will see how to fill out a 1040A form, how to use tax tables to find their tax if they do not itemize deductions, and how to calculate either the tax refund or the amount of tax due.

It is an old cliché that you can't avoid death or taxes, and for many people the most complicated mathematics problem that must be faced is the annual filing of their income tax form. Remember that even if you have your taxes prepared by a professional tax preparer, you must supply all the necessary information and you are still responsible for the accuracy of the return. In fact, even if the IRS (Internal Revenue Service) makes an error in its advice to you, you are still responsible for the accuracy of your return.

To obtain a job you must have a Social Security number and you are also required to fill out a form called a W-4 form. This determines your filing status (single or married) as well as the number of exemptions you claim for purposes of withholding. However, don't confuse withholding with taxes. A certain amount is withheld from your salary to be applied to your tax when you fill out the forms the following year—hopefully, before April 15. The amount of withholding has to be a fairly accurate estimate of the amount of tax due. If too much is withheld, you will receive a refund. If too little is withheld, you will owe some tax plus a possible penalty if it is too far under your actual tax.

The first step in filling out your income tax form is finding your adjusted gross income. All your employers are required to give you a W-2 form by January 31. You will obtain a W-2 form from each employer you had during the year. These forms will tell you your total wages for the year as well as all the amounts deducted from your salary during the year. A sample W-2 form is shown in Figure 8.7.

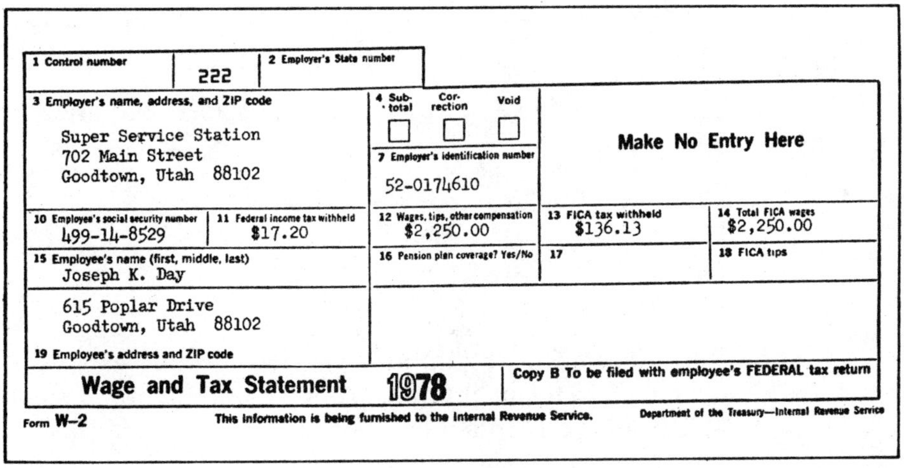
1 Control number 222
2 Employer's State number
3 Employer's name, address, and ZIP code
Super Service Station
702 Main Street
Goodtown, Utah 88102
4 Sub-total Cor-rection Void
Make No Entry Here
7 Employer's identification number
52-0174610
10 Employee's social security number 499-14-8529
11 Federal income tax withheld $17.20
12 Wages, tips, other compensation $2,250.00
13 FICA tax withheld $136.13
14 Total FICA wages $2,250.00
15 Employee's name (first, middle, last)
Joseph K. Day
16 Pension plan coverage? Yes/No
17
18 FICA tips
615 Poplar Drive
Goodtown, Utah 88102
19 Employee's address and ZIP code
Wage and Tax Statement 1978
Copy B To be filed with employee's FEDERAL tax return
Form W-2
This information is being furnished to the Internal Revenue Service.
Department of the Treasury—Internal Revenue Service

FIGURE 8.7 Sample W-2 form

Add up all your wages plus any other income, such as tips or other amounts received as income. Figure 8.8 shows a portion of form 1040A. On this form, total income is entered on Line 7, interest and dividends (less exclusion) on Line 8, unemployment on Line 9, and the total of these amounts on

FIGURE 8.8   Portion of form 1040A

Line 10. This amount is called the **adjusted gross income**. Example 1 illustrates how to correctly determine adjusted gross income.

EXAMPLE 1

Find the adjusted gross income and tell what amounts would be entered on each of the Lines 7–10 on form 1040A.

a. Sara earned $14,350 as a waitress last year, and another $983 in tips.

Line 7:	15,333	(14,350 + 983)
Line 8e:	0	
Line 9b:	0	
Line 10:	15,333	

b. Rita earned $6,418 as a grocery clerk, $7,260 as a salesperson, $48.96 in interest from her credit union, and $23.10 interest from a bank passbook.

Line 7:	13,678	(6,418 + 7,260)
Line 8a:	72	You can round amounts to the nearest dollar.
8b:	0	
8c:	72	
8d:	0	
8e:	72	
Line 9b:	0	
Line 10:	13,750	

c. Joe earned $11,350 working part-time while he was a student and also received $968 in stock dividends. Notice that Line 8d, on form 1040A, allows an exclusion for dividend income. This means that the first $200 of dividends from most stocks is excluded from tax ($400 for a married couple filing a joint return). Thus, Line 8b for Joe is $968 and Line 8d is $200.

Line 7:		11,350
Line 8a:	0	
8b:	968	
8c:	968	
8d:	200	
8e:		768
Line 9b:		0
Line 10:		12,118

∎

Almost 80% of the taxpayers are finished at this point because these people choose not to itemize deductions. All they need to do now is deduct their standard exemptions and use the Tax Table, which is designed for people with taxable incomes of less than $50,000. Part of this table is shown in Table 8.3 on pages 308–309.

EXAMPLE 2

Calculate the income tax for the people in Example 1 by using the tax tables in Table 8.3.

a. Sara's adjusted gross income is $15,333. Assume she is single with one child. The number of exemptions is 2, so in 1981 she would deduct $1,000 for each exemption:

$$\begin{aligned}&\$15,333\\-&\quad 2,000\\\hline&\$13,333 \leftarrow\text{Taxable income}\end{aligned}$$

Find the line of the 1981 Tax Table that says

At least	But less than	Your tax is
13,300	13,350	2,142

The tax is $2,142.

TABLE 8.3   1981 Tax Table

1981 Tax Table
Based on Taxable Income
For persons with taxable incomes of less than $50,000.

Example: Mr. and Mrs. Brown are filing a joint return. Their taxable income on line 34 is $23,270. First, they find the $23,250-23,300 income line. Next, they find the column for married filing jointly and read down the column. The amount shown where the income line and filing status column meet is $4,082. This is the tax amount they must write on line 35 of their return.

At least	But less than	Single	Married filing jointly *	Married filing separately	Head of a household
23,200	23,250	5,208	4,069	6,438	4,805
23,250	23,300	5,224	(4,082)	6,462	4,820
23,300	23,350	5,241	4,096	6,486	4,836

If line 34 (taxable income) is— / And you are— (Your tax is—)

At least	But less than	Single	Married filing jointly *	Married filing separately	Head of a household
8,000					
8,000	8,050	969	698	1,124	893
8,050	8,100	979	707	1,136	902
8,100	8,150	988	715	1,148	911
8,150	8,200	998	724	1,160	920
8,200	8,250	1,007	733	1,172	929
8,250	8,300	1,016	742	1,184	938
8,300	8,350	1,026	751	1,196	947
8,350	8,400	1,035	760	1,207	955
8,400	8,450	1,045	769	1,219	964
8,450	8,500	1,054	778	1,231	973
8,500	8,550	1,064	787	1,243	982
8,550	8,600	1,074	795	1,255	991
8,600	8,650	1,085	804	1,266	1,000
8,650	8,700	1,095	813	1,278	1,009
8,700	8,750	1,105	822	1,290	1,019
8,750	8,800	1,116	831	1,302	1,029
8,800	8,850	1,126	840	1,314	1,040
8,850	8,900	1,136	849	1,326	1,051
8,900	8,950	1,147	858	1,338	1,062
8,950	9,000	1,157	867	1,349	1,073
9,000					
9,000	9,050	1,167	875	1,361	1,084
9,050	9,100	1,178	884	1,373	1,095
9,100	9,150	1,188	893	1,385	1,106
9,150	9,200	1,199	902	1,397	1,116
9,200	9,250	1,209	911	1,409	1,127
9,250	9,300	1,219	920	1,421	1,138
9,300	9,350	1,230	929	1,432	1,149
9,350	9,400	1,240	938	1,444	1,160
9,400	9,450	1,250	947	1,456	1,171
9,450	9,500	1,261	955	1,468	1,182
9,500	9,550	1,271	964	1,480	1,192
9,550	9,600	1,282	973	1,492	1,203
9,600	9,650	1,292	982	1,503	1,214
9,650	9,700	1,302	991	1,515	1,225
9,700	9,750	1,313	1,000	1,527	1,236
9,750	9,800	1,323	1,009	1,539	1,247
9,800	9,850	1,333	1,018	1,551	1,258
9,850	9,900	1,344	1,027	1,563	1,268
9,900	9,950	1,354	1,035	1,575	1,279
9,950	10,000	1,364	1,044	1,586	1,290
10,000					
10,000	10,050	1,375	1,053	1,598	1,301
10,050	10,100	1,385	1,062	1,610	1,312
10,100	10,150	1,396	1,071	1,623	1,323
10,150	10,200	1,406	1,080	1,637	1,334
10,200	10,250	1,416	1,089	1,651	1,344
10,250	10,300	1,427	1,098	1,664	1,355
10,300	10,350	1,437	1,106	1,678	1,366
10,350	10,400	1,447	1,115	1,692	1,377
10,400	10,450	1,458	1,124	1,706	1,388
10,450	10,500	1,468	1,133	1,720	1,399
10,500	10,550	1,479	1,142	1,734	1,410
10,550	10,600	1,489	1,151	1,747	1,421
10,600	10,650	1,499	1,160	1,761	1,431
10,650	10,700	1,510	1,169	1,775	1,442
10,700	10,750	1,520	1,178	1,789	1,453
10,750	10,800	1,530	1,186	1,803	1,464
10,800	10,850	1,541	1,195	1,817	1,475
10,850	10,900	1,553	1,204	1,830	1,486
10,900	10,950	1,565	1,213	1,844	1,497
10,950	11,000	1,577	1,222	1,858	1,507
11,000					
11,000	11,050	1,589	1,231	1,872	1,518
11,050	11,100	1,601	1,240	1,886	1,529
11,100	11,150	1,613	1,249	1,899	1,540
11,150	11,200	1,624	1,258	1,913	1,551
11,200	11,250	1,636	1,266	1,927	1,562
11,250	11,300	1,648	1,275	1,941	1,573
11,300	11,350	1,660	1,284	1,955	1,583
11,350	11,400	1,672	1,293	1,969	1,594
11,400	11,450	1,684	1,302	1,982	1,605
11,450	11,500	1,696	1,311	1,996	1,616
11,500	11,550	1,707	1,320	2,010	1,627
11,550	11,600	1,719	1,329	2,024	1,638
11,600	11,650	1,731	1,338	2,038	1,649
11,650	11,700	1,743	1,346	2,052	1,659
11,700	11,750	1,755	1,355	2,065	1,670
11,750	11,800	1,767	1,364	2,079	1,681
11,800	11,850	1,778	1,373	2,093	1,693
11,850	11,900	1,790	1,382	2,107	1,704
11,900	11,950	1,802	1,392	2,121	1,716
11,950	12,000	1,814	1,402	2,134	1,728
12,000					
12,000	12,050	1,826	1,412	2,148	1,740
12,050	12,100	1,838	1,423	2,162	1,752
12,100	12,150	1,850	1,433	2,176	1,764
12,150	12,200	1,861	1,443	2,190	1,776
12,200	12,250	1,873	1,454	2,204	1,787
12,250	12,300	1,885	1,464	2,217	1,799
12,300	12,350	1,897	1,475	2,232	1,811
12,350	12,400	1,909	1,485	2,248	1,823
12,400	12,450	1,921	1,495	2,264	1,835
12,450	12,500	1,933	1,506	2,280	1,847
12,500	12,550	1,944	1,516	2,295	1,858
12,550	12,600	1,956	1,526	2,311	1,870
12,600	12,650	1,968	1,537	2,327	1,882
12,650	12,700	1,980	1,547	2,343	1,894
12,700	12,750	1,992	1,558	2,359	1,906
12,750	12,800	2,004	1,568	2,374	1,918
12,800	12,850	2,015	1,578	2,390	1,930
12,850	12,900	2,027	1,589	2,406	1,941
12,900	12,950	2,040	1,599	2,422	1,953
12,950	13,000	2,053	1,609	2,438	1,965
13,000					
13,000	13,050	2,065	1,620	2,453	1,977
13,050	13,100	2,078	1,630	2,469	1,989
13,100	13,150	2,091	1,640	2,485	2,001
13,150	13,200	2,104	1,651	2,501	2,013
13,200	13,250	2,117	1,661	2,517	2,024
13,250	13,300	2,130	1,672	2,532	2,036
13,300	13,350	2,142	1,682	2,548	2,048
13,350	13,400	2,155	1,692	2,564	2,060
13,400	13,450	2,168	1,703	2,580	2,072
13,450	13,500	2,181	1,713	2,596	2,084
13,500	13,550	2,194	1,723	2,611	2,095
13,550	13,600	2,207	1,734	2,627	2,107
13,600	13,650	2,219	1,744	2,643	2,119
13,650	13,700	2,232	1,755	2,659	2,131
13,700	13,750	2,245	1,765	2,675	2,143
13,750	13,800	2,258	1,775	2,690	2,155
13,800	13,850	2,271	1,786	2,706	2,167
13,850	13,900	2,284	1,796	2,722	2,178
13,900	13,950	2,296	1,806	2,738	2,190
13,950	14,000	2,309	1,817	2,754	2,202
14,000					
14,000	14,050	2,322	1,827	2,769	2,214
14,050	14,100	2,335	1,837	2,785	2,226
14,100	14,150	2,348	1,848	2,801	2,238
14,150	14,200	2,361	1,858	2,817	2,250
14,200	14,250	2,373	1,869	2,833	2,261
14,250	14,300	2,386	1,879	2,848	2,273
14,300	14,350	2,399	1,889	2,864	2,285
14,350	14,400	2,412	1,900	2,880	2,297
14,400	14,450	2,425	1,910	2,896	2,309
14,450	14,500	2,438	1,920	2,912	2,321
14,500	14,550	2,450	1,931	2,927	2,332
14,550	14,600	2,463	1,941	2,943	2,344
14,600	14,650	2,476	1,952	2,959	2,356
14,650	14,700	2,489	1,962	2,975	2,368
14,700	14,750	2,502	1,972	2,991	2,380
14,750	14,800	2,515	1,983	3,006	2,392
14,800	14,850	2,528	1,993	3,022	2,404
14,850	14,900	2,540	2,003	3,038	2,415
14,900	14,950	2,553	2,014	3,054	2,427
14,950	15,000	2,566	2,024	3,071	2,439
15,000					
15,000	15,050	2,580	2,034	3,089	2,451
15,050	15,100	2,595	2,045	3,107	2,464
15,100	15,150	2,609	2,055	3,126	2,477
15,150	15,200	2,624	2,066	3,144	2,490
15,200	15,250	2,639	2,076	3,162	2,503
15,250	15,300	2,654	2,086	3,180	2,516
15,300	15,350	2,669	2,097	3,199	2,528
15,350	15,400	2,684	2,107	3,217	2,541
15,400	15,450	2,698	2,117	3,235	2,554
15,450	15,500	2,713	2,128	3,254	2,567
15,500	15,550	2,728	2,138	3,272	2,580
15,550	15,600	2,743	2,149	3,290	2,593
15,600	15,650	2,758	2,159	3,308	2,606
15,650	15,700	2,772	2,169	3,327	2,618
15,700	15,750	2,787	2,180	3,345	2,631
15,750	15,800	2,802	2,190	3,363	2,644
15,800	15,850	2,817	2,200	3,381	2,657
15,850	15,900	2,832	2,211	3,400	2,670
15,900	15,950	2,846	2,221	3,418	2,683
15,950	16,000	2,861	2,232	3,436	2,695
16,000					
16,000	16,050	2,876	2,243	3,455	2,708
16,050	16,100	2,891	2,254	3,473	2,721
16,100	16,150	2,906	2,266	3,491	2,734
16,150	16,200	2,921	2,278	3,509	2,747
16,200	16,250	2,935	2,290	3,528	2,760

1981 Tax Table (Continued)

If line 34 (taxable income) is— At least	But less than	Single	Married filing jointly *	Married filing separately	Head of a household
16,250	16,300	2,950	2,302	3,546	2,772
16,300	16,350	2,965	2,314	3,564	2,785
16,350	16,400	2,980	2,326	3,582	2,798
16,400	16,450	2,995	2,337	3,601	2,811
16,450	16,500	3,009	2,349	3,619	2,824
16,500	16,550	3,024	2,361	3,637	2,837
16,550	16,600	3,039	2,373	3,655	2,849
16,600	16,650	3,054	2,385	3,674	2,862
16,650	16,700	3,069	2,397	3,692	2,875
16,700	16,750	3,083	2,409	3,710	2,888
16,750	16,800	3,098	2,420	3,729	2,901
16,800	16,850	3,113	2,432	3,747	2,914
16,850	16,900	3,128	2,444	3,765	2,926
16,900	16,950	3,143	2,456	3,783	2,939
16,950	17,000	3,158	2,468	3,802	2,952
17,000					
17,000	17,050	3,172	2,480	3,820	2,965
17,050	17,100	3,187	2,491	3,838	2,978
17,100	17,150	3,202	2,503	3,856	2,991
17,150	17,200	3,217	2,515	3,875	3,003
17,200	17,250	3,232	2,527	3,893	3,016
17,250	17,300	3,246	2,539	3,911	3,029
17,300	17,350	3,261	2,551	3,930	3,042
17,350	17,400	3,276	2,563	3,948	3,055
17,400	17,450	3,291	2,574	3,966	3,068
17,450	17,500	3,306	2,586	3,984	3,081
17,500	17,550	3,320	2,598	4,003	3,093
17,550	17,600	3,335	2,610	4,021	3,106
17,600	17,650	3,350	2,622	4,041	3,119
17,650	17,700	3,365	2,634	4,062	3,132
17,700	17,750	3,380	2,646	4,083	3,145
17,750	17,800	3,395	2,657	4,104	3,158
17,800	17,850	3,409	2,669	4,126	3,170
17,850	17,900	3,424	2,681	4,147	3,183
17,900	17,950	3,439	2,693	4,168	3,196
17,950	18,000	3,454	2,705	4,189	3,209
18,000					
18,000	18,050	3,469	2,717	4,210	3,222
18,050	18,100	3,483	2,728	4,232	3,235
18,100	18,150	3,498	2,740	4,253	3,247
18,150	18,200	3,513	2,752	4,274	3,260
18,200	18,250	3,529	2,764	4,295	3,274
18,250	18,300	3,546	2,776	4,317	3,290
18,300	18,350	3,562	2,788	4,338	3,305
18,350	18,400	3,579	2,800	4,359	3,320
18,400	18,450	3,596	2,811	4,380	3,336
18,450	18,500	3,613	2,823	4,402	3,351
18,500	18,550	3,630	2,835	4,423	3,366
18,550	18,600	3,646	2,847	4,444	3,381
18,600	18,650	3,663	2,859	4,465	3,397
18,650	18,700	3,680	2,871	4,486	3,412
18,700	18,750	3,697	2,883	4,508	3,427
18,750	18,800	3,713	2,894	4,529	3,443
18,800	18,850	3,730	2,906	4,550	3,458
18,850	18,900	3,747	2,918	4,571	3,473
18,900	18,950	3,764	2,930	4,593	3,489
18,950	19,000	3,781	2,942	4,614	3,504

If line 34 (taxable income) is— At least	But less than	Single	Married filing jointly *	Married filing separately	Head of a household
19,000					
19,000	19,050	3,797	2,954	4,635	3,519
19,050	19,100	3,814	2,965	4,656	3,535
19,100	19,150	3,831	2,977	4,678	3,550
19,150	19,200	3,848	2,989	4,699	3,565
19,200	19,250	3,865	3,001	4,720	3,580
19,250	19,300	3,881	3,013	4,741	3,596
19,300	19,350	3,898	3,025	4,762	3,611
19,350	19,400	3,915	3,037	4,784	3,626
19,400	19,450	3,932	3,048	4,805	3,642
19,450	19,500	3,949	3,060	4,826	3,657
19,500	19,550	3,965	3,072	4,847	3,672
19,550	19,600	3,982	3,084	4,869	3,688
19,600	19,650	3,999	3,096	4,890	3,703
19,650	19,700	4,016	3,108	4,911	3,718
19,700	19,750	4,032	3,120	4,932	3,733
19,750	19,800	4,049	3,131	4,954	3,749
19,800	19,850	4,066	3,143	4,975	3,764
19,850	19,900	4,083	3,155	4,996	3,779
19,900	19,950	4,100	3,167	5,017	3,795
19,950	20,000	4,116	3,179	5,038	3,810
20,000					
20,000	20,050	4,133	3,191	5,060	3,825
20,050	20,100	4,150	3,202	5,081	3,841
20,100	20,150	4,167	3,214	5,102	3,856
20,150	20,200	4,184	3,226	5,123	3,871
20,200	20,250	4,200	3,239	5,145	3,887
20,250	20,300	4,217	3,253	5,166	3,902
20,300	20,350	4,234	3,267	5,187	3,917
20,350	20,400	4,251	3,280	5,208	3,932
20,400	20,450	4,267	3,294	5,230	3,948
20,450	20,500	4,284	3,308	5,251	3,963
20,500	20,550	4,301	3,322	5,272	3,978
20,550	20,600	4,318	3,336	5,293	3,994
20,600	20,650	4,335	3,350	5,314	4,009
20,650	20,700	4,351	3,363	5,336	4,024
20,700	20,750	4,368	3,377	5,357	4,040
20,750	20,800	4,385	3,391	5,378	4,055
20,800	20,850	4,402	3,405	5,399	4,070
20,850	20,900	4,419	3,419	5,421	4,086
20,900	20,950	4,435	3,433	5,442	4,101
20,950	21,000	4,452	3,446	5,463	4,116
21,000					
21,000	21,050	4,469	3,460	5,484	4,131
21,050	21,100	4,486	3,474	5,506	4,147
21,100	21,150	4,503	3,488	5,527	4,162
21,150	21,200	4,519	3,502	5,548	4,177
21,200	21,250	4,536	3,516	5,569	4,193
21,250	21,300	4,553	3,529	5,590	4,208
21,300	21,350	4,570	3,543	5,612	4,223
21,350	21,400	4,586	3,557	5,633	4,239
21,400	21,450	4,603	3,571	5,654	4,254
21,450	21,500	4,620	3,585	5,675	4,269
21,500	21,550	4,637	3,598	5,697	4,285
21,550	21,600	4,654	3,612	5,718	4,300
21,600	21,650	4,670	3,626	5,739	4,315
21,650	21,700	4,687	3,640	5,760	4,330
21,700	21,750	4,704	3,654	5,782	4,346

If line 34 (taxable income) is— At least	But less than	Single	Married filing jointly *	Married filing separately	Head of a household
21,750	21,800	4,721	3,668	5,803	4,361
21,800	21,850	4,738	3,681	5,824	4,376
21,850	21,900	4,754	3,695	5,845	4,392
21,900	21,950	4,771	3,709	5,866	4,407
21,950	22,000	4,788	3,723	5,888	4,422
22,000					
22,000	22,050	4,805	3,737	5,909	4,438
22,050	22,100	4,821	3,751	5,930	4,453
22,100	22,150	4,838	3,764	5,951	4,468
22,150	22,200	4,855	3,778	5,973	4,483
22,200	22,250	4,872	3,792	5,994	4,499
22,250	22,300	4,889	3,806	6,015	4,514
22,300	22,350	4,905	3,820	6,036	4,529
22,350	22,400	4,922	3,833	6,058	4,545
22,400	22,450	4,939	3,847	6,079	4,560
22,450	22,500	4,956	3,861	6,100	4,575
22,500	22,550	4,973	3,875	6,121	4,591
22,550	22,600	4,989	3,889	6,142	4,606
22,600	22,650	5,006	3,903	6,164	4,621
22,650	22,700	5,023	3,916	6,185	4,637
22,700	22,750	5,040	3,930	6,206	4,652
22,750	22,800	5,056	3,944	6,227	4,667
22,800	22,850	5,073	3,958	6,249	4,682
22,850	22,900	5,090	3,972	6,270	4,698
22,900	22,950	5,107	3,986	6,293	4,713
22,950	23,000	5,124	3,999	6,317	4,728
23,000					
23,000	23,050	5,140	4,013	6,341	4,744
23,050	23,100	5,157	4,027	6,365	4,759
23,100	23,150	5,174	4,041	6,389	4,774
23,150	23,200	5,191	4,055	6,414	4,790
23,200	23,250	5,208	4,069	6,438	4,805
23,250	23,300	5,224	4,082	6,462	4,820
23,300	23,350	5,241	4,096	6,486	4,836
23,350	23,400	5,258	4,110	6,510	4,851
23,400	23,450	5,275	4,124	6,535	4,866
23,450	23,500	5,292	4,138	6,559	4,881
23,500	23,550	5,310	4,151	6,583	4,898
23,550	23,600	5,329	4,165	6,607	4,916
23,600	23,650	5,348	4,179	6,631	4,934
23,650	23,700	5,367	4,193	6,656	4,951
23,700	23,750	5,387	4,207	6,680	4,969
23,750	23,800	5,406	4,221	6,704	4,987
23,800	23,850	5,425	4,234	6,728	5,005
23,850	23,900	5,444	4,248	6,752	5,022
23,900	23,950	5,464	4,262	6,776	5,040
23,950	24,000	5,483	4,276	6,801	5,058
24,000					
24,000	24,050	5,502	4,290	6,825	5,076
24,050	24,100	5,521	4,304	6,849	5,094
24,100	24,150	5,541	4,317	6,873	5,111
24,150	24,200	5,560	4,331	6,897	5,129
24,200	24,250	5,579	4,345	6,922	5,147
24,250	24,300	5,598	4,359	6,946	5,165
24,300	24,350	5,618	4,373	6,970	5,182
24,350	24,400	5,637	4,386	6,994	5,200
24,400	24,450	5,656	4,400	7,018	5,218
24,450	24,500	5,675	4,414	7,043	5,236

b. Rita's adjusted gross income is $13,750. Assume she is single. The number of exemptions is 1, so deduct $1,000: $13,750 − $1,000 = $12,750. Now use the 1981 Tax Table:

At least	But less than	Your tax is
12,750	12,800	2,004

Notice that you use this line and *not* 12,700–12,750.

The tax is $2,004.

c. Joe's adjusted gross income is $12,118. Assume he is married with a wife and one child. The number of exemptions is 3, so deduct $3,000: $12,118 − $3,000 = $9,118. Find the line of the 1981 Tax Table that says

At least	But less than	Your tax is
9,100	9,150	893

The tax is $893.   ∎

Tax tables are based on what is called a **graduated income tax**. This means that the percent of tax paid goes up as income goes up. In 1981 if the taxable income for a single taxpayer is $10,000, then the tax rate is 21%. At $20,000 it is 34%, and at $50,000 it is 55%. The range is from 14% ($2,300) to 70% ($108,300). Now, due to the current rate of inflation, a $10,000 a year job would pay $26,000 in 10 years. At this rate, eventually, everyone would be in the highest income tax bracket. Notice that the government not only takes *more taxes* but also a *greater percentage* in the higher brackets. The idea that the brackets should be increased at the same rate as inflation is called **indexing**. Due to the tax reform of 1981, indexing will go into effect in 1985. (This same tax reform equalized the taxes paid by two income earners whether they were married or not.)

The final step in figuring your income tax is to compare the tax due with the amount paid through withholding. If the amount withheld is larger than the tax due, you are entitled to a refund. If the amount due is larger than the amount withheld, then you must send that amount along with your tax return by April 15.

EXAMPLE 3

a. Suppose Sara's withholding was $2,450 and her tax was $2,142. Calculate her refund or amount due.

b. Suppose Joe's withholding was $750 and his tax was $893. Calculate his refund or amount due.

Solution

a. Since her withholding was greater, she is due a refund of $2,450 − $2,142 = $308.

b. Since his withholding is less than the tax, he has a balance due of $143. ■

Problem Set 8.4

In Problems 1–10 find the adjusted gross income and tell what amount would be entered on each of Lines 7–10 on form 1040A (see Figure 8.8).

1. Carol earned $9,420 plus $1,960 in tips.

2. Bob earned $9,940 plus $1,560 in tips.

3. James earned $14,312 in salary plus $16.23 in interest from his credit union and $92.16 in interest from a bank passbook.

4. John, who is single, earned $17,412 in salary plus $92.63 in interest from a bank passbook and $625 in stock dividends.

5. Jane, who is single, earned $17,216 in salary plus $213 in stock dividends. She also had $28.92 in interest from her credit union.

6. Mary earned $13,416 and Dean earned $18,612 in salary. They also had $512 in stock dividends.

7. Jim earned $24,312 and Betty earned $6,810 in salary. They also had dividends of $928 from a stock account.

8. Jack earned $15,310 and Jill earned $16,525. They also had $148 in stock dividends.

9. Joe earned $25,400 and Josie earned $1,535. They also had $185 in stock dividends.

10. Ralph earned $14,325 and Fawn earned $9,890. They also had $2,950 in stock dividends as well as interest of $5,790.

In Problems 11–20 calculate the income tax from the adjusted gross income using Table 8.3.

11. Carol from Problem 1 is a single person.

12. Bob from Problem 2 is a single person.

13. James from Problem 3 is a single head of household with one child.

14. John from Problem 4 is a single head of household with one child.

15. Jane from Problem 5 is a single head of household with two children.

16. Bill and Laura have an adjusted gross income of $28,352 and are filing jointly; they have three children.

17. Karen and Steve have an adjusted gross income of $27,125 and are filing jointly; they have two children.

18. Phil earned $21,350 and Carolyn earned $23,475. They have no other income and no children. They are married and are filing separately. What are their combined income taxes?

19. Tom earned $25,490 and Sheila earned $17,208. They have no other income and no children. They are married and are filing separately. What are their combined income taxes?

20. Jim earned $16,306 and Betty earned $10,152. They have no other income and no children. Calculate their income taxes if they file jointly and if they file separately.

In Problems 21–30 compare the tax due with the amount paid through withholding. Give the difference and tell whether it is a refund or an amount due.

21. Carol from Problem 11 had a withholding of $1,560.

22. Bob from Problem 12 had $1,440 withheld from his salary.

23. James from Problem 13 had $2,280 withheld from his salary.

24. John from Problem 14 had a withholding of $2,750.

25. Jane from Problem 15 had a withholding of $2,750.

26. Laura from Problem 16 had $1,920 withheld from her salary, and Bill had $1,360 withheld from his salary.

27. Steve from Problem 17 had $2,840 withheld from his salary, and Karen had $1,920 withheld from her salary.

28. Phil from Problem 18 had $5,280 withheld from his salary, and Carolyn had $6,120 withheld from her salary.

29. Tom from Problem 19 had $7,080 withheld from his salary, and Sheila had $3,840 withheld from her salary.

30. Jim and Betty from Problem 20 decide to file a joint return. Jim had $265 withheld from his monthly check and Betty had $105 withheld from hers.

8.5 Review Problems

For Problems 1–4, assume that your beginning balance is $143.36. Fill out the given forms.

Section 8.1

1. Write a check for $32.95 to Doug's Shoes dated Sept. 19, 1983.

```
                                        NO. 182
                 _____ 19 ___        90-46
                                           1211
  PAY
  TO THE
  ORDER OF _____ $ _____

  _____ DOLLARS

      Exchange
        Bank
       Rockford, Ill.
  MEMO _____     _____
  ⑆0⑉4⑉00310⑆  903⑈391⑈
```

2. Enter the necessary information in the check register for the check written in Problem 1. Also, record check #183 for $57.14 to Safeway dated Sept. 19, 1983.

			BE SURE TO DEDUCT ANY PER CHECK CHARGES OR MAINTENANCE CHARGES THAT MAY APPLY					
DATE		CHECK NUMBER	CHECKS ISSUED TO OR DEPOSIT RECEIVED FROM	AMOUNT OF DEPOSIT	✔	AMOUNT OF CHECK	BALANCE 143	36
9	19	182	Doug's Shoes					

3. Suppose you deposit the following checks on 9/28/83:

Check 09-187 $827.46   Check 90-198 $100.07
Check 01-12 $ .10

Complete the deposit slip and then show this deposit on the check register of Problem 2. Next, record check #184 for $12.29 to Long's Drugs dated 10/1/83.

4. You receive the bank statement shown on p. 314. Notice that the balance shown is not the same as the balance shown in the check register. On the back of this bank statement is a reconciliation form which is also reproduced on p. 314. Follow the directions on this form for reconciling the check register and the bank statement. Notice that the only outstanding check is the one written in Problem 3 to Long's Drugs.

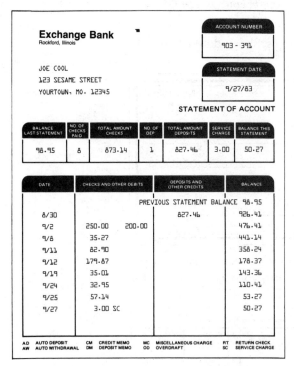

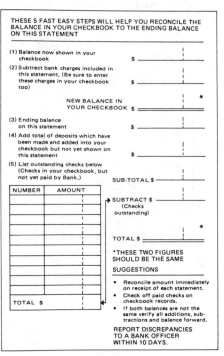

Section 8.2   **5.** If a coat you have been wanting costs $245 and the store has a special 20% off sale, what is the sale price for this coat?

Section 8.2   **6.** What is the discount on a $2,000 stereo system marked 15% off?

Section 8.3   **7.** What is the average amount per month spent for water if the bimonthly bills last year were $21, $25, $35, $43, $52, and $28?

Section 8.3   **8.** What is the amount of monthly reserve (to the nearest dollar) necessary if the total nonmonthly expenses last year were: Gifts, $125; Insurance, $425; Christmas, $360; Other, $725?

Section 8.4   **9.** Raymond earned $18,241 and Rita earned $8,421. They received $358 in stock dividends and $2,816 in other income during the year. What is their adjusted gross income? (See Figure 8.8 on page 306.)

Section 8.4   **10.** Raymond and Rita from Problem 9 have three children and Ray had $2,760 withheld from his salary, while Rita had $1,992 withheld from her salary. Use Table 8.3 on pages 308–309 to determine the amount of the balance due or the amount of refund they should expect.

Interest

9.1 Simple Interest

SITUATION   Blondie saw a mink coat that she would like very much to have. The cost of the coat is $3,650. Unfortunately, she doesn't have enough money to buy the coat, so she looks at her budget and decides that she can save ten cents a day toward the coat. Dagwood tells her that it will take her 100 years to save enough to buy the coat at the rate of ten cents per day. But wait a minute, Blondie! Banks will pay you interest on that money, which means that it will take you much less time to save the money than you think. In this section Dagwood and Blondie will learn about interest, beginning with something called *simple interest*. They will find out how to calculate simple interest in a variety of situations. Later in this chapter they will even see how they can earn interest on their accumulated interest because of something called *compound interest*. But, alas, they will also see in this chapter the effects of inflation on the price of that mink coat during the time they are saving to buy it.

There are certain arithmetical skills that enable us to make intelligent decisions about how we spend the money we earn. One of the most fundamental mathematical concepts that consumers, as well as business people, must understand is the notion of *interest*. Simply stated, **interest** is money paid for the use of money. That is, we receive interest when we let others use our money (when we deposit money in a savings account, for example), and we pay interest when we use the money of others (for example, borrowing from a bank).

The amount of the deposit or loan is called the **principal**, and the interest is stated as a percent of the principal, called the **interest rate**. The **time** is the length of time the money is borrowed or lent. The interest rate is usually an *annual interest rate*, and the time is stated in terms of a year unless otherwise stated. The fundamental interest formula is given below.

Simple Interest Formula

$$\text{INTEREST} = (\text{PRINCIPAL}) \times (\text{RATE}) \times (\text{TIME})$$
$$I = \quad\quad P \quad\quad \times \quad R \quad \times \quad T$$

or simply,

$$I = PRT$$

EXAMPLE 1

Look at the Blondie cartoon. Suppose Blondie saves 20¢ per day, but only for a year. At the end of a year she will have saved $73. If she then puts the money into a savings account at $8\frac{1}{2}\%$ interest, how much interest does the bank pay her after one year? After three years?

Solution

$$\text{INTEREST} = (\text{PRINCIPAL}) \times (\text{RATE}) \times (\text{TIME})$$

$$= \quad \$73 \quad \times .085 \times \quad 1$$

$\approx \$6.21$ Round your answer to the nearest cent.

For three years,

$\text{INTEREST} = \$73 \times .085 \times 3$ Do the arithmetic first; then round your
$\approx \$18.62$ answer. Notice that we didn't take the
                         rounded answer and multiply by 3.
                         This doesn't make much difference
                         here, but it will make a big difference
                         later in this chapter.

Many of the calculations in this chapter will be simplified if you have a calculator. You can do this problem on a calculator by pressing:

$$\boxed{73} \; \boxed{\times} \; \boxed{.085} \; \boxed{\times} \; \boxed{3} \; \boxed{=}$$

Display: 18.615   ■

There is a difference between asking for the amount of interest, as illustrated by Example 1, and asking for the **amount present**. The amount present is the amount you will have after the interest is added to the principal.

AMOUNT PRESENT = PRINCIPAL + INTEREST $A = P + I$

Amount Present

EXAMPLE 2

If $10,000 is deposited in an account earning 15% simple interest for 5 years, how much is in the account at the end of the 5 year period?

Solution

First, find the interest:

$$I = PRT$$
$$= \$10,000 \times .15 \times 5$$
$$= \$7,500$$

Next, find the amount present:

$$A = \$10,000 + \$7,500$$
$$= \$17,500 \qquad \blacksquare$$

Money is often borrowed or lent for periods of less than a year. This causes two problems: (1) you are usually given dates but need to express the length of time as a number of days; and (2) after you know the number of days, you must write it as a part of a year (since times and rates in the interest formula are usually stated in terms of years). This complicates the time part of the simple interest formula because it is stated as a (usually ugly) fraction.

EXAMPLE 3

Find the number of days between March 25 and September 15.

Solution

One method is to remember the number of days in each month.

Remember this elementary school poem?
Thirty days hath September, April, June, and November. All the rest have thirty-one.

February 28 alone, Except in leap years at which time, February's days are twenty-nine.

March:	6 days	Since there are 31 days in March,
April:	30 days	there are 6 days left in March after March 25.
May:	31 days	
June:	30 days	
July:	31 days	
August:	31 days	
September:	15 days	
Total:	174 days	

This is an awkward way of doing it, so most banks and businesses use a table of days of the year. In Table 9.1, find March 25 and September 15:

TABLE 9.1 The Number of Each Day of the Year*

Day of Mo.	Jan.	Feb.	Mar.	Apr.	May	June	July	Aug.	Sept.	Oct.	Nov.	Dec.	Day of Mo.
1	1	32	60	91	121	152	182	213	244	274	305	335	1
2	2	33	61	92	122	153	183	214	245	275	306	336	2
3	3	34	62	93	123	154	184	215	246	276	307	337	3
4	4	35	63	94	124	155	185	216	247	277	308	338	4
5	5	36	64	95	125	156	186	217	248	278	309	339	5
6	6	37	65	96	126	157	187	218	249	279	310	340	6
7	7	38	66	97	127	158	188	219	250	280	311	341	7
8	8	39	67	98	128	159	189	220	251	281	312	342	8
9	9	40	68	99	129	160	190	221	252	282	313	343	9
10	10	41	69	100	130	161	191	222	253	283	314	344	10
11	11	42	70	101	131	162	192	223	254	284	315	345	11
12	12	43	71	102	132	163	193	224	255	285	316	346	12
13	13	44	72	103	133	164	194	225	256	286	317	347	13
14	14	45	73	104	134	165	195	226	257	287	318	348	14
15	15	46	74	105	135	166	196	227	258	288	319	349	15
16	16	47	75	106	136	167	197	228	259	289	320	350	16
17	17	48	76	107	137	168	198	229	260	290	321	351	17
18	18	49	77	108	138	169	199	230	261	291	322	352	18
19	19	50	78	109	139	170	200	231	262	292	323	353	19
20	20	51	79	110	140	171	201	232	263	293	324	354	20
21	21	52	80	111	141	172	202	233	264	294	325	355	21
22	22	53	81	112	142	173	203	234	265	295	326	356	22
23	23	54	82	113	143	174	204	235	266	296	327	357	23
24	24	55	83	114	144	175	205	236	267	297	328	358	24
25	25	56	84	115	145	176	206	237	268	298	329	359	25
26	26	57	85	116	146	177	207	238	269	299	330	360	26
27	27	58	86	117	147	178	208	239	270	300	331	361	27
28	28	59	87	118	148	179	209	240	271	301	332	362	28
29	29	*	88	119	149	180	210	241	272	302	333	363	29
30	30		89	120	150	181	211	242	273	303	334	364	30
31	31		90		151		212	243		304		365	31

*In leap years, after February 28, add 1 to the tabulated number.

Sept. 15 is the   258th day
Mar. 25 is the   84th day
Subtract: 174 days   ■

EXAMPLE 4

Find the number of days between the given dates using Table 9.1.

a. April 15 to November 1

$$
\begin{array}{ll}
305 & \text{(Nov. 1)} \\
-\ 105 & \text{(Apr. 15)} \\
\hline
200 \text{ days}
\end{array}
$$

b. February 5, 1984 to October 12, 1984

$$
\begin{array}{ll}
286 & \text{(Oct. 12—add 1 to tabulated date because 1984 is a} \\
-\ 36 & \text{leap year)} \\
\hline
250 \text{ days}
\end{array}
$$

c. August 3, 1982 to May 15, 1983
First, find the number of days left in 1982:

$$
\begin{array}{r}
365 \quad \text{(Dec. 31)} \\
-\ 215 \quad \text{(Aug. 3)} \\
\hline
150 \ \text{days}
\end{array}
$$

Then, add the number of days in the following year:

$$
\begin{array}{r}
150 \\
+\ 135 \quad \text{(May 15)} \\
\hline
285 \ \text{days}
\end{array}
$$
■

Now, if the length of time for calculating interest is stated in days, then it is necessary to convert those days into part of a year so that it can be used in the simple interest formula. This can be done two ways:

1. Exact interest: 365 days per year
2. Ordinary interest: 360 days per year

Most applications and businesses use ordinary interest. So in this book, unless it is otherwise stated, assume ordinary interest. That is, use 360 for the number of days in a year.

EXAMPLE 5

Suppose you borrow $1,200 on March 25 at 21% simple interest. How much must be repaid on September 15?

Solution

From Example 3, there are 174 days between these dates.

$$
\begin{aligned}
\text{INTEREST} &= P \times R \times T \\
&= \$\cancel{1,200}^{12} \times \frac{21}{\cancel{100}} \times \frac{\cancel{174}^{58}}{\cancel{360}_{10}} \\
&= \frac{\$1,218}{10} \\
&= \$121.80
\end{aligned}
$$

The amount to be repaid means the interest plus the principal:

$$\text{AMOUNT} = \$1,200 + \$121.80$$
$$= \$1,321.80$$

This problem is greatly simplified if you have access to a calculator:

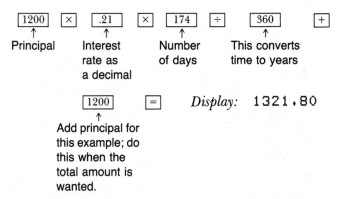

If you wanted exact simple interest in Example 5 instead of ordinary interest, the process would be the same except you would use 365 instead of 360.

Banks frequently sell certificates of deposit for periods of 30 days, 60 days, 90 days, or 180 days. In these cases the fractional component for time can easily be reduced:

Time	Time fraction
30 days	$\dfrac{30}{360} = \dfrac{1}{12}$
60 days	$\dfrac{60}{360} = \dfrac{1}{6}$
90 days	$\dfrac{90}{360} = \dfrac{1}{4}$
180 days	$\dfrac{180}{360} = \dfrac{1}{2}$

EXAMPLE 6

What is the interest for a ten thousand dollar 180 day certificate paying 18% interest?

Solution

$$\text{INTEREST} = P \times R \times T$$
$$= \$10,000 \times \frac{18}{100} \times \frac{1}{2}$$
$$= \$900$$ ■

Problem Set 9.1

Use the simple interest formula to find the interest in Problems 1–18.

1. $1,000 at 8% for 1 year
2. $5,000 at 9% for 2 years
3. $800 at 12% for 3 years
4. $1,200 at 8% for 3 years
5. $900 at 8% for 20 years
6. $1,600 at 6% for 12 years
7. $400 at 6% for 7 years
8. $5,000 at 8% for 6 years
9. $10,000 at 12% for 6 years
10. $500 at 15% for 5 years
11. $500 at $8\frac{1}{2}$% for 1 year
12. $300 at $8\frac{1}{2}$% for 5 years
13. $400 at $12\frac{1}{2}$% for 3 years
14. $800 at $12\frac{1}{2}$% for 4 years
15. $1,000 at 12% for 40 years
16. $10,000 at 15% for 40 years
17. $10,000 at 18% for 40 years
18. $100,000 at 20% for 40 years

Find the number of days between the given dates in Problems 19–24.

19. April 3, 1984 to August 21, 1984
20. July 16, 1983 to November 5, 1983
21. February 22, 1982 to October 30, 1982
22. November 17, 1982 to June 30, 1983
23. October 12, 1981 to April 26, 1983
24. January 5, 1984 to September 5, 1985

Find the amount to be repaid for the loans described in Problems 25–28.

25. $1,000 at 20% simple interest from June 4, 1983
to October 9, 1983

26. $5,000 at 15% simple interest from August 3, 1983
to October 12, 1983

27. $2,500 at 15% simple interest from February 5, 1984
to April 19, 1984

28. $12,000 at 19% simple interest from December 9, 1982
to June 5, 1983

29. Find the simple interest on a ten thousand dollar 180 day
certificate paying $19\frac{1}{2}\%$ interest.

30. Find the simple interest on a thirty-five hundred dollar 30
day certificate paying $14\frac{1}{2}\%$ interest.

9.2 Working with Simple Interest

SITUATION   Jerry has just received an inheritance of
$25,000 and he would like to use it to help him with his
retirement. Since Jerry is 25 years old, he figures that the
$25,000 can be invested for 40 years before he needs to use
it for retirement. Since he would like to have a monthly
income of $5,000, how much money will he need to have in
his retirement account to provide this income from interest
only? He also wants to know what interest rate would be
necessary in order for the $25,000 to grow to this amount.
Both of Jerry's questions are answered by working with the
simple interest formula introduced in the last section. In this
section, Jerry will see how to find the principal, the rate, or
the time, given the other components of the interest formula.

In the last section we used the simple interest formula to
find the interest, but it is frequently the case that we need to
find the principal, or the rate, or the time. These can be
easily found, but the calculations necessary to work out the
problems are often very difficult without a calculator. For
this reason, this section is optional, and should be studied
only if you have access to a calculator.

**Variations of the
Simple Interest Formula**

To find interest:

$$\text{INTEREST} = \text{PRINCIPAL} \times \text{RATE} \times \text{TIME}$$

To find principal:

$$\text{PRINCIPAL} = \frac{\text{INTEREST}}{\text{RATE} \times \text{TIME}}$$

To find rate:

$$\text{RATE} = \frac{\text{INTEREST}}{\text{PRINCIPAL} \times \text{TIME}}$$

To find time (in years):

$$\text{TIME} = \frac{\text{INTEREST}}{\text{PRINCIPAL} \times \text{RATE}}$$

EXAMPLE 1

Fill in the blanks.

	Interest	*Principal*	*Rate*	*Time*
a.	_____	$1,000	12%	2 yr
b.	$225	_____	9%	1 yr
c.	$112	$800	_____	2 yr
d.	$150	$500	6%	_____

Solution

a. $\text{INTEREST} = \$1,000 \times .12 \times 2$
$\qquad\qquad\quad = \$240$

b. $\text{PRINCIPAL} = \dfrac{\$225}{.09 \times 1}$
$\qquad\qquad\qquad = \$2,500$

c. $\text{RATE} = \dfrac{\$112}{\$800 \times 2}$
$\qquad\quad = .07 \text{ or } 7\%$

d. $\text{TIME} = \dfrac{\$150}{\$500 \times .06}$
$\qquad\quad = 5 \text{ yr}$ ∎

Study Example 1 to learn the *procedure* for working the different types of interest problems. Then study the rest of this section to learn the different types of applications you might encounter.

EXAMPLE 2

If a business borrows $18,000 and repays $26,100 in 3 years, what is the simple interest rate?

Solution

The time is 3 years; the principal is $18,000, and the interest is: $26,100 − $18,000 = $8,100.   So,

$$\text{RATE} = \frac{\text{INTEREST}}{\text{PRINCIPAL} \times \text{TIME}}$$

$$= \frac{\$8,100}{\$18,000 \times 3}$$

If your calculator has parentheses press:

$$\boxed{8100}\ \boxed{\div}\ \boxed{(}\ \boxed{18000}\ \boxed{\times}\ \boxed{3}\ \boxed{)}\ \boxed{=}$$

Display: ,15

If your calculator doesn't have parentheses, first calculate the value of the denominator and then divide it into the value of the numerator. You can also carry out two divisions to accomplish the same result:

$$\boxed{8100}\ \boxed{\div}\ \boxed{18000}\ \boxed{\div}\ \boxed{3}\ \boxed{=}$$

The simple interest is 15%.   ■

EXAMPLE 3

Suppose Blondie wants to save $3,720. If she has $3,000 and invests it at 8% simple interest, how long will it take her to obtain $3,720?

Solution

The interest she needs is: $3,720 − $3,000 = $720.   So,

$$\text{TIME} = \frac{\text{INTEREST}}{\text{PRINCIPAL} \times \text{RATE}}$$

$$= \frac{720}{3,000 \times .08}$$

Press: $\boxed{720}\ \boxed{\div}\ \boxed{3000}\ \boxed{\div}\ \boxed{.08}\ \boxed{=}$

Display: 3,

The time is 3 years.   ■

EXAMPLE 4

Example 3 was simplified to keep the arithmetic easy. But let's consider the cost of Blondie's mink coat in the cartoon in the last section—it cost $3,650. Now, how long (to the nearest day) would it take to save enough money for the purchase if Blondie invests her $3,000 at 8% simple interest?

Solution

Interest is: $3,650 − $3,000 = $650

$$\text{TIME} = \frac{\text{INTEREST}}{\text{PRINCIPAL} \times \text{RATE}}$$

$$= \frac{650}{3,000 \times .08}$$

On a calculator, the display for this calculation shows

$$2.7083333$$

To convert this to days, do not clear your calculator, but multiply the result by 360:

$$\boxed{\times}\ \boxed{360}\ \boxed{=}\ 975.$$

This is 975 (to the nearest day), which is 2 years and 255 days. ∎

EXAMPLE 5

In the opening situation, Jerry wanted to earn $5,000 per month interest. Suppose he figures the rate of interest at his retirement will be 15%. How much money does he need to have on deposit?

Solution

$$\text{PRINCIPAL} = \frac{\text{INTEREST}}{\text{RATE} \times \text{TIME}}$$

$$= \frac{5,000}{.15 \times \frac{1}{12}} \qquad \text{Time is 1 month, or } \tfrac{1}{12} \text{ year.}$$

Press: $\boxed{5000}\ \boxed{\div}\ \boxed{.15}\ \boxed{\div}\ \boxed{12}\ \boxed{1/x}\ \boxed{=}$

or $\boxed{(}\ \boxed{1}\ \boxed{\div}\ \boxed{12}\ \boxed{)}$

Display: $400000.$

He will need $400,000 on deposit to earn $5,000 per month interest. ∎

EXAMPLE 6

The $400,000 Jerry needs for retirement seems almost like an impossible amount! But Jerry does have a $25,000 inheritance. Using the simple interest formula, how long would it take to earn $375,000 interest at 15%?

Solution

$$\text{TIME} = \frac{\text{INTEREST}}{\text{PRINCIPAL} \times \text{RATE}}$$

$$= \frac{375,000}{25,000 \times .15}$$

Press: $\boxed{375000}$ $\boxed{\div}$ $\boxed{25000}$ $\boxed{\div}$ $\boxed{.15}$
Display: 100.

Jerry needs to wait 100 years for his $25,000 to grow to $400,000 at simple interest. ∎

Poor Jerry, since it is "only" 40 years until retirement! All is not lost, however, since it is possible that some of the interest Jerry earns also begins to earn interest. This is what is meant by *compound interest*, which we consider in the next section. *Preview:* Jerry will not only make enough to earn $5,000 per month, but can increase his monthly income upon retirement to over $80,000 per month! Wait for the next exciting installment to find out how.

Problem Set 9.2

Fill in the blanks in Problems 1–18.

	Interest	Principal	Rate	Time
1.	_____	$2,000	5%	1 yr
2.	$350	$2,500	7%	_____
3.	$66	_____	11%	6 yr
4.	$588	$700	_____	7 yr

	Interest	*Principal*	*Rate*	*Time*
5.	$432	$600	_____	8 yr
6.	$180	_____	15%	6 yr
7.	_____	$3,000	14%	3 yr
8.	$960	$1,500	16%	_____
9.	$432	_____	6%	8 yr
10.	$270	$300	_____	9 yr
11.	$624	$1,200	_____	4 yr
12.	$1,800	_____	9%	5 yr
13.	$5,850	$6,500	_____	5 yr
14.	$3,240	_____	12%	3 yr
15.	$320	$400	8%	_____
16.	$512.50	_____	$10\frac{1}{4}$%	180 days
17.	_____	$236.50	8.125%	255 days
18.	$58.03	$814.90	_____	213 days

19. If a business borrows $12,500 and repays $23,125 in five years, what is the simple interest rate?

20. Jerry wants to save $1,000. If he has $750 and invests it at 12% simple interest, how long (to the nearest day) will it take him to obtain $1,000?

21. If a friend tells you she earned $15,075 interest for the year on a 5 year certificate of deposit paying 15% simple interest, what is the amount of deposit?

22. If Rita receives $45.33 interest for a deposit earning 8% simple interest for 240 days, what is the amount of her deposit?

23. If John wants to retire with $10,000 per month, how much principal is necessary in order to generate this amount of monthly income if the interest rate is 15%?

24. If Melissa wants to retire with $50,000 per month, how much principal is necessary in order to generate this amount of monthly income if the interest rate is 12%?

25. If Jack wants to retire with $1,000 per month, how much principal is necessary in order to generate this amount of monthly income if the interest rate is 16%?

26. If John from Problem 23 has 30 years before retirement and has $100,000 to invest today, what interest rate does he need between now and then to achieve the amount of principal needed in Problem 23?

27. If Melissa from Problem 24 has 40 years before retirement and has $100,000 to invest today, what interest rate does she need to achieve the amount of principal needed in Problem 24?

28. If Jack from Problem 25 has 5 years before retirement and has $25,000 to invest today, what interest rate does he need to achieve the amount of principal needed in Problem 25?

9.3 Compound Interest

SITUATION Ron and Lorraine purchased a home some time ago and they still owe $20,000. Their payments are $195 per month and they have 10 years left to pay. They want to free themselves of the monthly payments by paying off the loan. They go to the bank and are told that to pay off the home loan they must pay $20,000 plus a "prepayment penalty" of 3%. (Older home loans often had prepayment clauses ranging from 1% to 5% of the amount to be paid off. Today, most lenders will waive this penalty because of the low interest rate of older loans. However, you must request the bank to do this.) This means that to own their home outright Ron and Lorraine would have to pay $20,600 ($20,000 + 3% of $20,000). They want to know if this is a wise financial move. In this section, Ron and Lorraine will learn about the effects of compound interest, and they will see that with their $20,600 they could not only pay off the home loan, but also end up with over $35,000! By the way, Jerry from the last section, will also find out how he can earn over $80,000 per month from his $25,000 inheritance.

Most banks do not pay interest according to the simple interest formula, but instead, after some period of time, they add the interest to the principal and then pay interest on this new, larger amount. When this is done, it is called **compound interest**.

EXAMPLE 1

Compare simple and compound interest for a $1,000 deposit at 8% for 3 years.

Solution

First, calculate the simple interest:

$$\text{INTEREST} = \$1,000 \times .08 \times 3$$
$$= \$240$$
$$\text{AMOUNT} = \$1,000 + \$240$$
$$= \$1,240$$

Next, assume that the interest is **compounded annually**. This means that the interest is added to the principal after one year has passed. This new amount then becomes the principal for the following year.

$$\textit{First year:} \quad \text{INTEREST} = \$1,000 \times .08 \times 1$$
$$= \$80$$
$$\text{AMOUNT} = \$1,000 + \$80$$
$$= \$1,080$$

So far, this is the same as simple interest. Now, however, in the second year the principal is $1,080 instead of $1,000.

$$\textit{Second year:} \quad \text{INTEREST} = \$1,080 \times .08 \times 1$$
$$= \$86.40$$
$$\text{AMOUNT} = \$1,080 + \$86.40$$
$$= \$1,166.40$$
$$\textit{Third year:} \quad \text{INTEREST} = \$1,166.40 \times .08 \times 1$$
$$\approx \$93.31$$
$$\text{AMOUNT} = \$1,166.40 + \$93.31$$
$$= \$1,259.71$$

Compound interest provides an additional $19.71 in the account. ∎

"The boy that by addition grows
And suffers no subtraction
Who multiplies the thing he knows
And carries every fraction
Who well divides the precious time
The due proportion giving
To sure success aloft will climb
Interest compound receiving."

The problem with compound interest is the difficulty in calculation. Notice that in order to simplify the calculations in Example 1, the variable representing time T was 1, and the process was repeated three times. Also notice that after the interest was found, it was added to the principal to be used in the next step. *This means that you can find the amount by adding 1 to the interest rate and then multiplying by the principal.*

EXAMPLE 2

If you deposit $5,000 in an account paying 10% interest for 5 years, compare the simple interest and the interest compounded annually.

Solution

Simple interest: INTEREST = $5,000 × .10 × 5 = $2,500
                      AMOUNT = $5,000 + $2,500 = $7,500

Compound interest:

$5,000	Original principal
× 1.10	1 + Interest rate
$5,500	Amount after 1 yr
× 1.10	
$6,050	Amount after 2 yr
× 1.10	
$6,655	Amount after 3 yr
× 1.10	
$7,320.50	Amount after 4 yr
× 1.10	
$8,052.55	Amount after 5 yr

Most calculators allow for what is called *constant multiplication*. This means the first number entered is automatically a constant multiplier. The constant multiplier in this example is 1.10. Enter the constant multiplier first; then multiply by the principal. Finally, press the equal sign for the number of years.

Press: 1.1   ×   5000   =   =   =
Display: 1.1 1.1 5000. 5500. 6050. 6655.
Press: =   =
Display: 7320.5 8052.55

The compound interest provides an additional $552.55 in the account.   ∎

If you analyze the steps for finding compound interest as shown in Example 2, you'll see that the work could be simplified by using an exponent.

Compound Interest Formula

$$\text{AMOUNT} = \text{PRINCIPAL}(1 + \text{RATE})^{\text{NUMBER OF YEARS}}$$

or in symbols,

$$A = P(1 + R)^t$$

where t is the time in years.

EXAMPLE 3

How much will you have if you invest \$2,000 for 20 years at 15% interest?

Solution

$$\text{AMOUNT} = 2{,}000(1 + .15)^{20}$$
$$= 2{,}000(1.15)^{20}$$

If your calculator has a button marked $\boxed{Y^x}$, then you can easily work this problem:

$$\boxed{2000}\;\boxed{\times}\;\boxed{1.15}\;\boxed{Y^x}\;\boxed{20}\;\boxed{=}$$

Display: 32733.075

You would have \$32,733.08. ∎

Many calculators don't have exponent keys, and even pressing the equal sign or repeat key twenty times is very tedious. For this reason, we use the compound interest table given in Table 9.2 (pages 334–335) for problems like Example 3.

EXAMPLE 4

Use Table 9.2 to answer Example 3.

Solution

First find the column labeled 15%. Then look down the column labeled N for $N = 20$ (the number of years), and look across that row. The entry is 16.366537. Multiply this number by the principal (this is easy on any four-function calculator): $16.366537 \times \$2{,}000 = \$32{,}733.07$ (rounded to the nearest cent). ∎

The next example illustrates the differences between simple interest and compound interest.

EXAMPLE 5

Consider the situation of Ron and Lorraine described at the beginning of this section. The first question is how much principal is necessary for Ron and Lorraine to generate $195 per month income (to make their mortgage payments) if they are able to invest it at 15% interest? This is an example of simple interest because the interest is withdrawn each month and is not left to accumulate:

$$\text{PRINCIPAL} = \frac{\text{INTEREST}}{\text{RATE} \times \text{TIME}}$$
$$= \frac{195}{.15 \times \frac{1}{12}}$$
$$= 15{,}600$$

This means that a deposit of $15,600 will be sufficient to pay off their home loan, with the added advantage that when their home is paid for they will still have their $15,600! Now, since they have $20,600 to invest, they can allow the left over $5,000 to grow at 15% interest. This is an example of compound interest because the interest is not withdrawn but accumulates. Since there are 10 years left, look at Table 9.2 for $N = 10$ under 15%:

$$\text{AMOUNT} = \$5{,}000 \times 4.045558$$
$$= \$20{,}227.79$$

This means that Ron and Lorraine have two options:

1. Use their $20,600 to pay off the loan.
2. Use their $20,600 as described in this example to pay off their loan. In 10 years they will have:

$20,227.79 + $15,600 = $35,827.79   ∎

EXAMPLE 6

In the last section, we described a situation where Jerry needed $400,000, and we found that at simple interest his

TABLE 9.2   Compound Interest Table
Compounded amount of $1.00 for *N* periods

N	1%	1½%	2%	2½%	3%	3½%	4%	N
1	1.010000	1.015000	1.020000	1.025000	1.030000	1.035000	1.040000	1
2	1.020100	1.030225	1.040400	1.050625	1.060900	1.071225	1.081600	2
3	1.030301	1.045678	1.061208	1.076891	1.092727	1.108718	1.124864	3
4	1.040604	1.061364	1.082432	1.103813	1.125509	1.147523	1.169859	4
5	1.051010	1.077284	1.104081	1.131408	1.159274	1.187686	1.216653	5
6	1.061520	1.093443	1.126162	1.159693	1.194052	1.229255	1.265319	6
7	1.072135	1.109845	1.148686	1.188686	1.229874	1.272279	1.315932	7
8	1.082857	1.126493	1.171659	1.218403	1.266770	1.316809	1.368569	8
9	1.093685	1.143390	1.195093	1.248863	1.304773	1.362897	1.423312	9
10	1.104622	1.160541	1.218994	1.280085	1.343916	1.410599	1.480244	10
11	1.115668	1.177949	1.243374	1.312087	1.384234	1.459970	1.539454	11
12	1.126825	1.195618	1.268242	1.344889	1.425761	1.511069	1.601032	12
13	1.138093	1.213552	1.293607	1.378511	1.468534	1.563956	1.665074	13
14	1.149474	1.231756	1.319479	1.412974	1.512590	1.618695	1.731676	14
15	1.160969	1.250232	1.345868	1.448298	1.557967	1.675349	1.800944	15
16	1.172579	1.268986	1.372786	1.484506	1.604706	1.733986	1.872981	16
17	1.164304	1.288020	1.400241	1.521618	1.652848	1.794676	1.947900	17
18	1.196147	1.307341	1.428246	1.559659	1.702433	1.857489	2.025817	18
19	1.208109	1.326951	1.456811	1.598650	1.753506	1.922501	2.106849	19
20	1.220190	1.346855	1.485947	1.638616	1.806111	1.989789	2.191123	20
25	1.282432	1.450945	1.640606	1.853944	2.093778	2.363245	2.665836	25
30	1.347849	1.563080	1.811362	2.097568	2.427262	2.806794	3.243398	30
35	1.416603	1.683881	1.999890	2.373205	2.813862	3.333590	3.946089	35
40	1.488864	1.814018	2.208040	2.685064	3.262038	3.959260	4.801021	40
45	1.564811	1.954213	2.437854	3.037903	3.781596	4.702359	5.841176	45
50	1.644632	2.105242	2.691588	3.437109	4.383906	5.584927	7.106683	50
55	1.728525	2.267944	2.971731	3.888773	5.082149	6.633141	8.646367	55
60	1.816697	2.443220	3.281031	4.399790	5.891603	7.878091	10.519627	60
65	1.909366	2.632042	3.622523	4.977958	6.829983	9.356701	12.798735	65
70	2.006763	2.835456	3.999558	5.632103	7.917822	11.112825	15.571618	70
75	2.109128	3.054592	4.415835	6.372207	9.178926	13.198550	18.945255	75
80	2.216715	3.290663	4.875439	7.209568	10.640891	15.675738	23.049799	80
85	2.329790	3.544978	5.382879	8.156964	12.335709	18.617859.	28.043605	85
90	2.448633	3.818949	5.943133	9.228856	14.300467	22.112176	34.119333	90
95	2.573538	4.114092	6.561699	10.441604	16.578161	26.262329	41.511386	95
100	2.704814	4.432046	7.244646	11.813716	19.218632	31.191408	50.504948	100

N	4½%	5%	5½%	6%	6½%	7%	7½%	N
1	1.045000	1.050000	1.055000	1.060000	1.065000	1.070000	1.075000	1
2	1.092025	1.102500	1.113025	1.123600	1.134225	1.144900	1.155625	2
3	1.141166	1.157625	1.174241	1.191016	1.207950	1.225043	1.242297	3
4	1.192519	1.215506	1.238825	1.262477	1.286466	1.310796	1.335469	4
5	1.246182	1.276282	1.306960	1.338226	1.370087	1.402552	1.435629	5
6	1.302260	1.340096	1.378843	1.418519	1.459142	1.500730	1.543302	6
7	1.360862	1.407100	1.454679	1.503630	1.553987	1.605781	1.659049	7
8	1.422101	1.477455	1.534687	1.593848	1.654996	1.718186	1.783478	8
9	1.486095	1.551328	1.619094	1.689479	1.762570	1.838459	1.917239	9
10	1.552969	1.628895	1.708144	1.790848	1.877137	1.967151	2.061032	10
11	1.622853	1.710339	1.802092	1.898299	1.999151	2.104852	2.215609	11
12	1.695881	1.795856	1.901207	2.012196	2.129096	2.252192	2.381780	12
13	1.772196	1.885649	2.005774	2.132928	2.267487	2.409845	2.560413	13
14	1.851945	1.979932	2.116091	2.260904	2.414874	2.578534	2.752444	14
15	1.935282	2.078928	2.232476	2.396558	2.571841	2.759032	2.958877	15
16	2.022370	2.182875	2.355263	2.540352	2.739011	2.952164	3.180793	16
17	2.113377	2.292018	2.484802	2.692773	2.917046	3.158815	3.419353	17
18	2.208479	2.406619	2.621466	2.854339	3.106654	3.379932	3.675804	18
19	2.307860	2.526950	2.765647	3.025600	3.308587	3.616528	3.951489	19
20	2.411714	2.653298	2.917757	3.207135	3.523645	3.869684	4.247851	20
25	3.005434	3.386355	3.813392	4.291871	4.827699	5.427433	6.098340	25
30	3.745318	4.321942	4.983951	5.743491	6.614366	7.612255	8.754955	30
35	4.667348	5.516015	6.513825	7.686087	9.062255	10.676581	12.568870	35
40	5.816365	7.039989	8.513309	10.285718	12.416075	14.974458	18.044239	40
45	7.248248	8.985008	11.126554	13.764611	17.011098	21.002452	25.904839	45
50	9.032636	11.467400	14.541961	18.420154	23.306679	29.457025	37.189746	50
55	11.256308	14.635631	19.005762	24.650322	31.932170	41.315001	53.390690	55
60	14.027408	18.679186	24.839770	32.987691	43.749840	57.946427	76.649240	60
65	17.480702	23.839701	32.464587	44.144972	59.941072	81.272861	110.039897	65
70	21.784136	30.426426	42.429916	59.075930	82.124463	113.989392	157.976504	70
75	27.146996	38.832686	55.454204	79.056921	112.517632	159.876019	226.795701	75
80	33.830096	49.561441	72.476426	105.795993	154.158907	224.234388	325.594560	80
85	42.158455	63.254353	94.723791	141.578904	211.211062	314.500328	467.433099	85
90	52.537105	80.730365	123.800206	189.464511	289.377460	441.102980	671.060065	90
95	65.470792	103.034676	161.801918	253.546255	396.472198	618.669748	963.394370	95
100	81.588518	131.501258	211.468636	339.302084	543.201271	867.716326	1383.077210	100

N	8%	9%	10%	11%	12%	13%	14%	N
1	1.080000	1.090000	1.100000	1.110000	1.120000	1.130000	1.140000	1
2	1.166400	1.188100	1.210000	1.232100	1.254400	1.276900	1.299600	2
3	1.259712	1.295029	1.331000	1.367631	1.404928	1.442897	1.481544	3
4	1.360489	1.411582	1.464100	1.518070	1.573519	1.630474	1.688960	4
5	1.469328	1.538624	1.610510	1.685058	1.762342	1.842435	1.925415	5
6	1.586874	1.677100	1.771561	1.870415	1.973823	2.081952	2.194973	6
7	1.713824	1.828039	1.948717	2.076160	2.210681	2.352605	2.502269	7
8	1.850930	1.992563	2.143589	2.304538	2.475963	2.658444	2.852586	8
9	1.999005	2.171893	2.357948	2.558037	2.773079	3.004042	3.251949	9
10	2.158925	2.367364	2.593742	2.839421	3.105848	3.394567	3.707221	10
11	2.331639	2.580426	2.853117	3.151757	3.478550	3.835861	4.226232	11
12	2.518170	2.812665	3.138428	3.498451	3.895976	4.334523	4.817905	12
13	2.719624	3.065805	3.452271	3.883280	4.363493	4.898011	5.492411	13
14	2.937194	3.341727	3.797498	4.310441	4.887112	5.534753	6.261349	14
15	3.172169	3.642482	4.177248	4.784589	5.473566	6.254270	7.137938	15
16	3.425943	3.970306	4.594973	5.310894	6.130394	7.067326	8.137249	16
17	3.700169	4.327633	5.054470	5.895093	6.866041	7.986078	9.276464	17
18	3.996019	4.717120	5.559917	6.543553	7.689966	9.024268	10.575169	18
19	4.315701	5.141661	6.115909	7.263344	8.612762	10.197423	12.055693	19
20	4.660957	5.604411	6.727500	8.062312	9.646293	11.523088	13.743490	20
25	6.848475	8.623081	10.834706	13.585464	17.000064	21.230542	26.461916	25
30	10.062657	13.267678	17.449402	22.892297	29.959922	39.115898	50.950159	30
35	14.785344	20.413968	28.102437	38.574851	52.799620	72.068506	98.100178	35
40	21.724521	31.409420	45.259256	65.000867	93.050970	132.781552	188.883514	40
45	31.920449	48.327286	72.890484	109.530242	163.987604	244.641402	363.679072	45
50	46.901613	74.357520	117.390853	184.564827	289.002190	450.735925	700.232988	50
55	68.913856	114.408262	189.059142	311.002466	509.320606	830.451725	1348.238807	55
60	101.257064	176.031292	304.481640	524.057242	897.596933	1530.053473	2595.918660	60
65	148.779847	270.845963	490.370725	883.066930	1581.872491	2819.024345	4998.219642	65
70	218.606406	416.730086	789.746957	1488.019132	2787.799828	5193.869624	9623.644985	70
75	321.204530	641.190893	1271.895371	2507.398773	4913.055841	9569.368113	18529.506390	75
80	471.954834	986.551668	2048.400215	4225.112750	8658.483100	17630.940454	35676.981807	80
85	693.456489	1517.932029	3298.969030	7119.560696	15259.205681	32483.864937	68692.981028	85
90	1018.915089	2335.526613	5313.022612	11996.873812	26891.934223	59849.415520	132262.467379	90
95	1497.120549	3593.497147	8556.676047	20215.430053	47392.776624	110268.668614	254660.083396	95
100	2199.761256	5529.040792	13780.612340	34064.175270	83522.265727	203162.874228	490326.238126	100

N	15%	16%	17%	18%	19%	20%	N
1	1.150000	1.160000	1.170000	1.180000	1.190000	1.200000	1
2	1.322500	1.345600	1.368900	1.392400	1.416100	1.440000	2
3	1.520875	1.560896	1.601613	1.643032	1.685159	1.728000	3
4	1.749006	1.810639	1.873887	1.938778	2.005339	2.073600	4
5	2.011357	2.100342	2.192448	2.287758	2.386354	2.488320	5
6	2.313061	2.436396	2.565164	2.699554	2.839761	2.985984	6
7	2.660020	2.826220	3.001242	3.185474	3.379251	3.583181	7
8	3.059023	3.278415	3.511453	3.758859	4.021385	4.299817	8
9	3.517876	3.802961	4.108400	4.435454	4.785449	5.159780	9
10	4.045558	4.411435	4.806828	5.233836	5.694684	6.191736	10
11	4.652391	5.117265	5.623989	6.175926	6.776674	7.430084	11
12	5.350250	5.936027	6.580067	7.287593	8.064242	8.916100	12
13	6.152788	6.885791	7.698679	8.599359	9.596448	10.699321	13
14	7.075706	7.987518	9.007454	10.147244	11.419773	12.839185	14
15	8.137062	9.265521	10.538721	11.973748	13.589530	15.407022	15
16	9.357621	10.748004	12.330304	14.129023	16.171540	18.488426	16
17	10.761264	12.467685	14.426456	16.672247	19.244133	22.186111	17
18	12.375454	14.462514	16.878953	19.673251	22.900518	26.623333	18
19	14.231772	16.776517	19.748375	23.214436	27.251616	31.948000	19
20	16.366537	19.460759	23.105599	27.393035	32.429423	38.337600	20
25	32.918953	40.874244	50.657826	62.668627	77.388073	95.396217	25
30	66.211772	85.849877	111.064650	143.370638	184.675312	237.376314	30
35	133.175523	180.314073	243.503474	327.997290	440.700607	590.668229	35
40	267.863546	378.721158	533.868713	750.378345	1051.667507	1469.771568	40
45	538.769269	795.443826	1170.479411	1716.683879	2509.650603	3657.261988	45
50	1083.657442	1670.703804	2566.215284	3927.356860	5988.913902	9100.438150	50
55	2179.622184	3509.048796	5626.293659	8984.841120	14291.666609	22644.802257	55
60	4383.998746	7370.201365	12335.356482	20555.139966	34104.970919	56347.514353	60
65	8817.787387	15479.940952	27044.628088	47025.180900	81386.522174	140210.646915	65
70	17735.720039	32513.164839	59293.941729	107582.222368	194217.025056	348888.956932	70
75	35672.867976	68288.754533	129998.886072	246122.063716	463470.508558	868147.369314	75
80	71750.879401	143429.715890	285015.802412	563067.660386	1106004.544354	2160228.462010	80
85	144316.646694	301251.407222	624882.050417	1288162.407650	2639317.992285	5375339.686589	85
90	290272.325206	632730.879999	1370022.050417	2947003.540121	6298346.150529	13375565.248934	90
95	583841.327636	1328951.025313	3003702.153303	6742030.208228	15030081.387632	33282686.520228	95
100	1174313.450700	2791251.199375	6585460.885837	15424131.905453	35867089.727971	82817974.522015	100

$25,000 would need 100 years to grow to $400,000 at 15% interest. Let's ask the same question again, but this time we'll allow the money to compound. Using Table 9.2, find 15% and $N = 40$:

$$\text{AMOUNT} = \$25,000 \times 267.863546$$
$$\approx \$6,696,589$$

He will have over $6\frac{1}{2}$ million dollars! This will generate him a lifetime income of about $83,707 per month (not the $5,000 a month he had hoped for).   ■

Remember that Table 9.2 gives you the amount present, whereas the simple interest formula gives the amount of interest.

Comparison of Simple and Compound Interest

	Interest, I	*Amount, A*
Simple interest formula	$I = P \times R \times T$ This is found first.	$A = P + I$ This is found after you've used the simple interest formula.
Compound interest	$I = A - P$ This is found after you've used Table 9.2.	**Use Table 9.2** This is found first.

This means that if you are working with compound interest and are asked for the amount of interest, you first use Table 9.2. Then you subtract the principal from the total.

Most banks will compound interest more frequently than once a year. A bank may pay interest:

1. **Semiannually**—this means every 180 days
2. **Quarterly**—this means every 90 days
3. **Monthly**—this means every 30 days
4. **Daily**—this means every day (for 360 days per year)

Paying 8% interest per year is the same as paying 4% interest semiannually or 2% interest every quarter. For interest compounded annually, the number of periods that interest is compounded is the same as the number of years. But if it is compounded semiannually, the number of periods that interest is compounded is double the number of years. Study Example 7.

EXAMPLE 7

Find the number of periods that interest is compounded and give the interest rate per period.

Compounding period	Yearly rate	Time	Number of periods	Rate per period
a. Annual	12%	3 yr	3	12%
b. Semiannual	12%	3 yr	6	6%
c. Quarterly	12%	3 yr	12	3%
d. Monthly	12%	3 yr	36	1%
e. Daily	12%	3 yr	1,080	.0$\overline{3}$%

To find the number of periods:
Multiply the time by   2 if semiannual
          by   4 if quarterly
          by 12 if monthly
          by 360 if daily
For part e,   3 × 360 = 1,080

To find the rate per period:
Divide the yearly rate by   2 if semiannual
            by   4 if quarterly
            by 12 if monthly
            by 360 if daily
For part e,   12 ÷ 360 = .033333 . . . ■

EXAMPLE 8

Use Table 9.2 to find the amount you would have if you invested $1,000 for ten years at 8% interest.

a. Compounded annually: $N = 10$ and the rate is 8%, so

we have

$$\text{AMOUNT} = \$1{,}000 \times 2.158925$$
$$\approx \$2{,}158.93$$

b. Compounded semiannually: $N = 20$ and the rate per period is 4%, so we have

$$\text{AMOUNT} = \$1{,}000 \times 2.191123$$
$$\approx \$2{,}191.12$$

c. Compounded quarterly: $N = 40$ and the rate per period is 2%, so we have

$$\text{AMOUNT} = \$1{,}000 \times 2.208040$$
$$\approx \$2{,}208.04 \quad\blacksquare$$

Table 9.2 is not extensive enough to find the monthly or daily compounding of $1,000 at 8% annual interest, but it will be sufficient for the examples in this book.

Problem Set 9.3

In Problems 1–6 compare the amount of simple interest and the interest if the investment is compounded annually.

1. $1,000 at 8% for five years
2. $5,000 at 10% for three years
3. $2,000 at 12% for three years
4. $2,000 at 12% for five years
5. $5,000 at 12% for twenty years
6. $1,000 at 14% for thirty years

In Problems 7–12 compare the amount you would have if the money were invested at simple interest or invested so that it compounded annually.

7. $1,000 at 8% for five years
8. $5,000 at 10% for three years
9. $2,000 at 12% for three years
10. $2,000 at 12% for five years
11. $5,000 at 12% for twenty years
12. $1,000 at 14% for thirty years

Fill in the blanks for Problems 13–26.

	Compounding period	Principal	Yearly rate	Time	Period rate	Number of periods	Entry in Table 9.2	Total amount	Total interest
13.	Annual	$1,000	9%	5 yr					
14.	Semiannual	$1,000	9%	5 yr					
15.	Annual	$500	8%	3 yr					
16.	Semiannual	$500	8%	3 yr					
17.	Quarterly	$500	8%	3 yr					
18.	Semiannual	$3,000	18%	3 yr					
19.	Quarterly	$5,000	18%	10 yr					
20.	Quarterly	$624	16%	5 yr					
21.	Quarterly	$5,000	20%	10 yr					
22.	Monthly	$350	12%	5 yr					
23.	Monthly	$4,000	24%	5 yr					
24.	Quarterly	$800	12%	90 days					
25.	Quarterly	$1,250	16%	450 days					
26.	Quarterly	$1,000	12%	900 days					

In Problems 27–29 find the interest using simple and compound interest.

27. $1,000 for 5 years at 10%: Compare simple interest and semiannual compounding.

28. $450 for 10 years at 8%: Compare simple interest and quarterly compounding.

29. $580 for 5 years at 12%: Compare simple interest and monthly compounding.

30. How much would you have in 5 years if you purchased a one thousand dollar 5 year savings certificate that paid 16% compounded quarterly?

31. What is the interest on $1,500 for 5 years at 12% compounded monthly?

32. What is the interest on $2,400 for 5 years at 12% compounded monthly?

33. What is the total amount after 15 years if you deposited $1,000 for your child's education and the interest was guaranteed at 16% compounded quarterly?

34. Conduct a survey of banks, savings and loan companies, and credit unions in your area. Prepare a report on the different types of savings accounts available and the interest rates they pay. Include methods of payment as well as the interest rates.

9.4 Inflation

> **SITUATION**   Carol and Jim have read the first sections of this chapter with interest (no pun intended). The results of compounding interest are truly remarkable and Jim is very interested in the possibilities. However, Carol reminds Jim of two factors which he is overlooking. The first is the effect of inflation, and the second is the effect of taxes. Carol says that Jerry's $80,000 per month looks great, but asks Jim to consider what the cost of living will be 40 years from now. "Just how is it possible to project into the future, especially with all the uncertainties of inflation?" asks Carol. "In fact," she continues, "Just last week an insurance man offered us a policy that will provide us with $100,000 per year income when we retire in 30 years. I don't even know if that's good or not!" In this section, Carol and Jim will see how to make projections regarding inflation.

Any discussion of compound interest is not complete without a discussion of inflation. The same procedure that we used to calculate compound interest can also be used to calculate the effects of inflation. The government releases reports of monthly and annual inflation rates. In 1981 the inflation rate was nearly 9%, but in 1982 it was less than 3%. Keep in mind that inflation rates can vary tremendously and that the best we can do in this section is to assume different constant inflation rates.

EXAMPLE 1

Carol and Jim are earning $20,000 per year and would like to know what salary they could expect in 20 years if inflation continues at 9%.

Solution

Use Table 9.2 and 9% for $N = 20$. The $20,000 is considered the principal.

$$\text{AMOUNT} = \$20,000 \times 5.604411$$
$$= \$112,088.22 \qquad \blacksquare$$

The answer to Example 1 means that if inflation continues at a constant 9%, in 20 years an annual salary of $112,000 will have about the same purchasing power as a salary of $20,000 today.

EXAMPLE 2

Jerry is presently earning $1,500 per month. If inflation continues at 9%, what is his projected salary in 40 years?

Solution

AMOUNT $= \$1{,}500 \times 31.409420$

$= \$47{,}114.13$ ——————Entry in Table 9.2 for 9%, $N = 40$ ■

Table 9.2 can be used in reverse to compare future amounts with comparable amounts today, as illustrated in Example 3.

EXAMPLE 3

If Jerry will have $83,000 per month in 40 years, what is the comparable amount today, if you assume a constant 9% inflation rate?

Solution

$$\text{AMOUNT} = \text{PRINCIPAL} \times \left(\begin{array}{c} \text{TABLE ENTRY} \\ 9\%, N = 40 \end{array} \right)$$

$$\$83{,}000 = \text{PRINCIPAL} \times 31.409420$$

Divide both sides by 31.409420 (you will need a calculator):

$$\boxed{83000} \; \boxed{\div} \; \boxed{31.409420} \; \boxed{=}$$

Display: 2642.5193

This tells Jerry that $83,000 compared with today's wages is about $2,643 per month. ■

The process illustrated in Example 3 is called finding the **present value** of some future amount.

EXAMPLE 4

Assuming a 9% inflation rate, compare $100,000 per year in 30 years with an equivalent amount today.

Solution

Find the present value of $100,000 at 9% in 30 years:

$$\text{AMOUNT} = \text{PRINCIPAL} \times \left(\begin{array}{c} \text{TABLE ENTRY} \\ 9\%, N = 30 \end{array} \right)$$

$$\$100,000 = \text{PRINCIPAL} \times 13.267678$$

$$\boxed{100000} \boxed{\div} \boxed{13.267678} \boxed{=}$$

Display: 7537.1139

The comparable amount today is about $7,537 per year.

∎

EXAMPLE 5

Rework Example 4 assuming a 12% inflation rate.

Solution

$$\$100,000 = \text{PRINCIPAL} \times 29.959922$$

PRINCIPAL ≈ 3,337.792 ⌐──── Table 9.2 entry for $N = 30$, 12%

The comparable amount today is about $3,338 per year.

∎

Problem Set 9.4

Find the cost of each item in Problems 1–6 in five years if you assume an inflation rate of 9%.

1. Cup of coffee, $0.65
2. Sunday paper, $0.75
3. Big Mac, $1.25
4. Gallon of gas, $1.35
5. Movie admission, $3.50
6. Record album, $5.95

Find the cost of each item in Problems 7–12 in ten years if you assume an inflation rate of 12%.

7. Textbook, $20
8. Electric bill, $65
9. Phone bill, $25
10. Pair of shoes, $45
11. New suit, $125
12. Monthly rent for a one bedroom apartment, $300

Find the cost of each item in Problems 13–18 in twenty years if you assume an inflation rate of 6%.

13. TV set, $600
14. Small car, $7,000
15. Large car, $12,000
16. One year tuition at a private college, $4,000
17. Yearly salary, $20,000
18. Average house, $65,000

Find the cost of each item in Problems 19–24 in twenty years if you assume an inflation rate of 9%.

19. Textbook, $20
20. Monthly rent for a one bedroom apartment, $300
21. Small car, $7,000
22. One year tuition at a private college, $4,000
23. Yearly salary, $15,000
24. Condominium, $50,000

Find the cost of each item in Problems 25–30 in twenty years if you assume an inflation rate of 12%.

25. Big Mac, $1.25   26. Large pizza, $9.00
27. Trans Am, $18,000   28. Yearly salary, $25,000
29. College tuition, $4,000   30. House, $85,000

Suppose an insurance agent offered you a policy that will provide you with a yearly income of $50,000 in 30 years. What is the comparative salary today assuming the inflation rates given in Problems 31–36?

31. 6%   32. 8%   33. 10%
34. 12%   35. 15%   36. 20%

37. Do you expect to live long enough to be a millionaire? Suppose your annual salary today is $19,000. If inflation continues at 12%, how long will it be before $19,000 increases to an annual salary of a million dollars?

38. Consult an almanac or some government sources, and then write a report on the current inflation rate. Project some of these results to the year of your own expected retirement.

9.5 Review Problems

Section 9.1   **1.** If $5,000 is invested at 14% simple interest for 5 years, what is the amount of interest?

Section 9.1   **2.** How much do you have in 5 years if you invest $5,000 at 14% simple interest?

Section 9.3   **3.** How much do you have in 5 years if you invest $5,000 at 14% compounded semiannually?

Section 9.3   **4.** If $5,000 is invested at 14% compounded quarterly for 5 years, what is the amount of interest?

Section 9.2   **5.** Suppose you have a $10,000 second mortgage on your home which has monthly payments of $125. How much do you need to invest at 17% in order to make this payment from the interest on your 17% investment?

Section 9.2   **6.** How long will it take $2,000 to double if it is invested at 12% simple interest?

Section 9.3   **7.** How long will it take $2,000 to double if it is invested at 12% compounded semiannually?

Section 9.3   **8.** How much will you have if you invest $10,000 in a 5 year certificate paying 18% compounded monthly?

Section 9.4   **9.** Suppose you just purchased a home for $80,000. If homes continue to increase in value at a 20% rate, what will this home be worth when it is paid off in 30 years?

Section 9.4   **10.** Suppose an insurance agent offered you a policy that will provide you with a yearly income of $25,000 in 30 years. What is the comparative income today, if you assume an inflation rate of 9%?

Consumer Applications   CHAPTER 10

10.1 Installment Buying

> SITUATION   Karen and Wayne need to buy a refrigerator because theirs just broke. Unfortunately, their savings account is depleted, and they will need to borrow money in order to buy a new one. The bank offers them a personal loan at 10% (discount rate), and the furniture store offers them an installment loan at 15% (add-on rate). Karen says that she has heard something about APR rates, but doesn't know what that means. Wayne says he thinks it has something to do with the prime rate, but he isn't sure what. In this section, Karen and Wayne will learn about installment loans, and, in particular, about add-on interest, discount interest, and APR (annual percentage rate). In the next section, they will learn about using a credit card (such as VISA or Mastercard) to finance a purchase.

There are two types of consumer credit that allow you to make installment purchases. The first, called **closed-end**, is the traditional installment loan. An **installment loan** is an agreement to pay off a loan or a purchase by making equal payments at regular intervals for some specified period of time. In this book it is assumed that all installment payments are made monthly. The loan is said to be *amortized* if it is completely paid off by these payments, and the payments are called **installments**. If the loan is not amortized, then there will be a larger final payment, called a *balloon payment*. With an *interest-only* loan, there is a monthly payment equal to the interest, with a final payment equal to the amount received when the loan was obtained.

The second type of consumer credit is called **open-end**, or credit card loan (Mastercard, VISA, Sears, and so on). It is also called *revolving credit*. This type of loan allows for purchases or cash advances up to a specified **line of credit**, and it has a flexible repayment schedule. This type of loan will be discussed in the next section.

There are two common methods for figuring interest, the **add-on interest** method and the **discount** method.

Add-On Interest

$$\text{INTEREST} = \text{PRINCIPAL} \times \text{RATE} \times \text{TIME}$$
$$\text{AMOUNT DUE} = \text{PRINCIPAL} + \text{INTEREST}$$
$$\text{AMOUNT OF EACH PAYMENT} = \frac{\text{AMOUNT DUE}}{\text{NUMBER OF PAYMENTS}}$$

You will notice that add-on interest is nothing more than the simple interest formula discussed in the last chapter.

EXAMPLE 1

Wayne and Karen want to purchase a refrigerator for $1,000, and they want to pay for it with installments over 3 years, or 36 months. The store has told them the add-on interest rate is 15%. What is the amount of each monthly payment?

Solution

$$
\begin{aligned}
\text{INTEREST} &= \$1,000 \times .15 \times 3 \\
&= \$450 \\
\text{AMOUNT DUE} &= \$1,000 + \$450 \\
&= \$1,450 \\
\text{MONTHLY PAYMENT} &= \frac{\$1,450}{36} \\
&= \$40.28
\end{aligned}
$$

When figuring monthly payments, always *round up* to the next higher penny (not to the nearest penny). ∎

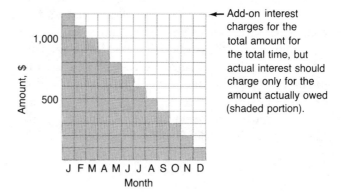

FIGURE 10.1   Amount owed on a
$1,200 loan repaid in 1 year by
making monthly payments of $100

The important thing to notice when working with add-on
interest is that you are paying more than the quoted interest
rate. The reason for this is that you are not keeping the
entire amount for the entire time. If the amount due is
$1,200 with monthly payments of $100, the amount owed as
compared to the amount on which you were charged interest
is compared in Figure 10.1.

In 1969 a *Truth-in-Lending Act* was passed by Congress; this
requires all lenders to state the true annual interest rate,
which is called the **annual percentage rate (APR)** and is
based on the actual amount owed (the shaded portion of
Figure 10.1). Regardless of the rate quoted, when you ask a
salesperson what the APR is, the law requires that you must
be told this rate. This regulation enables you to compare
interest rates *before* you sign a contract, which must state the
APR even if you haven't asked for it.

The actual formula for calculating the APR is too complicated to be useful to most business people, bankers, and
consumers. Instead, extensive tables are available from the
Federal Reserve System. A portion of one of these tables is
shown in Table 10.1.

**Procedure for Finding
the APR**

$$\left(\frac{\text{ACTUAL INTEREST}}{\text{AMOUNT BORROWED}} \right) \times 100 = \text{TABLE } 10.1 \text{ NUMBER}$$

TABLE 10.1 Annual Percentage Rates

Number of payments	10%	12%	14%	15%	16%	17%	18%	20%	25%	30%	35%	40%
6	2.94	3.53	4.12	4.42	4.72	5.02	5.32	5.91	7.42	8.93	10.45	11.99
12	5.50	6.62	7.74	8.31	8.88	9.45	10.02	11.16	14.05	16.98	19.96	22.97
18	8.10	9.77	11.45	12.29	13.14	13.99	14.85	16.52	20.95	25.41	29.96	34.59
24	10.75	12.98	15.23	16.37	17.51	18.66	19.82	22.15	28.09	34.19	40.44	46.85
30	13.43	16.24	19.10	20.54	21.99	23.45	24.92	27.89	35.49	43.33	51.41	59.73
36	16.16	19.57	23.04	24.80	26.57	28.35	30.15	33.79	43.14	52.83	62.85	73.20
42	18.93	22.96	27.06	29.15	31.25	33.37	35.51	39.85	51.03	62.66	74.74	87.24
48	21.74	26.40	31.17	33.59	36.03	38.50	41.00	46.07	59.15	72.83	87.06	101.82
60	27.48	33.47	39.61	42.74	45.91	49.12	52.36	58.96	76.11	94.12	112.94	132.51

The procedure for finding the APR using Table 10.1 is illustrated by Example 2.

EXAMPLE 2

Find the APR for the refrigerator loan described earlier in Example 1.

Solution

The actual interest is $450 and the amount borrowed is $1,000:

$$\left(\frac{\text{ACTUAL INTEREST}}{\text{AMOUNT BORROWED}}\right) \times 100 = \left(\frac{450}{1,000}\right) \times 100 = 45$$

Now, 45 is the number to be found in Table 10.1. Look across the row of the table corresponding to the number of payments (36 in this example): It is between 43.14 and 52.83, so the APR is between 25% and 30%. Since 45 is a little closer to 43.14, you could estimate the APR at 26%–27%. Remember, if we had the room to provide more extensive tables you could find the APR exactly, but for our purposes an estimate will be sufficient. ■

The second method for figuring the interest on installment purchases is called the **discount** method. This method is used to calculate interest on some personal loans.

With this method, the interest is subtracted from the amount you wish to borrow before you receive the money.

Discount Interest

> INTEREST = PRINCIPAL × RATE × TIME
>
> AMOUNT RECEIVED = PRINCIPAL − INTEREST
>
> $$\text{AMOUNT OF EACH PAYMENT} = \frac{\text{PRINCIPAL}}{\text{NUMBER OF PAYMENTS}}$$

Notice that this method is also based on the simple interest formula, and like the add-on method, the stated interest rate is much less than the actual APR.

EXAMPLE 3

Find the amount of payment and the APR for a 36 month loan of $1,400 discounted at 10%.

Solution

$$\text{INTEREST} = \$1,400 \times .10 \times 3$$
$$= \$420$$
$$\text{AMOUNT RECEIVED} = \$1,400 - \$420$$
$$= \$980$$
$$\text{AMOUNT OF PAYMENT} = \frac{\$1,400}{36}$$
$$= \$38.89$$

APR: $\left(\dfrac{\text{ACTUAL INTEREST}}{\text{AMOUNT BORROWED}}\right) \times 100 = \left(\dfrac{420}{980}\right) \times 100$
$$= 42.86$$

From Table 10.1, you can find this to be an APR of just under 25%.   ∎

Problem Set 10.1

Find the amount of interest and the monthly payment for each of the loans described in Problems 1–10.

1. Purchase a living room set for $1,200 at 12% add-on interest for 3 years.

2. Purchase a wood-burning stove for $650 at 11% add-on interest for 2 years.

3.   A 12 month loan of $1,200 with a discount rate of 12%

4.   A 4 year loan of $5,000 with a discount rate of 10%

5.   A $1,500 loan at 11% add-on rate for 2 years

6.   A $1,000 loan at 12% discount rate for 2 years

7.   A $1,000 loan at 12% add-on rate for 2 years

8.   A $2,400 loan at 15% add-on rate for 2 years

9.   A $2,400 loan at 15% discount rate for 2 years

10. A $4,500 loan at 18% add-on rate for 5 years

Find the APR for each loan listed in Problems 11–20.

11. See Problem 1.   **12.** See Problem 2.

13. See Problem 3.   **14.** See Problem 4.

15. See Problem 5.   **16.** See Problem 6.

17. See Problem 7.   **18.** See Problem 8.

19. See Problem 9.   **20.** See Problem 10.

21. A business needs to borrow $185,000 for one year and is offered the following loans:

Bank A: 14% add-on rate with monthly payments of $17,575
Bank B: 14% interest-only loan with monthly payments of $2,159 and a payment of $185,000 in 12 months
Bank C: 14% discount rate with monthly payments of $17,927

Which loan offers the lowest APR?

10.2 Buying with a Credit Card

SITUATION   Susan and Brad have a Mastercard, two VISA cards, three department store credit cards, and several oil company credit cards. "Enough!" says Brad, "We're through with all this plastic. From now on we'll limit ourselves to one card each." But which cards? Aren't all credit cards the same? In this section, Susan and Brad will see how to convert daily or monthly credit card rates into the APR. They will also learn about the three generally accepted methods for calculating the interest charges on credit cards. Even if the stated APR is the same, the charges can vary considerably. We'll also review how to select the cards that will cost you the least money for the way that you intend to use your own credit cards.

The most common types of open-end credit used today are through credit cards issued by VISA, Mastercard, department stores, and oil companies. Because it isn't necessary for you to apply for credit each time you want to charge an item, this type of credit is very convenient.

When comparing the interest rates on loans, you should always use the APR. In the last section we looked at add-on and discount rates. For comparison purposes you should convert these to their equivalent APR. Credit card rates, while based on the APR, are often stated as monthly or daily—rather than yearly—rates. Also, note that when dealing with credit cards, a 365 day year is used rather than a 360 day year.

EXAMPLE 1

Convert the given credit card rate to the APR.

a. $1\frac{1}{2}\%$ per month   Since there are 12 months per year, multiply by 12 to get: APR is 18%

b. Daily rate of .05753%   Multiply by 365:

$$\boxed{.05753} \;\boxed{\times}\; \boxed{365} \;\boxed{=}\; 20.99845$$

Rounding to the nearest tenth, this is equivalent to 21% APR.

■

In 1980 credit card rates changed considerably, and many cards introduced a yearly fee. Some are still free, others charge $1 every billing period the card is used by the consumer, but most now charge a $12 yearly fee. This affects the APR differently, depending on the amount the credit card is used during the year and the balance. If you always pay your credit card bills in full as soon as you receive them, then the card with no yearly fee would obviously be the best for you. On the other hand, if you use your card to stretch out your payments, then the APR is more important than this flat fee. For our purposes, we won't use this yearly fee in our calculations of APR on credit cards. In addition to fees, the interest rates or APRs for credit cards vary greatly. In May 1979 the average APR for credit cards was 17.06% and by

May 1980 the average rate had risen to 17.31%, but you could still find rates as low as 8% or as high as 24%. Remember that both VISA and Mastercard are issued by different banks, so the terms even in one locality can vary greatly. For example, I have the following credit cards:

VISA, Ranier National Bank	12% or 4% above the discount rate, whichever is greater (at this writing the discount rate is around 20%)
VISA, Citibank	18% on the first $500; 12% on amounts over $500
VISA, BankAmericard	18%
Mastercard, Wells Fargo Bank	20%

The finance charges can vary greatly even on credit cards showing the *same* APR, depending on the way the interest is calculated. There are three generally accepted methods for calculating these charges: **average daily balance, adjusted balance,** and **previous balance.**

Methods of Calculating Interest on Credit Cards

1. **Average daily balance:** Add the outstanding balance *each day* in the billing period. Divide by the number of days in the billing period. Multiply by the daily interest rate and then multiply by the number of days in the billing period.
2. **Adjusted balance:** The interest is calculated on the previous month's balance *less* credits and payments. This number is called the *adjusted balance*. Multiply the adjusted balance by the monthly interest rate.
3. **Previous balance:** Interest is calculated on the previous month's balance. Multiply the previous balance by the monthly interest rate.

In Examples 2–4 we will compare the finance charges on a $1,000 credit card purchase using these three different methods. Assume that a bill for $1,000 is received on April 1, and a payment is made. Then another bill is received on May 1, and this bill shows some finance charges. This finance charge is what we are calculating in Examples 2–4.

EXAMPLE 2

Calculate the interest on a $1,000 credit card bill showing an 18% APR using the average daily balance method. Assume that you sent a payment of $50 on April 1, and that it takes 10 days for this payment to be received and recorded.

Solution

For the first ten days the balance is $1,000; then the balance drops to $950. Add the balance for *each day*:

$$
\begin{aligned}
\text{10 days @ \$1,000} &= \$10,000 \\
\text{20 days @ \$950} &= \underline{\$19,000} \\
\text{Total:} & \quad\quad \$29,000
\end{aligned}
$$

Divide by the number of days (30 in April):

$$\$29,000 \div 30 = \$966.67$$

This is the average daily balance. Multiply the average daily balance by the daily interest rate and the number of days:

INTEREST = AVERAGE DAILY BALANCE × DAILY RATE × NUMBER OF DAYS

$$= \quad \$966.67 \quad\quad \times \quad \frac{.18}{365} \quad \times \quad 30$$

$$\boxed{966.67}\ \boxed{\times}\ \boxed{.18}\ \boxed{\div}\ \boxed{365}\ \boxed{\times}\ \boxed{30}\ \boxed{=}$$

Display: 14.301419

The interest is $14.30.   ∎

EXAMPLE 3

Calculate the interest on a $1,000 credit card bill showing an 18% APR using the adjusted balance method, and assume that $50 is sent and recorded by the due date.

Solution

First, find the adjusted balance:

Previous balance: $1,000
Less credits and payments: −   50
Adjusted balance: $950

INTEREST = ADJUSTED BALANCE × MONTHLY RATE
= $950   ×   .015
= $14.25   ↑

$1\frac{1}{2}$% per month = 18% APR

The interest is \$14.25. ■

EXAMPLE 4

Calculate the interest on a $1,000 credit card bill showing an 18% APR using the previous balance method and assuming $50 is sent and recorded by the due date.

Solution

The payment doesn't affect the finance charge for this month unless the bill is paid in full, so

INTEREST = PREVIOUS BALANCE × MONTHLY RATE
= $1,000   ×   .015
= $15

The interest is \$15. ■

You can sometimes make good use of credit cards by taking advantage of the period in which no finance charges are levied. That is, many credit cards charge no interest if you pay in full within a certain period of time (usually 20 or 30 days). On the other hand, if you borrow cash on your credit card, many credit cards have an additional charge for cash advances—and these can be as high as 4%. This 4% is *in addition to* the normal finance charges.

Problem Set 10.2

Convert each credit card rate in Problems 1–6 to the APR. (These rates were the listed finance charges on purchases under $500 on a Citibank VISA statement.)

1. Oregon, $1\frac{1}{4}$% per month
2. Arizona, $1\frac{1}{3}$% per month

3. New York, $1\frac{1}{2}\%$ per month
4. Tennessee, 0.02740% daily rate
5. Ohio, 0.02192% daily rate
6. Nebraska, 0.03014% daily rate
7. Repeat Example 2, but assume a $500 payment instead of a $50 payment.
8. Repeat Example 3, but assume a $500 payment instead of a $50 payment.
9. Repeat Example 4, but assume a $500 payment instead of a $50 payment.
10. Which method of calculating interest is most advantageous to the consumer?
11. Which method of calculating interest is least advantageous to the consumer?

Calculate the monthly finance charge for each credit card transaction in Problems 12–23. Assume that it takes 10 days for a payment to be received and recorded and that the month is 30 days long.

	Balance	Rate	Payment	Method
12.	$300	18%	$50	Previous balance
13.	$300	18%	$50	Adjusted balance
14.	$300	18%	$50	Average daily balance
15.	$300	18%	$250	Previous balance
16.	$300	18%	$250	Adjusted balance
17.	$300	18%	$250	Average daily balance
18.	$3,000	21%	$150	Previous balance
19.	$3,000	21%	$150	Adjusted balance
20.	$3,000	21%	$150	Average daily balance
21.	$3,000	21%	$1,500	Previous balance
22.	$3,000	21%	$1,500	Adjusted balance
23.	$3,000	21%	$1,500	Average daily balance

24. Most credit cards provide for a minimum finance charge of 50¢ per month. Suppose you buy a $30 item and make five payments of $5 and then pay the remaining balance. What is the APR for this purchase?
25. The finance charge statement on a Sears Revolving Charge Card statement is shown at the top of the next page. Why do you suppose the limitation on the 50¢ finance charge is from $1.00 to $33.00?

FINANCE CHARGE is based upon account activity during the billing period preceding the current billing period, and is computed upon the "previous balance" ("new balance" outstanding at the end of the preceding billing period) before deducting payments and credits or adding purchases made during the current billing period. The FINANCE CHARGE will be the greater of 50¢ (applied to previous balances of $1.00 through $33.00) or an amount determined as follows:

PREVIOUS BALANCE	PERIODIC RATE	ANNUAL PERCENTAGE RATE
$1,000 or under	1.5% per month	18%
Excess over **$1,000**	1.0% per month	12%

26. Suppose you wish to purchase a home air conditioner from Sears that costs $800. You have three credit plans from which to choose: (1) **revolving:** payments of $80 per month until paid (APR 18%); (2) **easy payment:** payments of $29.02 per month for 36 months; (3) **modernizing credit plan:** payments of $38.69 per month for 24 months. What is the APR for each of these choices, and which is the least expensive?

10.3 Buying a New Automobile

SITUATION   "When I get a full-time job, the first thing I'm going to do is buy a Trans Am," said Rod.

"You've got to be kidding," interrupted Steve, "the insurance will be exorbitant! Anyway, my Uncle Sid is a car dealer and he'll give you the best deal."

In this section, Rod will learn how to price an automobile, how to deal with a dealer, and finally, how to finance an automobile.

One of the things you are most likely to buy in your lifetime is an automobile. Unless you are paying cash, there are two prices that must be added together to give the true cost of your car—the cost of the car and the cost of the financing. The dealer with the best price for the car may not have the best price for financing, and you should shop for the two items separately.

You can do a great deal of "shopping around," but before you talk to a salesperson about price you should decide on the size, style, and accessories you want. An excellent source for helping you make these decisions is *Consumer Reports* (available in most libraries). Every April an issue of this magazine is devoted to new cars and car buying.

TABLE 10.2   Dealer's Cost for 1982 American Automobiles

Type of car	Dealer's cost factor	
	Average	Range
Small	.86	.84–.95
Midsize	.85	.79–.91
Large	.84	.77–.85

TABLE 10.3   Prices for Selected New 1982 Cars

Automobile	List	List factor	Options	Option factor	Destination charges
Chevrolet Chevette 4-door diesel hatchback	$6,727	.90	$570	.95	$250
Datsun 210 2-door hatchback	$6,149	.86	$910	.86	$250
Honda Civic 4-door	$6,749	.87	$200	.87	$250
Plymouth Gran Fury	$7,750	.91	$731	.85	$200
Dodge Mirada	$8,519	.85	$801	.85	$400
Volkswagen Jetta 2-door diesel	$9,020	.87	$1,095	.87	$300
Datsun Maxima wagon	$11,859	.86	$330	.86	$400
Oldsmobile 98 2-door	$12,117	.85	$1,520	.85	$400
Cadillac Seville	$23,433	.85	$1,760	.77	$400
Mercedes-Benz 300D	$28,483	.80	$1,250	.80	$400

When you're buying a car from a dealer, it's important to remember that the salesperson is a professional and you are an amateur. The price that appears on the window is not the price that is paid by most people who buy a car. It is necessary to negotiate a fair price. Table 10.2 will help you approximate the price the dealer paid for the car.

Specific prices and dealer cost factors are given in Table 10.3. The information in the box "How to Read an Automobile Price Sticker" (page 360) will help you decide on the true value of the car. Most dealers will sell the car for 10% over dealer cost, and high-volume dealers may settle for as little as 5% over their cost. You should be familiar with the following terminology regarding the price of a new automobile:

The **sticker price** is the price posted on the window of the car. This is the *maximum* price you would pay for the car, and it is the price the dealer would like to obtain for the car.

The **list price** is the same as the sticker price.

The **dealer cost** is the price that the dealer actually paid for the car. This is the *minimum* price you would pay for the car. However, since the dealer must cover overhead and profit to stay in business, a certain amount must be added to the cost.

5% price is the dealer cost plus 5% of the dealer's cost. To this amount you must add dealer preparation charges, destination charges, taxes, and license fees. This is the lowest price you should realistically expect to pay for a car.

10% price is the dealer cost plus 10% of the dealer's cost. To this amount you must add dealer preparation charges, destination charges, taxes, and license fees. This is the highest price the dealer should realistically expect you to pay for the car, but remember, the dealer is a professional negotiator. Don't give up easily in trying to get this price for the car you want to buy.

Pricing an Automobile

How to Read an Automobile Price Sticker
Example for a 1982 Monte Carlo

```
1982 MONTE CARLO 2DR SP     8,177.10
     AUTOMATIC TRANSMISSI  NO CHARGE
     POWER BRAKES          NO CHARGE
     POWER STEERING        NO CHARGE
     STEEL BELTED RADIAL   NO CHARGE
     ACOUSTIC INSULATION   NO CHARGE
     ROOF DRIP MOLDINGS    NO CHARGE
     WHEEL OPENING MOLDIN  NO CHARGE
     COLOR KEYED SEAT&SHO  NO CHARGE
     DUAL HORNS            NO CHARGE
     FRONT STABILIZER BAR  NO CHARGE
     ROCKER PANEL MOLDING  NO CHARGE
     BUMPER GUARDS         NO CHARGE
     COMPUTER COMMAND CON  NO CHARGE
      (GASOLINE ENGINE ON  NO CHARGE
AU3  POWER DOOR LOCK SYST     106.00
A01  TINTED GLASS              88.00
A31  POWER WINDOWS            165.00
BW2  DELUXE BODY SIDE MOL      57.00
B85  SIDE WINDOW SILL/HCC      45.00
B93  DOOR EDGE GUARDS          15.00
CD4  INTERMITTENT WINDSHI      47.00
C60  AIR CONDITIONING         675.00
D35  SPORT MIRRORS LH REM      48.00
K35  AUTO SPEED CONTROL W     155.00
LD5  3.8 LITER 2-BBL - V6  NO CHARGE
MX1  AUTOMATIC TRANSMISSI  NO CHARGE
N33  COMFORTILT STEERING       95.00
N95  WIRE WHEEL COVERS        153.00
P42  PUNCTURE SEALANT TIR     105.48
CXW  P195/75R-14 S/B RAD       62.00
TR9  AUXILIARY LIGHTING        38.00
U35  QUARTZ ELECTRIC CLOC      32.00
U69  AM/FM RADIO              165.00
YF5  CALIFORNIA EMISSION       65.00
77D  CFF3 REDWOOD CLOTH 5     133.00
77L  REDWOOD METALLIC      NO CHARGE
                            2,249.48
TRANSPORTATION AND HANDLING   399.00
TOTAL                      10,825.58
```

Manufacturer's suggested retail price: This is the **list price**, or **base price**, for the car. To find the **dealer's cost**, multiply this number by a **cost factor**. Since this is a midsize car, Table 10.2 tells you that the dealer's cost will vary between

$$\$8,177 \times .79 \approx \$6,460$$

and

$$\$8,177 \times .91 \approx \$7,441$$

However, if you don't know the exact cost factor (see Table 10.3), you can use the average given for the appropriate size in Table 10.2. For the Monte Carlo, this is .85:

$$\$8,177 \times .85 \approx \$6,950$$

Options: The **cost factor** for the options is often different from the cost factor for the car (see Table 10.3). For this Monte Carlo the options total $2,249. This puts the **dealer's cost** at about

$$\$2,249 \times .85 \approx \$1,912$$

Total of dealer's base price and options: For this Monte Carlo the total is about $8,862 (the sum of $6,950 + $1,912).

Add **destination charge**.

Dealer's cost is about $9,261. You can use this as a basis for making the dealer an offer on this car.

EXAMPLE 1

Determine the dealer's cost for a car with a sticker price of $8,350 and options of $1,685, destination charges of $200, and dealer preparation charges of $200. Suppose the cost factor is .81 for the car and .78 for the options.

Solution

$$\text{DEALER'S COST FOR THE CAR} = \$8,350 \times .81 = \$6,763.50$$
$$\text{DEALER'S COST FOR THE OPTIONS} = \$1,685 \times .78 = \$1,314.30$$
$$\text{DEALER'S COST} = \$6,763.50 + \$1,314.30$$
$$= \mathbf{\$8,077.80} \qquad \blacksquare$$

EXAMPLE 2

Make both a 5% and a 10% offer on the car described in Example 1.

Solution

From Example 1, the dealer's cost is $8,077.80.

5% offer: $8,077.80 × .05 = $403.89;
            $8,077.80 + $403.89 = **$8,481.69**

10% offer: $8,077.80 × .10 = $807.78;
            $8,077.80 + $807.78 = **$8,885.58**   ∎

If you wanted the car described in Example 1, you would expect to pay between $8,482 and $8,886 for the car. You would then need to add sales tax, license, destination charges, and dealer preparation charges to this price.

The rules for determining the finance rate for a car are the same as the general rules we discussed in Section 10.1. Many people are very careful about negotiating a good price for the car but then take the dealer's first offer on financing. The finance rates vary as much as the price of the car. As you can see from Table 10.4, there are several sources for automobile loans. Even though these rates are subject to frequent change, you can see that they vary considerably. In addition, if you have the dealer arrange the loan through a bank or through a company like the General Motors Acceptance Corporation, the dealer will usually raise the rate to cover the costs of arranging the loan. However, during periods of high interest rates, such loans may be offered at lower rates to stimulate sales. The most economical sources are listed first on Table 10.4.

TABLE 10.4   Sources for New Car Loans*

Passbook loans: Generally 2% above the prevailing rate of the passbook account

Life insurance loans: 8%–14%

Credit union loans: 12%–20%

Commercial bank loans: Vary considerably, depending on location; 15%–23%

Dealer loans: 13%–25%

Finance company loans: 15%–35%

Average of all loans: 17.36%

*Keep in mind that interest rates are constantly changing (on almost a daily basis). This table was prepared in January 1982 and will not provide the most up-to-date interest rate information. It will, however, show you the comparisons among various types of loans. That is, you would expect all of these rates to fluctuate, but the differences shown here should remain about the same.

Problem Set 10.3

Use Table 10.3 to determine the dealer's cost (to the nearest dollar) for the list price of each car given in Problems 1–10.

1. Chevrolet Chevette
2. Datsun 210
3. Honda Civic
4. Plymouth Gran Fury
5. Dodge Mirada
6. Volkswagen Jetta
7. Datsun Maxima
8. Oldsmobile 98
9. Cadillac Seville
10. Mercedes-Benz 300D

Use Table 10.3 to determine the dealer's cost (to the nearest dollar) for the options of each car named in Problems 11–20.

11. Chevrolet Chevette
12. Datsun 210
13. Honda Civic
14. Plymouth Gran Fury
15. Dodge Mirada
16. Volkswagen Jetta
17. Datsun Maxima
18. Oldsmobile 98
19. Cadillac Seville
20. Mercedes-Benz 300D

To determine a range of prices to pay for a car:

a. *Find the dealer's cost for the car (Problems 1–10).*
b. *Find the dealer's cost for the options (Problems 11–20).*
c. *Total parts a and b and add 5% to the total.*
d. *Total parts a and b and add 10% to the total.*
e. *Add the destination charge (varies depending on where you live, but for these problems use the amount shown in Table 10.3) to your answers to parts c and d. Round your answer to the nearest dollar.*

The results of part e give you a range of prices that you should reasonably expect to pay for a car. Carry out these steps in Problems 21–30 for each car listed.

21. Chevrolet Chevette   22. Datsun 210
23. Honda Civic   24. Plymouth Gran Fury
25. Dodge Mirada   26. Volkswagen Jetta
27. Datsun Maxima   28. Oldsmobile 98
29. Cadillac Seville   30. Mercedes-Benz 300D

31. A newspaper advertisement offers a $4,000 car for nothing down and 36 easy monthly payments of $141.62. What is the annual percentage rate?

32. A car dealer will sell you the $6,798 car of your dreams for $798 down and payments of $168.51 per month for 48 months. What is the annual percentage rate?

33. A car dealer carries out the following calculations:

LIST PRICE	$5,368
OPTIONS	$1,625
DESTINATION CGS	$ 200
SUBTOTAL	$7,193
TAX	$ 432
LESS TRADE-IN	$2,932
AMOUNT TO BE FINANCED	$4,693
8% interest for 48 months	$1,501.76
TOTAL	$6,194.76
MONTHLY PAYMENT	$ 129.06

What is the annual percentage rate charged?

10.4 Buying a Home

SITUATION   Dorothy and Wes have found the "home of their dreams" after having saved enough money for the down payment. They have never purchased a home before, or anything else involving so much money, and they are very excited. There seems to be so much to do, and so many new words are being used by the realtor and the banker that they are also confused. In this section, Dorothy and Wes will be introduced to the terminology of buying a house, as well as the steps in negotiating a sales contract, obtaining a mortgage, and settling escrow.

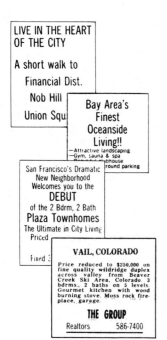

For many people, buying a home is the single most important financial step of a lifetime. There are three steps in buying and financing a home. The first is finding a home you would like to buy and then reaching an agreement with the seller on the price and terms. At this step you will need to negotiate a *purchase-and-sale agreement* or *sales contract*. The second step is finding a lender to finance the purchase. You will need to understand interest as you shop around to obtain the best terms. Finally, the third step is payment of certain *closing costs* in a process called *settlement* or **closing**, where the deal is finalized. The term **closing costs** refers to money exchanged at the settlement, above and beyond the down payment on the property. These costs can include an attorney's fee, the lender's administration fee, taxes to be held in escrow, and *points*, which we will define later in this section.

The first step in buying a home is finding a house you can afford and then coming to an agreement with the seller. A real estate agent can help you find the type of home you want for the money you can afford to pay. A great deal depends on the amount of down payment you can make. A useful rule of thumb to determine the monthly payment you can afford is given below:

1. Subtract any monthly bills (not paid off in the next 6 months) from your gross monthly income.
2. Multiply by 36%.

Your house payment should not exceed this amount. Another way of determining whether you can afford a house is to multiply your annual salary by 4; the purchase price should not exceed this amount. Today, the 36% factor is more common and is the one we will use in this book.

EXAMPLE 1

If Dorothy and Wes' gross monthly income is $2,500 and their current monthly payments on bills are $285, what is the maximum amount they should plan to spend for house payments?

Solution

Step 1. $2,500 - $285 = $2,215
Step 2. $2,215 × .36 = $797.40

The maximum house payment should be under $800 per month. ∎

After you have found a home that you can afford and that you would like to buy, you will be asked to sign a *sales contract* and make a deposit. The deposit is sometimes called *earnest money*, which simply shows you are serious about buying the house. The next step is to find a lender.

The amount you will need to borrow depends on the price of the home and the amount of your down payment. The down payment can be as large as you wish, but most lenders have some minimum down payment requirements depending on the appraised value of the property. Although it is sometimes possible to buy a home with a 5% down payment, the amount usually runs between 10% and 30% of the appraised value.

EXAMPLE 2

Determine the down payment for these homes.

a. Purchase price: $76,000
  10% down:

  DOWN PAYMENT = $76,000 × .10
         = **$7,600**

b. Purchase price: $120,000
  20% down:

  DOWN PAYMENT = $120,000 × .20
         = **$24,000** ∎

Next, a lender agrees to provide the money you need to buy a specific home. You, in turn, promise to repay the money based on terms set forth in an agreement, or loan contract, called a **mortgage**. As the borrower, you pledge your home as security. It remains pledged until the loan is paid off. If you fail to meet the terms of the contract, the lender has the right to **foreclose**, which means that the lender may take possession of the property.

There are three types of mortgage loans: (1) conventional loans made between you and a private lender; (2) VA loans made to eligible veterans (these are guaranteed by the Veteran's Administration, so they cost less than the other types of loans); and (3) FHA loans made by private lenders and insured by the Federal Housing Administration. Regardless of the type of loan you obtain, you will pay certain lender costs. By lender costs, we mean all the charges required by the lender: closing costs plus interest.

When you shop around for a loan, certain rates will be quoted:

1. **Interest rate:** This is the annual interest rate for the loan; it fluctuates on a daily basis. The APR, as stated on the loan agreement, will generally be just a little higher than the quoted interest rate. This is because the quoted interest rate is usually based on ordinary interest (360 day year) and the APR is based on exact interest (365 day year).
2. **Origination fee:** This is a one-time charge to cover the lender's administrative costs in processing the loan. It may be a flat $100–$300 fee, or it may be expressed as a percentage of the loan.
3. **Points:** This refers to discount points. This is a one-time charge used to adjust the yield on the loan to what the market conditions demand. It is used to offset constraints placed on the yield by state and federal regulators. Each point is equal to 1% of the amount of the loan.

EXAMPLE 3

If you are obtaining a $55,000 loan and the bank charges $5\frac{1}{2}$ points, what is the fee charged?

Solution

$$\text{FEE} = \$55,000 \times .055 = \mathbf{\$3,025}$$ ∎

It is desirable to include these three charges—interest rate, origination fee, and points—in one formula so that you can decide which lending institution is giving you the best terms on your loan. This could save you thousands of dollars over the life of your loan. The **comparison rate** formula shown in the box can be used to calculate the combined effects of these fees. Even though it is not perfectly accurate, it is usually close enough for meaningful comparisons among lenders.

$$\begin{pmatrix} \text{COMPARISON} \\ \text{RATE} \end{pmatrix} = \begin{pmatrix} \text{INTEREST} \\ \text{RATE} \end{pmatrix} + .125 \begin{pmatrix} \text{POINTS} + \dfrac{\text{ORIGINATION FEE}}{\text{AMOUNT OF LOAN}} \end{pmatrix}$$

Comparison Rate for Home Loans

EXAMPLE 4

Dorothy and Wes want to borrow $60,000. Lender *A* quotes 14.5% + 3 points + $250, and lender *B* quotes 14% + 5 points + $150. Which lender is making the better offer from Dorothy and Wes' viewpoint?

Solution

Calculate the comparison rate for both lenders (a calculator is helpful for this calculation).

Lender A:

$$\begin{pmatrix} \text{COMPARISON} \\ \text{RATE} \end{pmatrix} = \begin{pmatrix} \text{INTEREST} \\ \text{RATE} \end{pmatrix} + .125 \begin{pmatrix} \text{POINTS} + \dfrac{\text{ORIGINATION FEE}}{\text{AMOUNT OF LOAN}} \end{pmatrix}$$

$$= \quad .145 \quad + .125 \left(.03 + \frac{250}{60,000} \right)$$

Press: $\boxed{.145}$ $\boxed{+}$ $\boxed{.125}$ $\boxed{\times}$ $\boxed{(}$ $\boxed{.03}$ $\boxed{+}$ $\boxed{250}$ $\boxed{\div}$ $\boxed{60000}$ $\boxed{)}$ $\boxed{=}$

Display: .14927083

This is a comparative rate of 14.9%.

Lender B: COMPARISON RATE $= .14 + .125 \left(.05 + \dfrac{150}{60,000} \right)$

Press: $\boxed{.14}$ $\boxed{+}$ $\boxed{.125}$ $\boxed{\times}$ $\boxed{(}$ $\boxed{.05}$ $\boxed{+}$ $\boxed{150}$ $\boxed{\div}$ $\boxed{60000}$ $\boxed{)}$ $\boxed{=}$

Display: .1465625

This is a comparative rate of 14.7%.
Lender *B* is giving the better offer. ■

Now that you can find the comparable interest rates, the next step is to determine the amount of your mortgage payments. Most lenders assume a 30 year period, but you might be able to afford a 20 or 25 year loan, which would greatly reduce the total finance charge.

The amount of money you borrow, of course, depends on the cost of the home and the size of the down payment. Some financial counselors suggest making as large a down payment as you can afford, and others suggest making as small a down payment as is allowed. There are many factors, such as your tax bracket and your investment potential, that you will have to assess in order to determine how large a down payment you want to make. Table 10.5 may help you with this decision.

EXAMPLE 5

The home you select costs $68,500, and you pay 20% down. What is your down payment? How much will be financed?

Solution

$$\text{DOWN PAYMENT} = \$68,500 \times .20 = \mathbf{\$13,700}$$
$$\text{AMOUNT TO BE FINANCED} = \text{TOTAL AMOUNT} - \text{DOWN PAYMENT}$$
$$= \$68,500 - \$13,700$$
$$= \mathbf{\$54,800}$$ ■

We will use Table 10.6 to calculate the amount of monthly payments for a home loan. You will need to know the amount to be financed, the interest rate, and the length of time the loan is to be financed.

TABLE 10.5   Effect of Down Payment on the Cost of a $120,000 Home with Interest at 12%

Down payment	Percent	Monthly payment (principal and interest)			Total interest		
		20 years	*25 years*	*30 years*	*20 years*	*25 years*	*30 years*
$ 0	0%	$1,321.20	$1,263.60	$1,234.80	$317,160	$379,200	$444,480
6,000	5%	1,255.14	1,200.42	1,173.06	301,300	360,240	422,260
12,000	10%	1,189.08	1,126.44	1,111.32	285,440	341,280	400,030
24,000	20%	1,056.96	1,010.88	987.84	253,730	303,360	355,580
31,000	25%	979.89	937.17	915.81	235,230	281,240	329,660
36,000	30%	924.84	884.52	864.36	222,010	265,440	311,140

Note: Monthly payments are rounded to the nearest $1 and total interest to the nearest $10.

TABLE 10.6 Cost to Finance $1,000 for Selected Numbers of Years and Rates of Interest

Rate of interest	Financed for 20 years		Financed for 25 years		Financed for 30 years	
	Monthly cost	Total cost	Monthly cost	Total cost	Monthly cost	Total cost
6%	$ 7.17	$1,721	$ 6.45	$1,935	$ 6.00	$2,160
7%	7.76	1,862	7.07	2,121	6.66	2,398
8%	8.37	2,009	7.72	2,316	7.34	2,642
9%	9.00	2,160	8.40	2,520	8.05	2,898
10%	9.66	2,318	9.09	2,727	8.78	3,161
$10\frac{1}{2}$%	9.98	2,397	9.44	2,833	9.15	3,294
11%	10.32	2,477	9.80	2,940	9.52	3,427
$11\frac{1}{2}$%	10.66	2,559	10.16	3,049	9.90	3,564
12%	11.01	2,643	10.53	3,160	10.29	3,704
$12\frac{1}{2}$%	11.37	2,729	10.91	3,273	10.68	3,845
13%	11.52	2,765	11.28	3,384	11.07	3,985
$13\frac{1}{2}$%	12.04	2,890	11.66	3,498	11.46	4,126
14%	12.44	2,986	12.04	3,612	11.85	4,266
$14\frac{1}{2}$%	12.80	3,072	12.42	3,726	12.25	4,410
15%	13.17	3,161	12.81	3,843	12.65	4,554
$15\frac{1}{2}$%	13.54	3,250	13.20	3,960	13.05	4,698
16%	13.92	3,341	13.59	4,077	13.45	4,842
17%	14.67	3,521	14.38	4,314	14.26	5,134
18%	15.44	3,706	15.18	4,554	15.07	5,425
19%	16.21	3,890	15.98	4,794	15.89	5,720
20%	16.99	4,078	16.79	5,037	16.71	6,016

EXAMPLE 6

What is the monthly payment for a home loan of $54,800 if the interest rate is $14\frac{1}{2}$% financed for 30 years?

Solution

Notice that Table 10.6 is in terms of thousands of dollars. This means that you need to divide the amount to be financed by 1,000 (you can do this mentally). For this example, we get 54.8. Next, find the entry in Table 10.6 for $14\frac{1}{2}$% for 30 years; it shows a monthly cost of $12.25. This is $12.25 per thousand, so multiply by 54.8:

$$\$12.25 \times 54.8 = \$671.30$$

The monthly payments for this home are $671.30. ■

In 1982 the average price of a new home in the United States was $110,000 and the interest rate was 15%. There was a considerable amount of press coverage about how this was affecting the economy because most people couldn't afford a home. The concluding example of this section deals with these figures and asks several questions regarding these average prices and rates.

EXAMPLE 7

Suppose you want to purchase a home for $110,000 with a 30 year mortgage at 15% interest. Suppose that you can put 20% down.

a. What is the amount of the down payment?
b. What is the amount to be financed?
c. What are the monthly payments?
d. What is the total amount of interest paid on the 30 year loan?
e. What is the necessary monthly income for a family to be able to afford this home?

Solution

a. DOWN PAYMENT $= \$110,000 \times .20 = $ **$22,000**

b. AMOUNT TO BE FINANCED $=$ PRICE $-$ DOWN PAYMENT
$$= \$110,000 - \$22,000$$
$$= \mathbf{\$88,000}$$

c. MONTHLY PAYMENTS $= 88 \times \$12.65 = $ **$1,113.20**

d. The total interest can also be found from Table 10.6 as follows:

$$\text{TOTAL INTEREST} = \text{TOTAL COST} - \text{AMOUNT BORROWED}$$
$$= 88 \times \$4,554 - \$88,000$$
$$= \$400,752 - \$88,000$$
$$= \mathbf{\$312,752}$$

e. Remember, the amount of house payment should be no more than 36% of total monthly income. This means:

36% of WHAT NUMBER is $1,113.20?

$$\frac{36}{100} = \frac{1,113.20}{W}$$

Press: $\boxed{100}$ $\boxed{\times}$ $\boxed{1113.2}$ $\boxed{\div}$ $\boxed{36}$ $\boxed{=}$ 3092.2222

This means that the monthly salary would need to be about $3,092. (This is an annual salary of about $37,100.) ■

Problem Set 10.4

Determine the maximum monthly payment for a house (to the nearest dollar), given the information in Problems 1–6.

1. Gross monthly income $985; current monthly payments $147

2. Gross monthly income $1,240; current monthly payments $215

3. Gross monthly income $1,480; current monthly payments $520

4. Gross monthly income $2,300; current monthly payments $350

5. Gross monthly income $2,800; current monthly payments $540

6. Gross monthly income $3,600; current monthly payments $370

Determine the down payment for each home described in Problems 7–12.

7. $48,500;   5% down

8. $69,900;   20% down

9. $53,200;   10% down

10. $64,350;   10% down

11. $85,000;   20% down

12. $112,000;   30% down

What are the bank charges for the points indicated in Problems 13–18.

13. $48,500;   2 points

14. $69,900;   3 points

15. $53,200;   3 points

16. $64,350;   5 points

17. $85,000;   7 points

18. $112,000;   9 points

For Problems 19–24, determine the comparable interest rate (to two decimal places) for a $50,000 loan when the quoted information is given.

19. 11.5% + 2 pts + $450

20. 14.25% + 3 pts + $250

21. 13.7% + 7 pts + $250

22. 14.5% + 4 pts + $350

23. 15.3% + 5 pts + $150

24. 14.8% + 6 pts + $200

Use Table 10.6 to estimate the monthly payment for each loan described in Problems 25–36.

25. $48,500;   5% down;   20 years;   14%

26. $48,500;   5% down;   25 years;   14%

27. $48,500;   5% down;   30 years;   14%

28. $69,900;   20% down;   20 years; $14\frac{1}{2}$%

29. $69,900;   20% down;   25 years; $14\frac{1}{2}$%

30. $69,900;   20% down;   30 years; $14\frac{1}{2}$%

31. $85,000;   20% down;   20 years;   15%

32. $85,000;   20% down;   25 years;   15%

33. $85,000;   20% down;   30 years;   15%

34. $112,000;   30% down;   20 years;   15%

35. $112,000;   30% down;   25 years;   15%

36. $112,000;   30% down;   30 years;   15%

37. Suppose you want to purchase a home for $75,000 with a 30 year mortgage at 15% interest. Suppose that you can put 20% down.
  a. What is the amount of the down payment?
  b. What is the amount to be financed?
  c. What are the monthly payments?
  d. What is the total amount of interest paid on the 30 year loan?
  e. What is the necessary monthly income for you to be able to afford this home?

38. Suppose you want to purchase a home for $100,000 with a 30 year mortgage at 15% interest, and that you can put 20% down. Answer the questions asked in Problem 37.

39. Suppose you want to purchase a home for $150,000 with a 30 year mortgage at 15% interest, and that you can put 25% down. Answer the questions asked in Problem 37.

40. Do some research and write a report about buying a home in your community. Interview a real estate agent, various lenders, and some escrow agents. Be sure to include specific information about the local customs and interest rates.

Closing

Buying a home and getting a loan to finance it involve filling out a number of papers before the property officially becomes yours.

Between the signing of the sales contract and the closing of the loan, three things usually need to be done: (1) the property will have to be appraised, (2) evidence of title will have to be obtained, and (3) a survey of the land will need to be made. At the time the loan is closed, the note and mortgage will need to be signed and the deed transferring the title to the buyer will need to be executed and then recorded.

In order to carry out these tasks, you will need to select a settlement agent. Settlement practices vary from locality to locality and even within the same county or city. In various areas, settlements are conducted by lending institutions, title insurance companies, escrow companies, real estate brokers, or attorneys for the buyer or seller. By investigating and comparing practices and rates, you may find that the first suggested settlement agent may not be the least expensive. You might save money by taking the initiative in arranging for settlement and selecting the firm and location that meets your needs.

Figure 10.2 shows the form developed by HUD (U.S. Department of Housing and Urban Development) that must be filled out by the person who conducts the settlement meeting. This statement must be delivered or mailed to you before settlement, unless you waive that right.

HUD-1 REV. 5/76

FORM APPROVED
OMB NO. 63-R-1501

A.	B. TYPE OF LOAN
U. S. DEPARTMENT OF HOUSING AND URBAN DEVELOPMENT SETTLEMENT STATEMENT	1. ☐ FHA 2. ☐ FmHA 3. ☐ CONV. UNINS. 4. ☐ VA 5. ☐ CONV. INS. 6. FILE NUMBER: 7. LOAN NUMBER 8. MORTGAGE INSURANCE CASE NUMBER:

C. NOTE: *This form is furnished to give you a statement of actual settlement costs. Amounts paid to and by the settlement agent are shown. Items marked "(p.o.c.)" were paid outside the closing; they are shown here for informational purposes and are not included in the totals.*

D. NAME OF BORROWER:	E. NAME OF SELLER:	F. NAME OF LENDER:
G. PROPERTY LOCATION:	H. SETTLEMENT AGENT: PLACE OF SETTLEMENT	I. SETTLEMENT DATE:

J. SUMMARY OF BORROWER'S TRANSACTION		K. SUMMARY OF SELLER'S TRANSACTION	
100. GROSS AMOUNT DUE FROM BORROWER:		400. GROSS AMOUNT DUE TO SELLER:	
101. Contract sales price		401. Contract sales price	
102. Personal property		402. Personal property	
103. Settlement charges to borrower *(line 1400)*		403.	
104.		404.	
105.		405.	
Adjustments for items paid by seller in advance		*Adjustments for items paid by seller in advance*	
106. City/town taxes to		406. City/town taxes to	
107. County taxes to		407. County taxes to	
108. Assessments to		408. Assessments to	
109.		409.	
110.		410.	
111.		411.	
112.		412.	
120. GROSS AMOUNT DUE FROM BORROWER		420. GROSS AMOUNT DUE TO SELLER	
200. AMOUNTS PAID BY OR IN BEHALF OF BORROWER:		500. REDUCTIONS IN AMOUNT DUE TO SELLER:	
201. Deposit or earnest money		501. Excess deposit *(see instructions)*	
202. Principal amount of new loan(s)		502. Settlement charges to seller *(line 1400)*	
203. Existing loan(s) taken subject to		503. Existing loan(s) taken subject to	
204.		504. Payoff of first mortgage loan	
205.		505. Payoff of second mortgage loan	
206.		506.	
207.		507.	
208.		508.	
209.		509.	
Adjustments for items unpaid by seller		*Adjustments for items unpaid by seller*	
210. City/town taxes to		510. City/town taxes to	
211. County taxes to		511. County taxes to	
212. Assessments to		512. Assessments to	
213.		513.	
214.		514.	
215.		515.	
216.		516.	
217.		517.	
218.		518.	
219.		519.	
220. TOTAL PAID BY/FOR BORROWER		520. TOTAL REDUCTION AMOUNT DUE SELLER	
300. CASH AT SETTLEMENT FROM/TO BORROWER		600. CASH AT SETTLEMENT TO/FROM SELLER	
301. Gross amount due from borrower *(line 120)*		601. Gross amount due to seller *(line 420)*	
302. Less amounts paid by/for borrower *(line 220)*	( )	602. Less reductions in amount due seller *(line 520)*	( )
303. CASH (☐ FROM) (☐ TO) BORROWER		603. CASH (☐ TO) (☐ FROM) SELLER	

Page 2

L. SETTLEMENT CHARGES	PAID FROM BORROWER'S FUNDS AT SETTLEMENT	PAID FROM SELLER'S FUNDS AT SETTLEMENT
700. TOTAL SALES/BROKER'S COMMISSION based on price $ @ % =		
Division of Commission (line 700) as follows:		
701. $ to		
702. $ to		
703. Commission paid at Settlement		
704.		
800. ITEMS PAYABLE IN CONNECTION WITH LOAN		
801. Loan Origination Fee %		
802. Loan Discount %		
803. Appraisal Fee to		
804. Credit Report to		
805. Lender's Inspection Fee		
806. Mortgage Insurance Application Fee to		
807. Assumption Fee		
808.		
809.		
810.		
811.		
900. ITEMS REQUIRED BY LENDER TO BE PAID IN ADVANCE		
901. Interest from to @ $ /day		
902. Mortgage Insurance Premium for months to		
903. Hazard Insurance Premium for years to		
904. years to		
905.		
1000. RESERVES DEPOSITED WITH LENDER		
1001. Hazard insurance months @ $ per month		
1002. Mortgage insurance months @ $ per month		
1003. City property taxes months @ $ per month		
1004. County property taxes months @ $ per month		
1005. Annual assessments months @ $ per month		
1006. months @ $ per month		
1007. months @ $ per month		
1008. months @ $ per month		
1100. TITLE CHARGES		
1101. Settlement or closing fee to		
1102. Abstract or title search to		
1103. Title examination to		
1104. Title insurance binder to		
1105. Document preparation to		
1106. Notary fees to		
1107. Attorney's fees to		
(includes above items numbers;)		
1108. Title insurance to		
(includes above items numbers;)		
1109. Lender's coverage $		
1110. Owner's coverage $		
1111.		
1112.		
1113.		
1200. GOVERNMENT RECORDING AND TRANSFER CHARGES		
1201. Recording fees: Deed $; Mortgage $; Releases $		
1202. City/county tax/stamps: Deed $; Mortgage $		
1203. State tax/stamps: Deed $; Mortgage $		
1204.		
1205.		
1300. ADDITIONAL SETTLEMENT CHARGES		
1301. Survey to		
1302. Pest inspection to		
1303.		
1304.		
1305.		
1400. TOTAL SETTLEMENT CHARGES *(enter on lines 103, Section J and 502, Section K)*		

HUD-1 REV. 5/76

FIGURE 10.2 HUD settlement statement

10.5 Review Problems

Section 10.1

In Problems 1 and 2 assume you want to purchase an $1,800 ring with no money down and monthly payments for three years.

1. Use a simple add-on interest rate of 15% to calculate the monthly payments.

2. Use Table 10.1 on page 349 to estimate the APR for the loan described in Problem 1.

Section 10.2

Suppose you purchase a $550 TV set on a VISA credit card with an APR of 18% and receive the bill. You then make the payment indicated in Problems 3–5. Show the total interest charged on your next statement if your credit card uses the indicated method of calculating interest.

3. Average daily balance (allow 10 days for payment and a 30 day month):
   a. $50 payment   b. $500 payment

4. Adjusted balance:
   a. $50 payment   b. $500 payment

5. Previous balance:
   a. $50 payment   b. $500 payment

Section 10.3

6. Suppose the car you want has a list price of $11,640 with options of $1,350. Assume that the cost factor for the car is .79 and for the options is .81. What is the dealer's cost for the car with options?

Section 10.3

7. a. Make a 5% offer for the car described in Problem 6.
   b. Make a 10% offer for the car described in Problem 6.

Section 10.4

For Problems 8–10 assume that you select a home that costs $85,000 and you will finance the home for 30 years.

8. Determine the following down payments:
   a. 10% down   b. 20% down   c. 25% down

9. Use Table 10.6 on page 369 and Problem 8 to estimate the monthly payments for each 30 year loan described below:
   a. 10% down; 15% interest
   b. 20% down; 14% interest
   c. 25% down; 14% interest

10. Suppose you decide to pay 30% down. Lender *A* quotes 14% + 5 points + $200 and lender *B* quotes $14\frac{1}{4}\%$ + 3 points + $300. Which lender is giving you the better offer, and by how much?

Probability

CHAPTER 11

11.1 Probability Experiments

SITUATION   We see examples of probability every day. Weather forecasts, stock market analysis, contests, children's games, television game shows, and gambling all involve the ideas of probability. Probability is the mathematics of uncertainty. "Wait a minute," interrupts Sammy, "I thought mathematics was absolute and that there was no uncertainty about it." Well, Sammy, suppose you are playing a game of Monopoly. Do you know what you'll land on in your next turn? "No, but neither do you!" That's right, but mathematics can tell us what property you are *most likely* to land on. There is nothing uncertain about the probability, but rather probability is a way of describing the uncertain. For example, what is the probability of tossing a coin and obtaining a head? "That's easy, it's one-half," answers Sammy. Right, but *why* do you say it is one-half? In the following section, we'll investigate the definition of probability as well as some of the procedures in dealing with probability. But first, in this section, we'll describe probability in terms of actual physical experiments. Then you'll be asked to make some guesses (conjectures) about probability.

Six experiments are described in this section. With these experiments we'll be able to introduce you to some of the terminology of probability. No special knowledge is assumed. You are simply asked to carry out these experiments honestly and make some observations. You will need three coins, a pair of dice, and three 3″ × 5″ cards (or any other convenient size) to complete these experiments.

Experiment 1: Tossing a Coin

In this experiment, we flip a coin 50 times. Make sure that, each time the coin is flipped, it rotates several times in the air and lands on a table or on the floor. Keep a record similar to that in Table 11.1. The figures in each table of this section are the result of an actual experiment.

Repeat the experiment. The results are called **data**. We will study the analysis of data in the next chapter. There are many ways to represent data, and you should choose the one most convenient for your purposes. For example, the findings in Table 11.1 could be recorded more conveniently by using tally marks, as in Table 11.2.

TABLE 11.1   Outcomes of Tossing a Single Coin

Number of throw	Outcome	Number of throw	Outcome	Number of throw	Outcome	Number of throw	Outcome	Number of throw	Outcome
1	H	11	H	21	H	31	H	41	T
2	T	12	T	22	T	32	T	42	H
3	H	13	T	23	T	33	H	43	H
4	H	14	T	24	H	34	H	44	T
5	T	15	H	25	H	35	H	45	H
6	T	16	T	26	H	36	T	46	H
7	T	17	T	27	T	37	T	47	H
8	T	18	H	28	T	38	T	48	T
9	H	19	H	29	H	39	T	49	H
10	H	20	T	30	T	40	T	50	T

Totals: Heads, 24;   Tails, 26

TABLE 11.2   Outcomes of Tossing a Single Coin—Tally Method

Outcome	Number of occurrences	Total	Percent
Heads	~~HHT~~ ~~HHT~~ ~~HHT~~ ~~HHT~~ IIII	24	48%
Tails	~~HHT~~ ~~HHT~~ ~~HHT~~ ~~HHT~~ ~~HHT~~ I	26	52%

Can you formulate any conjectures concerning these data? Find the percent of heads and tails for your data. For our example, we have:

$$\frac{24}{50} \text{ heads} \qquad \frac{26}{50} \text{ tails}$$

$$.48 \text{ heads} \qquad .52 \text{ tails}$$

$$48\% \text{ heads} \qquad 52\% \text{ tails}$$

From our experience, we see that this is what we would expect; that is, heads have about a "50–50 chance" of occurrence. If we write the percent as a fraction, we say that the **probability** of heads occurring is about $\frac{1}{2}$. This means that *if we repeated the experiment a large number of times,* we could expect heads to occur about $\frac{1}{2}$ the time.

Experiment 2: Tossing Three Coins

In this experiment, we flip three coins simultaneously 50 times. We are interested in the probability of obtaining 3

heads, 2 heads and 1 tail, 1 head and 2 tails, and 3 tails. The procedure is to flip the coins simultaneously and let them fall to the floor (it might be easier to flip them if you place them in a cup). Record the result in each case, as in Table 11.3.

TABLE 11.3   Outcomes of Flipping Three Coins Simultaneously

Outcome	First trial	Second trial	Third trial	Total	Percent of occurrence
Three heads	7	5	6	18	12%
Two heads and one tail	15	19	22	56	$37\frac{1}{3}\%$
Two tails and one head	24	22	13	59	$39\frac{1}{3}\%$
Three tails	4	4	9	17	$11\frac{1}{3}\%$
Totals	50	50	50	150	100%

Repeat the experiment three times, and record the totals. Next, find the percent of occurrence by dividing the number of times the event occurred by the total number of occurrences. Can you make any conjectures? It appears that these outcomes are *not* equally likely to occur. Can you make a conjecture about the probability of three heads occurring?

Notice some properties of probability. If we specify the **probability** of three heads as the **percent of occurrence** (or as a fraction), we see that this percent must be between 0% and 100%. (Why?) Therefore, if we express the probability as a fraction, the fraction will be between 0 and 1. The closer the fraction is to 0, the less likely the event is to occur; the closer it is to 1, the more likely the event is to occur. An event that can never occur has probability 0, and an event that is certain to occur has probability 1.

Experiment 3: Rolling One Die*

In this experiment, we roll a single die 50 times and record the results, as in Table 11.4.

Repeat the experiment three times. Find the frequency of each outcome, and then compute the percent of occurrence of each.

*The word *die* is the singular form of *dice*. All dice used in this text are fair; that is, they are not loaded unless so designated.

TABLE 11.4 Outcomes of Rolling One Die

Outcome	First trial	Second trial	Third trial	Total	Percent of occurrence
1	10	4	10	24	16%
2	8	5	6	19	$12\frac{2}{3}\%$
3	9	12	7	28	$18\frac{2}{3}\%$
4	6	11	11	28	$18\frac{2}{3}\%$
5	5	7	11	23	$15\frac{1}{3}\%$
6	12	11	5	28	$18\frac{2}{3}\%$
Totals	50	50	50	150	100%

Can you make a conjecture about the probability of each? Notice that the average of each is close to $\frac{1}{6} = 16\frac{2}{3}\%$. That is, for one die, it appears that all outcomes are equally likely.

Experiment 4: Rolling a Pair of Dice

In this experiment, we roll two dice. Here, the possible sums are 2, 3, 4, 5, 6, 7, 8, 9, 10, 11, and 12. In each of these examples, we have made a list of all possible outcomes. This list of outcomes is called the **sample space** of the experiment. Roll the pair of dice 50 times.

TABLE 11.5 Outcomes of Rolling a Pair of Dice

Outcome	First trial	Second trial	Third trial	Total	Percent of occurrence
2	1	0	1	2	$1\frac{1}{3}\%$
3	3	2	2	7	$4\frac{2}{3}\%$
4	5	4	5	14	$9\frac{1}{3}\%$
5	2	7	6	15	10%
6	9	8	6	23	$15\frac{1}{3}\%$
7	8	11	7	26	$17\frac{1}{3}\%$
8	6	8	9	23	$15\frac{1}{3}\%$
9	6	3	7	16	$10\frac{2}{3}\%$
10	4	4	2	10	$6\frac{2}{3}\%$
11	5	2	5	12	8%
12	1	1	0	2	$1\frac{1}{3}\%$
Totals	50	50	50	150	100%

Repeat the experiment three times. Now total to find the frequency of each outcome. Next, compute the percent of occurrence for each.

Can you make a conjecture about the probability of each? In this problem, it is clear that there is no tendency for each outcome to be equally likely, as in the previous experiment. In fact, it seems as if there is a strong tendency for the middle outcomes to be more probable. Can you explain why?

Experiment 5: Tossing a Coin and Rolling a Die

In this experiment, we simultaneously toss a coin and roll a die. Repeat the experiment 50 times and tabulate the results, as shown in Table 11.6.

TABLE 11.6 Outcomes of Tossing a Coin and Rolling a Die

Outcome	First trial	Second trial	Third trial	Total	Percent of occurrence
1H	5	4	3	12	8%
1T	6	3	3	12	8%
2H	1	8	2	11	$7\frac{1}{3}\%$
2T	1	3	7	11	$7\frac{1}{3}\%$
3H	4	5	4	13	$8\frac{2}{3}\%$
3T	3	2	7	12	8%
4H	8	5	1	14	$9\frac{1}{3}\%$
4T	7	3	1	11	$7\frac{1}{3}\%$
5H	1	1	9	11	$7\frac{1}{3}\%$
5T	8	5	3	16	$10\frac{2}{3}\%$
6H	4	5	6	15	10%
6T	2	6	4	12	8%
Totals	50	50	50	150	100%

After performing this experiment, total and find the frequency and percent of each outcome. Can you make a conjecture about the probabilities involved? We saw that the probability of heads should be $\frac{1}{2}$, and the probability of rolling a 3 on a die is $\frac{1}{6}$. How does the probability of obtaining 3H compare with the product of these probabilities?

(Probability of 3) × (Probability of H)

$$\frac{1}{6} \quad \times \quad \frac{1}{2} \quad = \frac{1}{12}$$

This result follows when the events are **independent**—that is, when they are completely unrelated (the occurrence of one event does not affect the occurrence of the other event).

Suppose we reclassify this problem to determine the probability of obtaining an even number or a 3. This result can be tabulated as shown in Table 11.7. If we obtained an even number or a 3 in Table 11.6, we classify the result as a Success in Table 11.7; otherwise, we call it a Failure. Do the same for your data. Compute the total percent of each.

TABLE 11.7 Outcomes of Obtaining an Even Number or a 3 on Tossing a Coin and Rolling a Die

Outcome	First trial	Second trial	Third trial	Total	Percent of occurrence
Success	30	37	32	99	66%
Failure	20	13	18	51	34%

Notice that the event of rolling an even number and the event of obtaining a 3 are **mutually exclusive**. That is, if either one occurs, the other one cannot occur. In our example, if a 3 appears, then an even number cannot appear on that *same* roll. Also, if an even number comes up, then a 3 can't come up. Compare the result you obtain if you compute the following:

(Probability of an even number) + (Probability of 3)

$$\frac{1}{2} \quad + \quad \frac{1}{6} \quad = \frac{2}{3}$$

Do you think this will always be true? To find the probability of mutually exclusive occurrences, we *add* the probabilities of each.

On the other hand, suppose we again reclassify the problem to determine the probability of obtaining an even number or a head. The result could be tabulated as in Table 11.8.

TABLE 11.8   Outcomes of Obtaining an Even Number or a Head
on Tossing a Coin and Rolling a Die

Outcome	First trial	Second trial	Third trial	Total	Percent of occurrence
Success	33	40	37	110	$73\frac{1}{3}\%$
Failure	17	10	13	40	$26\frac{2}{3}\%$

Do the same for your data. If we obtain an even number or a head in Table 11.6, we call the result a Success; otherwise, it is a Failure.

$$(\text{Probability of even}) + (\text{Probability of head})$$
$$\frac{1}{2} \qquad + \qquad \frac{1}{2} \qquad = 1$$

Does this result fit the data obtained? Can you explain how the experiment of Table 11.7 differs from the experiment of Table 11.8? That is, for the probability of "an even or a 3," we can add the respective probabilities; but for the probability of "an even or a head," we apparently cannot. Can you make any conjectures concerning this experiment?

Experiment 6: Three Card Problem

In this experiment, you need to prepare three cards that are identical except for color. One card is black on both sides, one is white on both sides, and one is black on one side and white on the other. One card is selected at random and placed flat on the table. You will see either a black card or a white card; record the color. This is not the probability with which we are concerned; rather, we are interested in predicting the probability of the *other* side being black or white. Record the color of the second side, as shown in Table 11.9. Repeat the experiment 50 times, and find the percent of occurrence of black or white with respect to the known color.

Consider the following argument. The probability of a white on the underside of the shown card is $\frac{1}{2}$, since the chosen card must be one of two possibilities. Assume that the visible side is white. Then it cannot be the black–black card;

Black on
both sides

Black on
one side.
white on
the other

White on
both sides

TABLE 11.9 Outcomes of Three Card Experiment

Color of chosen card	Number of times	Outcome (color of second side)	First trial	Percent of occurrence*
White	26	White	17	65.38%
		Black	9	34.62%
Black	24	White	9	37.5%
		Black	15	62.5%

*This is with respect to the known color. For example, the first entry is found by dividing: $17 \div 26 \approx .6538$, or 65.38%.

it must be the white–white or the white–black. Thus, the probability should be $\frac{1}{2}$. Does this argument match the data? If not, then either the data are biased, or the argument isn't correct. You might wish to repeat the experiment.

The probabilities obtained in this section are **empirical probabilities**. In the next section, we will find **theoretical probabilities**. Our mathematical model will be good only insofar as these two probabilities are "about the same." That is, if we perform the experiment enough times, the difference between the empirical and the theoretical probability can be made as small as we please.

Problem Set 11.1

1–6. Complete the six experiments described in this section, and state any appropriate conclusions or conjectures.

*Problems 7–10 deal with **random numbers**. We are all familiar with the idea of "picking a number at random." What do we mean by this? Can you really pick random numbers? Problems 7–9 ask you to pick random numbers using different methods. You are then asked to check the "randomness" of the numbers you selected. **After** you have selected the numbers, look at page 385 and apply the tests of randomness. **Do not read these tests until after** you have selected your "random numbers."*

7. Pick random numbers "from your head." Try writing down 100 random digits between 0 and 9 inclusive. Remember to write down the numbers as you think of them. You may not go back and change the numbers once they have been put down.

8. Pick random numbers using cards. From an ordinary deck of cards, remove all jacks, queens, and kings. Consider the ten as 0, the ace as 1, and the other cards as their face values. Shuffle the cards thoroughly. Select one card, note the result, and return it to the deck. Shuffle, and repeat 100 times.

9. Repeat Problem 8, except pick 20 cards, note the result, and return the cards to the deck. Shuffle, and repeat 5 times.

10. Pick random numbers using a table of random numbers. Consult a table of random numbers such as Table 11.10.

TABLE 11.10
A Table of 3,500 Random Digits
The random digits in this table are arranged in groups of five for ease of reading

Line/Col.	(1)	(2)	(3)	(4)	(5)	(6)	(7)	(8)	(9)	(10)	(11)	(12)	(13)	(14)
1	10480	15011	01536	02011	81647	91646	69179	14194	62590	36207	20969	99570	91291	90700
2	22368	46573	25595	85393	30995	89198	27982	53402	93965	34095	52666	19174	39615	99505
3	24130	48360	22527	97265	76393	64809	15179	24830	49340	32081	30680	19655	63348	58629
4	42167	93093	06243	61680	07856	16376	39440	53537	71341	57004	00849	74917	97758	16379
5	37570	39975	81837	16656	06121	91782	60468	81305	49684	60672	14110	06927	01263	54613
6	77921	06907	11008	42751	27756	53498	18602	70659	90655	15053	21916	81825	44394	42880
7	99562	72905	56420	69994	98872	31016	71194	18738	44013	48840	63213	21069	10634	12952
8	96301	91977	05463	07972	18876	20922	94595	56869	69014	60045	18425	84903	42508	32307
9	89579	14342	63661	10281	17453	18103	57740	84378	25331	12566	58678	44947	05585	56941
10	85475	36857	43342	53988	53060	59533	38867	62300	08158	17983	16439	11458	18593	64952
11	28918	69578	88231	33276	70997	79936	56865	05859	90106	31595	01547	85590	91610	78188
12	63553	40961	48235	03427	49626	69445	18663	72695	52180	20847	12234	90511	33703	90322
13	09429	93969	52636	92737	88974	33488	36320	17617	30015	08272	84115	27156	30613	74952
14	10365	61129	87529	85689	48237	52267	67689	93394	01511	26358	85104	20285	29975	89868
15	07119	97336	71048	08178	77233	13916	47564	81056	97735	85977	29372	74461	28551	90707
16	51085	12765	51821	51259	77452	16308	60756	92144	49442	53900	70960	63990	75601	40719
17	02368	21382	52404	60268	89368	19885	55322	44819	01188	65255	64835	44919	05944	55157
18	01011	54092	33362	94904	31273	04146	18594	29852	71585	85030	51132	01915	92747	64951
19	52162	53916	46369	58586	23216	14513	83149	98736	23495	64350	94738	17752	35156	35749
20	07056	97628	33787	09998	42698	06691	76988	13602	51851	46104	88916	19509	25625	58104
21	48663	91245	85828	14346	09172	30168	90229	04734	59193	22178	30421	61666	99904	32812
22	54164	58492	22421	74103	47070	25306	76468	26384	58151	06646	21524	15227	96909	44592
23	32639	32363	05597	24200	13363	38005	94342	28728	35806	06912	17012	64161	18296	22851
24	29334	27001	87637	87308	58731	00256	45834	15398	46557	41135	10367	07684	36188	18510
25	02488	33062	28834	07351	19731	92420	60952	61280	50001	67658	32586	86679	50720	94953
26	81525	72295	04839	96423	24878	82651	66566	14778	76797	14780	13300	87074	79666	95725
27	29676	20591	68086	26432	46901	20849	89768	81536	86645	12659	92259	57102	80428	25280
28	00742	57392	39064	66432	84673	40027	32832	61362	98947	96067	64760	64584	96096	98253
29	05366	04213	25669	26422	44407	44048	37937	63904	45766	66134	75470	66520	34693	90449
30	91921	26418	64117	94305	26766	25940	39972	22209	71500	64568	91402	42416	07844	69618
31	00582	04711	87917	77341	42206	35126	74087	99547	81817	42607	43808	76655	62028	76630
32	00725	69884	62797	56170	86324	88072	76222	36086	84637	93161	76038	65855	77919	88006
33	69011	65797	95876	55293	18988	27354	26575	08625	40801	59920	29841	80150	12777	48501
34	25976	57948	29888	88604	67917	48708	18912	82271	65424	69774	33611	54262	85963	03547
35	09763	83473	73577	12908	30883	18317	28290	35797	05998	41688	34952	37888	38917	88050
36	91567	42595	27958	30134	04024	86385	29880	99730	55536	84855	29080	09250	79656	73211
37	17955	56349	90999	49127	20044	59931	06115	20542	18059	02008	73708	83517	36103	42791
38	46503	18584	18845	49618	02304	51038	20655	58727	28168	15475	56942	53389	20562	87338
39	92157	89634	94824	78171	84610	82834	09922	25417	44137	48413	25555	21246	35509	20468
40	14577	62765	35605	81263	39667	47358	56873	56307	61607	49518	89656	20103	77490	18062
41	98427	07523	33362	64270	01638	92477	66969	98420	04880	45585	46565	04102	46880	45709
42	34914	63976	88720	82765	34476	17032	87589	40836	32427	70002	70663	88863	77775	69348
43	70060	28277	39475	46473	23219	53416	94970	25832	69975	94884	19661	72828	00102	66794
44	53976	54914	06990	67245	68350	82948	11398	42878	80287	88267	47363	46634	06541	97809
45	76072	29515	40980	07391	58745	25774	22987	80059	39911	96189	41151	14222	60697	59583
46	90725	52210	83974	29992	65831	38857	50490	83765	55657	14361	31720	57375	56228	41546
47	64364	67412	33339	31926	14883	24413	59744	92351	97473	89286	35931	04110	23726	51900
48	08962	00358	31662	25388	61642	34072	81249	35648	56891	69352	48373	45578	78547	81788
49	95012	68379	93526	70765	10593	04542	76463	54328	02349	17247	28865	14777	62730	92277
50	15664	10493	20492	38391	91132	21999	59516	81652	27195	48223	46751	22923	32261	85653

[There are whole books of random numbers—for example, Rand Corporation's book entitled *A Million Random Digits with 100,000 Normal Deviants* (New York: The Free Press, 1955).] Go to any place on the table and, by reading across a row or down a column, pick 100 numbers.

11. **The game of WIN.** Construct a set of nonstandard dice as shown.

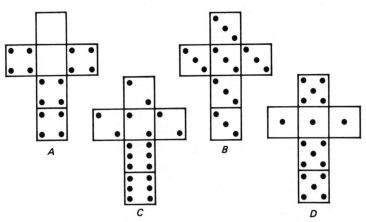

(Blanks are counted as zero.) Consider a game called WIN. The game is for two players. Each player is to choose one of the dice, and then the players roll the dice simultaneously. The player with the larger number wins the game. By experimenting with the dice, which one would you choose? If you were to play the game of WIN, would you like to choose your die first or second?

Tests of Randomness

Do you think you can pick numbers at random? Well, check it out by looking at Problems 7–10. There are several tests for randomness. Here are two of them.

1. Each digit should occur approximately the same number of times. This means that, given enough such digits, the percent of occurrence for each should be close to 10%.
2. Are there any sequences of numbers? In naming 100 digits, we could expect a random selection to have a sequence of several digits.

In many applications of science, it is important to obtain random numbers. Computers have been a great help in selecting random numbers for us.

12. A tournament is organized between Shannon's Stompers and Melissa's Marvels. The first team to win five games wins the tournament. Suppose the teams each have an equal chance of winning and the Marvels are leading the series three games to two when play is interrupted.

   a. How could you use Table 11.10 to simulate the finish of the series and decide the winner?

   b. Carry out the scheme you described in part a. Record the name of the winner of the series.

   c. Repeat the experiment 25 times (or more if necessary) and record the frequency with which the Marvels win the series.

11.2 Introduction to Probability

Engraving by Darcis: *Le Trente-et-un*

SITUATION   Chevalier was a heavy loser at cards, but he was certain he could turn his losses into wins if he only understood more about the game. He sent some questions about the game to his friend Blaise who in turn involved Pierre. Blaise and Pierre together were able to instruct

Chevalier how to play the game more intelligently than he had been playing. In this section, we'll introduce you to some of the secrets Blaise and Pierre were able to share with Chevalier. These include the definition of probability and its application to simple situations including dice, cards, and simple board games.

We will now formalize the results of the previous section and develop a mathematical model giving **theoretical probabilities**. That is, we'll formulate definitions, assume some properties, and thus create a model that will serve as a predictor of the actual percent of occurrence for a particular experiment. Our model will be good only insofar as it conforms to experimental data.

First we should settle on some terminology. We will consider experiments and events. The **sample space** of an experiment is the set of all possible outcomes. In each of the experiments of the previous section, we tabulated the sample space under the column headed "Outcomes." Now, an **event** is simply a set of possible outcomes (a subset of the sample space). For example, in Experiment 5, we tossed a coin and rolled a die. The sample space is

{1H, 1T, 2H, 2T, 3H, 3T, 4H, 4T, 5H, 5T, 6H, 6T}

and an event might be "Obtaining an even number or a head." That is, the following subset of the sample space is considered to be this event:

{1H, 2H, 2T, 3H, 4H, 4T, 5H, 6H, 6T}

The event "Obtaining a 5" is

{5H, 5T}

If the sample space can be divided into **mutually exclusive** and **equally likely** outcomes, we can define the probability of an event. Let's consider Experiment 1, where we tossed a single coin. We see here that a suitable sample space is

S = {Heads, Tails}

Suppose we wish to consider the event of obtaining heads; we'll call it event A:

A = {Heads}

We wish to define the probability of event A, which we denote by $P(A)$. Notice that the outcomes in the sample space are mutually exclusive. That is, if one occurs, the other cannot occur. If we flip a coin, there are two possible outcomes, and *one and only one* outcome can occur on a toss. If each outcome in the sample space is equally likely, we define the probability of A as

$$P(A) = \frac{\text{Number of successful results}}{\text{Number of possible results}}$$

A "successful" result is a result that corresponds to the probability we are seeking—in this case, {Heads}. Since we can obtain a head in only one way (Success), and the total number of possible outcomes is two, then the probability of heads is given by this definition as

$$P(\text{Heads}) = P(A) = \frac{1}{2}$$

This, of course, corresponds to our empirical results and to our experience of flipping coins. We now give a definition of probability.

Definition of Probability

> If an experiment can occur in any of n mutually exclusive and equally likely ways, and if s of these ways are considered favorable, then the probability of the event E, denoted by $P(E)$, is
>
> $$P(E) = \frac{s}{n} = \frac{\text{Number of outcomes favorable to } E}{\text{Number of all possible outcomes}}$$

EXAMPLE 1

Use the definition of probability to find the probabilities of obtaining white and black using the spinner shown and assuming that the arrow will never lie on a border line.

a. $P(\text{White}) = \dfrac{2}{3}$ ⟵ Two sections are white
⟵ Three sections altogether

b. $P(\text{Black}) = \dfrac{1}{3}$

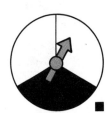

EXAMPLE 2

Consider a jar containing marbles as shown. Suppose each marble has an equal chance of being picked from the jar. Find the indicated probabilities.

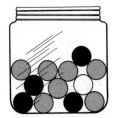

a. $P(\text{Black}) = \dfrac{4}{12}$ ←— 4 black marbles in jar
←— 12 marbles in jar

$= \dfrac{1}{3}$ Reduce fractions

b. $P(\text{Gray}) = \dfrac{7}{12}$

c. $P(\text{White}) = \dfrac{1}{12}$ ■

Reduced fractions are used to state probabilities when the fractions are fairly simple. If, however, the fractions are not simple, the probabilities are usually stated as decimals, as shown by Example 3.

EXAMPLE 3

Suppose that in a certain study, 46 out of 155 people showed a certain kind of behavior. Assign a probability to this behavior.

Solution

$$P = \dfrac{46}{155}$$ *Press:*   46 ÷ 155 =
$$\approx .30$$ *Display:*   .29677419   ■

Notice that Examples 1 and 2 are theoretical probabilities while Example 3 is an empirical probability. If, for example, we *actually* spun the dial in Example 1, or *actually* drew a marble from the jar in Example 2, and if we repeated the experiment a large number of times, we would expect the empirical probabilities and theoretical probabilities to be almost the same. The results obtained by applying this definition *must* be consistent with the results obtained by actually performing an experiment a large number of times.

EXAMPLE 4

Compare the empirical and theoretical probabilities for Experiment 3 (rolling a single die).

Solution

The sample space is {1, 2, 3, 4, 5, 6}, as shown in Figure 11.1.

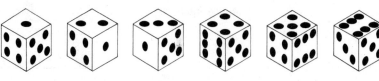

FIGURE 11.1   Sample space for tossing a single die

Let's find the theoretical probability for one of these, say the probability of tossing a two:

$$P(\text{Two}) = \frac{\text{Number of successful outcomes}}{\text{Number of all possible outcomes}} = \frac{1}{6}$$

The probability of any other single occurrence is also $\frac{1}{6}$. Since $\frac{1}{6} \approx .1667$, we can compare the empirical and theoretical probabilities as shown in Table 11.11. These seem to be consistent.

TABLE 11.11

Outcome	Theoretical probability	Empirical probability*
1	.1667	.1600
2	.1667	.1267
3	.1667	.1867
4	.1667	.1867
5	.1667	.1533
6	.1667	.1867

*These figures vary according to the particular experiment and the number of trials. The figures used here are our actual results from Experiment 3.   ■

EXAMPLE 5

Compare the empirical and theoretical probabilities for Experiment 4 (rolling a pair of dice).

Solution

Let's begin by reasoning in the same fashion as in Example 4. The sample space is {2, 3, 4, 5, 6, 7, 8, 9, 10, 11, 12}. Thus,

$$P(\text{Two}) = \frac{1}{11} \approx .0909$$

Suppose we say, as in Example 4, that the probability of any other single event will also be $\frac{1}{11}$. Table 11.12 compares this with the results we obtained for Experiment 4.

TABLE 11.12

Outcome	Theoretical probability	Empirical probability
2	.0909 ⟵⟶	.0133
3	.0909	.0467
4	.0909	.0933
5	.0909	.1000
6	.0909 ⎫	.1533 ⎫
7	.0909 ⎬	.1733 ⎬
8	.0909 ⎭	.1533 ⎭
9	.0909	.1067
10	.0909	.0667
11	.0909	.0800
12	.0909 ⟵⟶	.0133

The empirical and theoretical probabilities do *not* seem to be consistent. The difficulty is that the sample space we are using doesn't list equally likely possibilities. You must be careful to set up a sample space of equally likely possibilities. The correct sample space is shown in Figure 11.2.

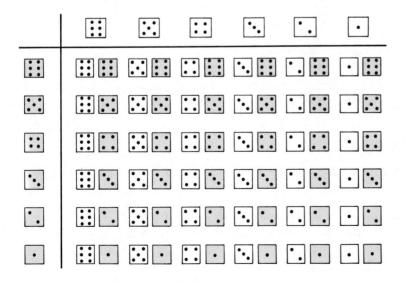

FIGURE 11.2 Sample space for tossing a pair of dice

Now,

$$P(\text{Two}) = \frac{1}{36} \longleftarrow \text{Number of equally likely possibilities in sample space}$$

$$P(\text{Three}) = \frac{2}{36} \longleftarrow \text{Do you see that there are two possibilities of obtaining a three in Figure 11.2?}$$

$$P(\text{Four}) = \frac{3}{36}$$

After calculating the others (you will be asked to do some of these in Problem Set 11.2), we compare these theoretical probabilities with the empirical probabilities in Table 11.13.

TABLE 11.13

Outcome	Theoretical probability	Empirical probability
2	.0278	.0133
3	.0556	.0467
4	.0833	.0933
5	.1111	.1000
6	.1389	.1533
7	.1667	.1733
8	.1389	.1533
9	.1111	.1067
10	.0833	.0667
11	.0556	.0800
12	.0278	.0133

The results are consistent.   ■

Example 6 illustrates the use of the words *and* and *or* in probability problems.

EXAMPLE 6

Suppose that a single card is selected from an ordinary deck of 52 cards.

a. What is the probability that it is a heart?
b. What is the probability that it is an ace?
c. What is the probability that it is a two *or* a king?

d. What is the probability that it is a two *or* a heart?

e. What is the probability that it is a two *and* a heart?

f. What is the probability that it is a two *and* a king?

Solution

The sample space is shown in Figure 11.3.

FIGURE 11.3   Sample space using a deck of cards

Spades (black cards)

Clubs (black cards)

Hearts (red cards)

Diamonds (red cards)

a. $P(\text{Heart}) = \dfrac{13}{52} = \dfrac{1}{4}$

b. $P(\text{Ace}) = \dfrac{4}{52} = \dfrac{1}{13}$

c. The word *or* is used in probability to mean that either the first event occurs *or* the second event occurs; that is, both events are counted as successes when applying the definition of probability. In this example, there are 4 twos and 4 kings in a deck, so

$$P(\text{Two } or \text{ King}) = \frac{8}{52} = \frac{2}{13}$$

d. This seems to be very similar to part c, but there is one important difference. Look at the sample space and notice that although there are 4 twos and 13 hearts, the total number of successes is *not* 4 + 13 = 17, *but rather* 16. The reason for this is that you cannot count the same card twice; the two of hearts should *not* be counted *once* as a two *and again* as a heart. Thus,

$$P(\text{Two } or \text{ Heart}) = \frac{16}{52} = \frac{4}{13}$$

e. The word *and* is used in probability to mean that both events occur. This means, for this example, that when you draw the single card it must be *both* a two *and* a heart. By looking at the sample space, you can see that there is only one way of success. Thus,

$$P(\text{Two } and \text{ Heart}) = \frac{1}{52}$$

f. There is no way of drawing both a two *and* a king with a single draw, so

$$P(\text{Two } and \text{ King}) = \frac{0}{52} = 0 \qquad \blacksquare$$

Notice from Example 6f that the probability of an event that cannot occur is 0. In Problem Set 11.2 you are asked to show that the probability of an event that must occur is 1. These are the two extremes. All other probabilities will fall somewhere in between. The closer the probability is to 1, the more likely the event; the closer it is to 0, the less likely the event.

Procedure for Finding Probabilities

1. Describe and identify the sample space, and then count the number of elements (these should be equally likely); call this number n.
2. Count the number of occurrences that interest us; call this the number of successes and denote it by s.
3. Compute the probability of the event: $P(E) = \dfrac{s}{n}$

The procedure just outlined will not always work. If it doesn't, you need a more complicated model, or else you must proceed experimentally, as in the last section. This model will, however, be sufficient for the problems you will find in this book.

Problem Set 11.2

1. What is the difference between empirical and theoretical probabilities?

2. Define probability.

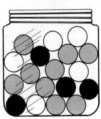

Use the spinner shown here for Problems 3–5 and find the requested probabilities. Assume that the pointer can never lie on a border line.

3. *P*(White)   4. *P*(Black)   5. *P*(Gray)

Consider the jar containing marbles shown here. Suppose each marble has an equal chance of being picked from the jar. Find the probabilities in Problems 6–8.

6. *P*(White)   7. *P*(Black)   8. *P*(Gray)

A single card is selected from an ordinary deck of cards. Find the probabilities in Problems 9–11.

9. *P*(Five of hearts)   10. *P*(Five)   11. *P*(Heart)

Give the probabilities in Problems 12–15 in decimal form (correct to two decimal places). A calculator may be helpful with these problems.

12. Last year, 1,485 calculators were returned to the manufacturer. If 85,000 were produced, assign a number to specify the probability that a particular calculator would be returned.

13. Last semester, a certain professor gave 13 As out of 285 grades. If grades were assigned randomly, what is the probability of an A?

14. Last year, in a certain city it rained on 85 days. What is the probability of rain on a day selected at random?

15. A certain campus club is having a raffle and they are selling 1,500 tickets. If the people on your floor of the dorm bought 285 of those tickets, what is the probability that someone on your floor will hold the winning ticket?

Suppose you toss a coin and roll a die. The sample space is shown here. Use this sample space to answer the questions in Problems 16–19.

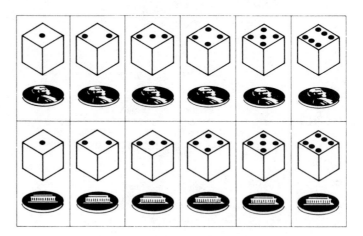

16. What is the probability of obtaining:
    a. A tail *and* a five?   b. A tail *or* a five?
    c. A head *and* a two?

17. What is the probability of obtaining:
    a. A tail?   b. One, two, three, *or* four?
    c. A head *or* a two?

18. What is the probability of obtaining:
    a. A head *and* an odd number?
    b. A head *or* an odd number?

19. What is the probability of obtaining:
    a. A head *and* a five?   b. A head *or* a five?

Use the sample space shown in Figure 11.2 to answer the questions in Problems 20–30. The requested probabilities in Problems 20–27 refer to the sum of the numbers shown on the tops of the dice.

20. P(Five)   21. P(Six)   22. P(Seven)   23. P(Eight)

24. P(Nine)   25. P(Four *or* Five)

26. P(Even number)   27. P(Eight *or* Ten)

28. Dice is a popular game in gambling casinos. Two dice are tossed, and various amounts are paid according to the outcome. If a seven or eleven occurs on the first roll, the player wins. What is the probability of winning on the first roll?

29. In dice, the player loses if the outcome of the first roll is a two, three, or twelve. What is the probability of losing on the first roll?

30. In dice, a pair of ones is called *snake eyes*. What is the probability of losing a dice game by rolling snake eyes?

31. Consider a die with only four sides, marked one, two, three, and four. Write out a sample space similar to the one shown in Figure 11.2 for rolling a pair of these dice.

32. Using the sample space you found in Problem 31, find the probability that the sum of the dice is the given number. Assume equally likely outcomes.
a. $P(\text{Two})$ **b.** $P(\text{Three})$ **c.** $P(\text{Four})$

33. Using the sample space you found in Problem 31, find the probability that the sum of the dice is the given number. Assume equally likely outcomes.
a. $P(\text{Five})$ **b.** $P(\text{Six})$ **c.** $P(\text{Seven})$

34. The game of Dungeons and Dragons uses nonstandard dice. Consider a die with eight sides marked one, two, three, four, five, six, seven, and eight. Write out a sample space similar to the one shown in Figure 11.2 for rolling a pair of these dice.

35. Consider the game of WIN described in Problem 11, page 385. Suppose your opponent picks die *A*. Suppose you pick die *B*. Then we can enumerate the sample space as shown here.

B \ A	0	0	4	4	4	4	
3	(3, 0)	(3, 0)	(3, 4)	(3, 4)	(3, 4)	(3, 4)	
3	(3, 0)	(3, 0)	(3, 4)	(3, 4)	(3, 4)	(3, 4)	
3	(3, 0)	(3, 0)	(3, 4)	(3, 4)	(3, 4)	(3, 4)	← A wins
3	(3, 0)	(3, 0)	(3, 4)	(3, 4)	(3, 4)	(3, 4)	
3	(3, 0)	(3, 0)	(3, 4)	(3, 4)	(3, 4)	(3, 4)	
3	(3, 0)	(3, 0)	(3, 4)	(3, 4)	(3, 4)	(3, 4)	

↑ B wins

We see that the probability of *A* winning is $\frac{24}{36}$, or $\frac{2}{3}$.

By enumeration of the sample space, find your probability of winning if you choose:
a. Die *C* **b.** Die *D*
c. If your opponent picks die *A*, which die would you pick?

36. In WIN (see Problem 35), if your opponent picks die *B*, which would you pick?

37. In WIN (see Problem 35), if your opponent picks die *C*, which would you pick?

38. In WIN (see Problem 35), if your opponent picks die *D*, which would you pick?

39. By considering Problems 35–38, see if you can write a strategy for playing WIN. If you follow this strategy, what is your probability of winning?

40. Show that the probability of some event E falls between 0 and 1. Also, show that $P(E) = 0$ if the event cannot occur and $P(E) = 1$ if the event must occur.

Probabilities of Poker Hands

Hand		Number of favorable events	Probability
Royal flush		4	.00000153908
Other straight flush		36	.00001385169
Four of a kind		624	.00024009604
Full house		3,744	.00144057623
Flush		5,108	.00196540155
Straight		10,200	.00392464682
Three of a kind		54,912	.02112845138
Two pair		123,552	.04753901561
One pair		1,098,240	.42256902761
Other hands		1,302,540	.50117739403
Totals		2,598,960	1.00000000000

11.3 Probability Models

SITUATION "Last week I won a free Big Mac at
McDonald's. I sure was lucky!" exclaimed Hal.
  "Do you go there often?" asked Pat.
  "Only twenty or thirty times a month. And the odds of
winning a Big Mac were only 20 to 1."
  "Don't you mean 1 to 20?" queried Pat.
  "Don't confuse me with details. I don't even understand
odds at the racetrack, and I go all the time. Why do I need to
know anything about odds anyway?"
  In this section, Hal will learn about odds, the calculation of
odds, and their relationship to probability. He will also learn
about two models to help in calculating probabilities—one
model to use in finding the probability that something does
not occur, and the other model to use for calculating the
probability when the sample space has been altered because
of some additional given or known information.

Suppose the odds against you in a certain game are 2 to 5;
what does this mean? There are two ways of stating odds,
odds in favor and **odds against**. If the odds against your
winning are 2 to 5, then the odds in favor of your winning
are 5 to 2. In general, we give the following definition of
odds:

Definition of Odds

The **odds in favor of an event E**, where $P(E)$ is the
probability that E will occur, is

$$P(E) \quad \text{to} \quad 1 - P(E) \qquad \text{or} \qquad \frac{P(E)}{1 - P(E)}$$

The **odds against an event E** is

$$1 - P(E) \quad \text{to} \quad P(E) \qquad \text{or} \qquad \frac{1 - P(E)}{P(E)}$$

EXAMPLE 1

What are the odds in favor of drawing a heart from an
ordinary deck of cards?

Solution

Let $H = \{$Draw a heart$\}$. Then $P(H) = \frac{1}{4}$ and $1 - P(H) = \frac{3}{4}$. When stating odds, it is not customary to state them using fractions. That is, we do not generally say the odds are $\frac{1}{4}$ to $\frac{3}{4}$. The procedure, then, is to first write the odds as a ratio, and then simplify:

$$\frac{\frac{1}{4}}{\frac{3}{4}} = \frac{1}{4} \times \frac{4}{3} = \frac{1}{3}$$

After the fraction is simplified, state the odds: **The odds in favor of drawing a heart are 1 to 3.** ∎

EXAMPLE 2

If a contest has 1,000 entries and you purchase 10 tickets, what are the odds against your winning?

Solution

$$P(\text{Winning}) = \frac{10}{1,000} = \frac{1}{100} \qquad 1 - P(\text{Winning}) = \frac{99}{100}$$

$$\textit{Odds against:} \quad \frac{\frac{99}{100}}{\frac{1}{100}} = \frac{99}{100} \times \frac{100}{1} = \frac{99}{1}$$

The odds against winning are 99 to 1. ∎

In Examples 1 and 2 we found the odds by first calculating the probability. Suppose, instead, you are given the odds and wish to calculate the probability.

Procedure for Finding the Probability, Given the Odds

> If the odds in favor of an event E are s to b, then the probability of E is given by
>
> $$P(E) = \frac{s}{b + s}$$

EXAMPLE 3

If the odds in favor of some event are 2 to 5, what is the probability?

Solution

In this example, $s = 2$ and $b = 5$, so

$$P(E) = \frac{2}{7}$$ ■

EXAMPLE 4

If the odds against you are 100 to 1, what is the probability?

Solution

First change (mentally) to odds in favor: 1 to 100. Then,

$$P(E) = \frac{1}{101}$$ ■

The number $1 - P(E)$ is related to the probability that an event E cannot occur. The **complement** of any event E is the event that E does not occur, and is denoted by $\bar{E}$. That is, since any event either *must occur* or *must not occur*, the sum of the probability of an event occurring and that of the same event not occurring is 1. In symbols,

$$P(E) + P(\bar{E}) = 1$$

This is usually written in a more useful form:

$$P(E) = 1 - P(\bar{E}) \qquad \text{or} \qquad P(\bar{E}) = 1 - P(E)$$

EXAMPLE 5

From the table of probabilities of poker hands at the end of Section 11.2, the probability of obtaining one pair is about .42. What is the probability of not obtaining one pair?

Solution

Since $P(\text{Pair}) \approx .42$,

$$
\begin{aligned}
P(\text{Not a pair}) &= 1 - P(\text{Pair}) \\
&\approx 1 - .42 \\
&= .58
\end{aligned}
$$ ■

Consider a portion of the letter to Dear Abby shown in the margin. Let's verify the mathematics mentioned in this letter and assume that each child is as likely to be a boy as a girl.*

EXAMPLE 6

What is the probability of a family having two children of opposite sexes?

Solution

The sample space might look like this:

> 2 boys
> a boy and a girl
> 2 girls

But this sample space presents problems in computing probabilities, since the outcomes are not equally likely. Consider the following **tree diagram**:

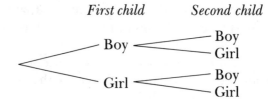

We see that there are four equally likely outcomes:

> Boy—Boy
> Boy—Girl
> Girl—Boy
> Girl—Girl

Thus, the probability of having a boy and a girl with two children is:

$$\frac{\text{Number of favorable outcomes}}{\text{Total number of all possible outcomes}} = \frac{2}{4} = \frac{1}{2}$$

There is a probability of $\frac{1}{2}$ of having a boy and a girl family.   ■

*There is some medical evidence to show that this is not true. However, for our purposes it is sufficiently accurate to make this simplifying assumption.

EXAMPLE 7

Suppose a family wanted to have four children. What is the probability of having two boys and two girls?

Solution

In this problem, there are 16 possibilities. (Can you list them?) Of these possibilities, there are 6 ways of obtaining 2 boys and 2 girls: BBGG, BGBG, BGGB, GBGB, GBBG, and GGBB. Thus, the probability is

$$\frac{6}{16} = \frac{3}{8}$$

A common fallacy is to say "$\frac{1}{2}$ boys and $\frac{1}{2}$ girls; thus, the probability is $\frac{1}{2}$." ∎

EXAMPLE 8

What is the probability of a family having eight girls in a row?

Solution

Consider a tree diagram: First child, 2 possibilities; Second child, 4 possibilities; Third child, 8 possibilities; . . . for the Eighth child, there must be 256 possibilities. Thus, the probability is $\frac{1}{256}$. ∎

The doctor, as reported in the Dear Abby letter, was incorrect in saying the odds were 100 to 1. If the probability is $\frac{1}{256}$, then what are the odds? (*Answer:* 1 to 255.) But even this is incorrect reasoning. Consider a couple with two girls who want their next child to be a boy. They know a little about probability and reason that, since the chance of having three girls is only $\frac{1}{8}$, they really have a good chance of having a girl the next time. This is a common misconception, even believed by the doctor in the Dear Abby letter. It is important to remember in independent outcomes that the probability of a future event is unaffected by past occurrences. The probability of a boy on the next birth is the same as always: $\frac{1}{2}$—*regardless of the previous record.*

Another example of this fallacy is given by Darrell Huff in his book *How to Take a Chance.* On August 18, 1913, at a casino in Monte Carlo, black came up 26 times in a row on a roulette wheel. If you had bet $1 on black and continued to let your bet ride for the entire run of blacks, you would have won $67,108,863. What actually happened confirms our statement that many people commit this fallacy. After about the 15th occurrence of black, there was a near-panic rush to bet on red. People kept doubling up on their bets in the belief that, after black came up the 20th time, there was not a chance in a million of another repeat. The casino came out ahead by several million francs. Remember: For each spin of the wheel, the probability of black remains the same, *regardless of past performance*.

Frequently, we wish to compute the probability of an event and we have additional information that will alter the sample space. For example, suppose a family has two children. What is the probability that the family has two boys? $P(2 \text{ boys}) = \frac{1}{4}$ (Why?) Let's complicate the problem a little. Suppose we know that the older child is a boy. We have altered the sample space as follows:

Sample space

B   B
B   G   } This is the altered sample space.
G̶   B̶
G̶   G̶

We see that the probability is now $\frac{1}{2}$.

If, instead, we know that at least one child is a boy, we alter the sample space as shown:

Sample space

B   B
B   G
G   B
G̶   G̶

Here we see that the probability is $\frac{1}{3}$!

These are problems involving *conditional probability*. We speak about the *probability of an event E, given that another event F has occurred*. We denote this by $P(E|F)$. This means that, rather than simply compute $P(E)$, we reevaluate $P(E)$ in light of the information that F has occurred. That is, we consider the altered sample space.

EXAMPLE 9

Suppose you toss two coins (or a single coin twice). What is the probability that two heads are obtained if you already know that at least one head is obtained?

Solution

Consider the altered sample space:

$$\textit{Sample space}$$

H H
H T
T H
~~T T~~

The probability is $\frac{1}{3}$.   ∎

Problem Set 11.3

1. What are the odds in favor of drawing an ace from an ordinary deck of cards?

2. What are the odds in favor of drawing a heart from an ordinary deck of cards?

3. What are the odds against a family of four children containing four boys?

4. Suppose the odds that a man will be bald by the time he is 60 are 9 to 1. State this as a probability.

5. Suppose the odds are 33 to 1 that someone will lie to you at least once in the next seven days. State this as a probability.

6. Racetracks quote the approximate odds for each race on a large display board called a *tote board*. Here's what it might say for a particular race:

Horse number	Odds
1	2 to 1
2	15 to 1
3	3 to 2
4	7 to 5
5	1 to 1

What would be the probability of winning for each of these horses? [*Note:* The odds stated are for the horse *losing*. Thus, $P(\text{Horse 1 losing}) = \frac{2}{2+1} = \frac{2}{3}$ so that $P(\text{Horse 1 winning}) = 1 - \frac{2}{3} = \frac{1}{3}$.]

Fill in the blanks for Problems 7–12.

	Experiment	Outcomes considered success	P(Success)	Outcomes considered failure	P(Failure)
7.	Tossing a coin	Heads	$\frac{1}{2}$	_____	_____
8.	Rolling a die	4 or 6	$\frac{1}{3}$	_____	_____
9.	Guessing an answer on a multiple-choice test	Correct guess	$\frac{1}{5}$	_____	_____
10.	Card game	Drawing a heart	.18	_____	_____
11.	Baseball game	White Sox win	.57	_____	_____
12.	Football game	Your school wins	.83	_____	_____

13. Three fair coins are tossed. What is the probability that at least one is a head?

14. A card is selected from an ordinary deck. What is the probability that it is not a face card?

15. Choose a natural number between 1 and 100, inclusive. What is the probability that the one chosen is not a multiple of five?

16. In your own words, explain what we mean by "conditional probability."

17. Suppose a family wants to have four children.
   a. What is the sample space?
   b. What is the probability of 4 girls? 4 boys?
   c. What is the probability of 1 girl and 3 boys? 1 boy and 3 girls?
   d. What is the probability of 2 boys and 2 girls?
   e. What is the sum of your answers in parts b through d?

18. What is the probability of a family of three children containing exactly two boys, given that at least one of them is a boy?

19. What is the probability of flipping a coin four times and obtaining exactly three heads, given that at least two are heads?

For Problems 20–22, see Figure 11.2 on page 391.

20. On a single roll of a pair of dice, what is the probability that the sum is 7, given that at least one die came up 2?

21. On a single roll of a pair of dice, suppose we are given that at least one die came up 2. What is the probability that the sum is:

a. 3? **b.** 4? **c.** 2?

22. What are the odds in favor of rolling a 7 or 11 on a single roll of a pair of dice?

23. **Experiment.** Consider the birthdates of some famous mathematicians:

Abel	August 5, 1802
Cardano	September 21, 1576 (died)
Descartes	March 31, 1596
Euler	April 15, 1707
Fermat	August 20, 1601 (baptized)
Galois	October 25, 1811
Gauss	April 30, 1777
Newton	December 25, 1642
Pascal	June 19, 1623
Riemann	September 31, 1815

Add to this list the birthdates of the members of your class. *But, first, what is your guess as to the probability that at least two people in this group will have exactly the same birthdate (not counting the year)?* Be sure to make your guess *before* finding out the birthdates of your classmates. The answer, of course, depends on the number of people on the list. We've listed 10 mathematicians, and you may have 20 people in your class, giving 30 names on the list.

24. **Birthday problem.** This problem is related to Problem 23. Suppose you select 23 people out of a crowd. The probability that two or more of them will have the same birthday is greater than 50%! This is a seemingly paradoxical situation that will fool most people. A chart showing the probabilities is given here.

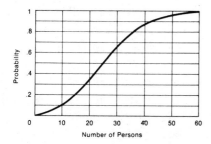

Probabilities for the birthday problem

a. Pick 23 names at random from a biological dictionary or a *Who's Who* and verify some of the probabilities of the table. The more you go past 23, the greater the probability of the same birthday.

b. The graph shown here is getting closer to a probability of 1 as the number of people increases. How many people are necessary for the probability to actually reach 1?

11.4 Mathematical Expectation

SITUATION   Suppose your friend George shows you three cards. One is white on both sides, one is black on both sides, and the last is black on one side and white on the other. (Remember Experiment 6?) He mixes the cards and lets you select one at random and place it on the table. Suppose the upper side turns out to be black. It is not the white–white card; it must be the black–black or black–white card. (Remember conditional probability.) "Thus," says George, "I'll bet you $1 that the other side is black." Would you play? Perhaps you hesitate. Now George says he feels generous. You need only pay him 75¢ if you lose, and he will still pay you $1 if he loses. Would you play now?

In this section, you'll see how to analyze a variety of gambling situations. Whether you enjoy gambling and games of chance, or are opposed to them and would never play a gambling game, you should find some valuable information in this section. Gambling situations range from dice, cards, and slot machines, to buying insurance and selling a home. By analyzing these games, you can show that the gambler is destined for financial ruin, given enough time and limited resources. You can also find situations that should not be considered gambling—situations in which you can't lose.

Smiles toothpaste is giving away $10,000. All you must do to have a chance to win is send in a postcard with your name on it (the fine print says you do not need to buy a tube of toothpaste). Is it worthwhile to enter?

Suppose the contest receives one million postcards (a conservative estimate). We wish to compute your **expected value** (or your expectation) in entering this contest. We denote the expectation by E and compute:

$E = (\text{AMOUNT TO WIN}) \times (\text{PROBABILITY OF WINNING})$

$\quad = \$10,000 \times \left(\dfrac{1}{1,000,000}\right)$

$\quad = \$0.01$

A game is **fair** if the expected value equals the cost of playing the game. Is this game fair? If the toothpaste company charges you 1¢ to play the game, it is fair. But how much does it really cost you to enter the contest? How much does the postcard cost? We see that it is not a fair game.

EXAMPLE 1

Suppose you are going to roll two dice. You will be paid $3.60 if you roll two 1s. You won't receive anything for any other outcome. How much should you be willing to pay for the privilege of rolling the dice?

Solution

You should be willing to pay the mathematical expectation of the game, which is computed as follows:

$E = (\text{AMOUNT OF WINNINGS}) \times (\text{PROBABILITY OF WINNING})$

$\quad = \$3.60 \times \left(\dfrac{1}{36}\right)$

$\quad = \$0.10$

You should be willing to pay 10¢ to play the game. ∎

When we say that the expectation is 10¢, we certainly do *not* mean that you will win 10¢ every time you play. (Indeed, you will *never* win 10¢; it will be either $3.60 or nothing.) We mean that, if you were to play this game a very large number of times, you could expect to win an *average* of 10¢ per game.

Most games of chance will not be fair (the "house" must make a living). The question often is "Which games come closest to being fair?" We will consider this question in some of the problems.

Sometimes there is more than one payoff, as illustrated by Example 2.

EXAMPLE 2

A recent contest offered one grand prize worth $10,000, two second prizes worth $5,000 each, and ten third prizes worth $1,000 each. What is the expectation if you assume that there were one million entries and that winners' names were replaced in the pool after being drawn?

Solution

$$P(\text{1ST PRIZE}) = \frac{1}{1,000,000} \qquad P(\text{2ND PRIZE}) = \frac{2}{1,000,000}$$

$$P(\text{3RD PRIZE}) = \frac{10}{1,000,000}$$

$$E = \begin{pmatrix} \text{AMOUNT OF} \\ \text{1ST PRIZE} \end{pmatrix} \times P(\text{1ST PRIZE}) + \begin{pmatrix} \text{AMOUNT OF} \\ \text{2ND PRIZE} \end{pmatrix}$$

$$\times P(\text{2ND PRIZE}) + \begin{pmatrix} \text{AMOUNT OF} \\ \text{3RD PRIZE} \end{pmatrix} \times P(\text{3RD PRIZE})$$

$$= \$10,000 \times \left(\frac{1}{1,000,000} \right) + \$5,000 \times \left(\frac{2}{1,000,000} \right)$$

$$+ \$1,000 \times \left(\frac{10}{1,000,000} \right)$$

$$= \$0.01 + \$0.01 + \$0.01$$
$$= \$0.03 \qquad\qquad\qquad \blacksquare$$

Now we give a formal definition of expectation.

Mathematical Expectation

> If an event has several possible outcomes with probabilities $p_1, p_2, p_3, \ldots,$ and for each of these outcomes the amount that can be won is $a_1, a_2, a_3, \ldots,$ the **mathematical expectation E** of the event is
>
> $$E = a_1 p_1 + a_2 p_2 + a_3 p_3 + \cdots$$

EXAMPLE 3

A contest offered the following prizes (as indicated in the fine print):

Prize	Value	Probability of winning
Grand prize trip	$1,500 = a_1$	$.000026 = p_1$
Weber Kettle	$110 = a_2$	$.000032 = p_2$
Magic Chef Range	$279 = a_3$	$.000016 = p_3$
Murray Bicycle	$191 = a_4$	$.000021 = p_4$
Lawn Boy Mower	$140 = a_5$	$.000026 = p_5$
Samsonite Luggage	$183 = a_6$	$.000016 = p_6$

What is the expected value for this contest?

Solution

$$E = a_1p_1 + a_2p_2 + a_3p_3 + a_4p_4 + a_5p_5 + a_6p_6$$
$$= 1,500(.000026) + 110(.000032) + 279(.000016)$$
$$\quad + 191(.000021) + 140(.000026) + 183(.000016)$$
$$= .057563$$

The expected value is a little less than 6¢. Suppose we mail in our entry as specified in the rules. What is the cost of the stamp and the envelope? ■

Life insurance is also a form of gambling. When you purchase a policy you are betting that you will die during the term of the policy, and the company is betting that you will live. The probability that you will die in any particular year of your life is an empirical probability and is listed in what is called a *mortality table* (see Table 11.14 on page 412).

EXAMPLE 4

Suppose you turned 21 years old today and you wish to take out a $10,000 insurance policy for one year. How much should you be willing to pay for the policy?

Solution

This is also an expectation problem, in which you must find the probability that a 21-year-old person will die during his or her 21st year. If we consult Table 11.14, we see that, for 96,478 persons of age 21, we could expect 177 of them to die within a year. Thus,

$$P(\text{Death during 21st year}) = .0018$$

TABLE 11.14   Mortality Table
Based on 100,000 living at age 0, giving: ℓ_x, NUMBER OF LIVING; d_x, NUMBER OF DEATHS; p_x, PROBABILITY OF LIVING; q_x, PROBABILITY OF DYING, for age x from 0 to 99.

x	ℓ_x	d_x	p_x	q_x	x	ℓ_x	d_x	p_x	q_x
0	100000	708	.9929	.0071	50	87624	729	.9917	.0083
1	99292	175	.9982	.0018	51	86895	792	.9909	.0091
2	99117	151	.9985	.0015	52	86103	858	.9900	.0100
3	98966	144	.9986	.0015	53	85245	928	.9891	.0109
4	98822	138	.9986	.0014	54	84317	1003	.9881	.0119
5	98684	133	.9987	.0014	55	83314	1083	.9870	.0130
6	98551	128	.9987	.0013	56	82231	1168	.9858	.0142
7	98423	124	.9987	.0013	57	81063	1260	.9845	.0156
8	98299	121	.9988	.0012	58	79803	1357	.9830	.0170
9	98178	119	.9988	.0012	59	78446	1458	.9814	.0186
10	98059	119	.9988	.0012	60	76988	1566	.9797	.0204
11	97940	120	.9988	.0012	61	75422	1677	.9778	.0222
12	97820	123	.9988	.0013	62	73745	1793	.9757	.0243
13	97697	129	.9987	.0013	63	71952	1912	.9734	.0266
14	97568	136	.9986	.0014	64	70040	2034	.9710	.0291
15	97432	142	.9986	.0015	65	68006	2159	.9683	.0318
16	97290	150	.9985	.0016	66	65847	2287	.9653	.0347
17	97140	157	.9984	.0016	67	63560	2418	.9620	.0381
18	96983	164	.9983	.0017	68	61142	2548	.9583	.0417
19	96819	168	.9983	.0017	69	58594	2672	.9544	.0456
20	96651	173'	.9982	.0018	70	55922	2784	.9502	.0498
21	96478	177	.9982	.0018	71	53138	2877	.9459	.0542
22	96301	179	.9982	.0019	72	50261	2948	.9414	.0587
23	96122	182	.9981	.0019	73	47313	2993	.9368	.0633
24	95940	183	.9981	.0019	74	44320	3019	.9319	.0681
25	95757	185	.9981	.0019	75	41301	3030	.9266	.0734
26	95572	187	.9981	.0020	76	38271	3030	.9208	.0792
27	95385	190	.9980	.0020	77	35241	3020	.9143	.0857
28	95195	193	.9980	.0020	78	32221	2998	.9070	.0931
29	95002	198	.9979	.0021	79	29223	2957	.8988	.1012
30	94804	202	.9979	.0021	80	26266	2888	.8901	.1100
31	94602	207	.9978	.0022	81	23378	2790	.8807	.1194
32	94395	212	.9978	.0023	82	20588	2659	.8709	.1292
33	94183	218	.9977	.0023	83	17929	2499	.8606	.1394
34	93965	226	.9976	.0024	84	15430	2314	.8500	.1500
35	93739	235	.9975	.0025	85	13116	2113	.8389	.1611
36	93504	247	.9974	.0027	86	11003	1901	.8272	.1728
37	93257	261	.9972	.0028	87	9102	1685	.8149	.1851
38	92996	280	.9970	.0030	88	7417	1470	.8018	.1982
39	92716	301	.9968	.0033	89	5947	1263	.7876	.2124
40	92415	326	.9965	.0035	90	4684	1068	.7720	.2280
41	92089	354	.9962	.0039	91	3616	888	.7544	.2456
42	91735	383	.9958	.0042	92	2728	725	.7342	.2658
43	91352	414	.9955	.0045	93	2003	579	.7109	.2891
44	90938	447	.9951	.0049	94	1424	450	.6840	.3160
45	90491	484	.9947	.0054	95	974	341	.6499	.3501
46	90007	525	.9942	.0058	96	633	253	.6003	.3997
47	89482	569	.9937	.0064	97	380	185	.5132	.4869
48	88913	618	.9931	.0070	98	195	129	.3385	.6615
49	88295	671	.9924	.0076	99	66	66	.0000	1.0000

Based on the 1958 CSO Mortality Table prepared in cooperation with the National Association of Insurance Commissioners. Courtesy of the Society of Actuaries, Chicago, Illinois.

The expected value is

$$E = \$10,000 \times .0018$$
$$= \$18.00$$

This means that you should be willing to pay $18. But if the insurance company charges $18 for such a policy, their gain (in the long run) will be zero. The actual premium, then, will be $18 plus administrative costs and profit. ∎

Sometimes there are situations in which the amount you must pay to play a game has a role in the amount you will win, so that it is not feasible to compare the expectation with the amount you must pay to play. In such situations, you calculate the expenses for playing as part of the amount to win, and then you can have a positive, negative, or zero expectation. In this case, if the expectation is positive, then the game is in your favor; if it is negative, then the game is in your opponent's favor; and if the expectation is zero, then the game is said to be fair.

EXAMPLE 5

Walt, who is a realtor, knows that if he takes a listing to sell a house, it will cost him $1,000. However, if he sells the house, he will receive 6% of the selling price. If another realtor sells the house, Walt will receive 3% of the selling price. If the house is unsold in 3 months, he will lose the listing and receive nothing. Suppose the probabilities for selling a particular $100,000 house are as follows: The probability that Walt will sell the house is .4; the probability that another agent will sell the house is .2; and the probability that the house will remain unsold is .4. What is Walt's expectation if he takes this listing?

Solution

First calculate

$$6\% \text{ of } \$100,000: \quad .06 \times \$100,000 = \$6,000$$
$$3\% \text{ of } \$100,000: \quad .03 \times \$100,000 = \$3,000$$

Probability that Walt sells the house

Probability that another
agent sells the house

Probability that
house is not sold

Notice that the expense is
subtracted from the profit.

$$E = (\$6,000 - \$1,000)(.4) + (\$3,000 - \$1,000)(.2) + (\$0 - \$1,000)(.4)$$
$$= \$5,000(.4) + \$2,000(.2) + (-\$1,000)(.4)$$
$$= \$2,000 + \$400 + (-\$400)$$
$$= \$2,000$$

Walt's expectation is $2,000.   ∎

EXAMPLE 6

Remember the situation described in the beginning of this section? Your friend George shows you three cards. One is white on both sides, one is black on both sides, and the last is black on one side and white on the other. He mixes the cards and lets you select one at random and place it on the table. If you select the right color as being on the underside, you win $1; if you don't, you lose 75¢. Should you play?

Solution

Both these propositions can be answered by finding the expectation for the problem.

$$E = (\text{AMOUNT IF BLACK}) \times P(\text{BLACK})$$
$$+ (\text{AMOUNT IF WHITE}) \times P(\text{WHITE})$$

Notice: An *incorrect* assumption is that $P(\text{BLACK}) = P(\text{WHITE}) = \frac{1}{2}$. Recall that the empirical results of Experiment 6 showed that $P(\text{BLACK}) \neq \frac{1}{2}$. What is it then? Consider the sample space. Let's distinguish between the front (side 1) and back (side 2) of each card. The sample space of *equally likely* events is

	Card 1		Card 2		Card 3	
Side showing:	B_1	B_2	B	W	W_1	W_2
Side not showing:	B_2	B_1	W	B	W_2	W_1

We see that

$$P(\text{BLACK FACE DOWN}) = \frac{3}{6} = \frac{1}{2}$$

But we have additional information. We wish to compute the *conditional probability* of black, given that a black card is face up. Alter the sample space to take into account the additional information:

	Card 1		Card 2		Card 3	
Side showing:	B_1	B_2	B	W̸	W̸₁	W̸₂
Side not showing:	B_2	B_1	W	B̸	W̸₂	W̸₁

The conditional probability is

$$P\left(\begin{array}{c}\text{A BLACK IS ON THE BOTTOM}\\ \text{GIVEN THAT A BLACK IS ON THE TOP}\end{array}\right) = \frac{2}{3}$$

Thus, the desired expectation is

$$E = \left(\begin{array}{c}\text{AMOUNT}\\ \text{TO LOSE}\end{array}\right)\left(\begin{array}{c}\text{PROBABILITY}\\ \text{OF LOSING}\end{array}\right) + \left(\begin{array}{c}\text{AMOUNT}\\ \text{TO WIN}\end{array}\right)\left(\begin{array}{c}\text{PROBABILITY}\\ \text{OF WINNING}\end{array}\right)$$

$$= (-.75)\left(\frac{2}{3}\right) + (1)\left(\frac{1}{3}\right)$$

$$\approx -.17$$

Since the expectation is negative, you should not play. What amount would make this a fair game?   ■

Problem Set 11.4

1. Suppose you roll two dice. You will be paid $5 if you roll a double. You will not receive anything for any other outcomes. How much should you be willing to pay for the privilege of rolling the dice?

2. A magazine subscription service is having a contest in which the prize is $80,000. If the company receives one million entries, what is the expectation of the contest?

3. Suppose you have 5 quarters, 5 dimes, 10 nickels, and 5 pennies in your pocket. You reach in and choose a coin at random so you can tip your barber. What is the barber's expectation? What tip is the barber most likely to receive?

4. A game involves tossing two coins and receiving 50¢ if they are both heads. What is a fair price to pay for the privilege of playing?

5. Krinkles potato chips is having a "Lucky Seven Sweepstakes." The one grand prize is $70,000; 7 second prizes each pay $7,000; 77 third prizes each pay $700; and 777 fourth prizes each pay $70. How much is the expectation of this contest if there are ten million entries?

6. A punch-out card contains 100 spaces. One space pays $100; five spaces pay $10; and the others pay nothing. How much should you pay to punch out one space?

Use the mortality table (Table 11.14) to answer the questions in Problems 7–12.

7. What is the expected value for a 1 year $10,000 policy issued at age 10?

8. What is the expected value for a 1 year $10,000 policy issued at age 38?

9. What is the expected value for a 1 year $10,000 policy issued at age 65?

10. What is the expected value for a 1 year $20,000 policy issued at age 18?

11. What is the expected value for a 1 year $25,000 policy issued at age 19?

12. What is the expected value for a 1 year $50,000 policy issued at age 23?

13. In old gangster movies on TV, you often hear of "numbers runners" or the "numbers racket." The game, which is still played today, involves betting $1 and picking three digits, such as 245. Then, the next day some procedure for randomly selecting numbers is used. For example, the last three digits of the number of stocks sold on a particular day as reported in *The Wall Street Journal*. If the payoff is $500, what is the expectation for this numbers game?

14. A realtor who takes the listing on a house to be sold knows that she will spend $800 trying to sell the house. If she sells it herself, she will earn 6% of the selling price. If another realtor sells a house from her list, our realtor will earn only 3% of the price. If the house is unsold in 6 months, she will lose the listing. Suppose the probabilities are as follows:

	Probability
Sell by herself	.50
Sell by another realtor	.30
Not sell in 6 months	.20

What is the expected profit for listing an $85,000 house?

State and national lotteries are often held to increase government income. The prizes are large, but, because of the large number of tickets sold, the expectation is low. The percentage returned to the gamblers is much smaller than with other forms of legal gambling.

15. An oil drilling company knows that it will cost $25,000 to sink a test well. If oil is hit, the income for the drilling company will be $425,000. If natural gas only is hit, the income will be $125,000. If nothing is hit, they will have no income. If the probability of hitting oil is $\frac{1}{40}$ and if the probability of hitting gas is $\frac{1}{20}$, what is the expectation for the drilling company? Should they sink the test well?

16. In Problem 15, suppose the income for hitting oil is changed to $825,000 and the income for gas to $225,000. Now, what is the expectation for the drilling company? Should they sink the test well?

17. Suppose you roll one die. You are paid $5 if you roll a 1, and you pay $1 otherwise. What is the expectation?

18. A game involves drawing a single card from an ordinary deck. If an ace is drawn, you receive 50¢; if a heart is drawn, you receive 25¢; if the queen of spades is drawn, you receive $1. If the cost of playing is 10¢, should you play?

19. Consider the following game where a player rolls a single die. If a prime (2, 3, or 5) is rolled, the player wins $2. If a square (1 or 4) is rolled, the player wins $1. However, if the player rolls a perfect number (6), then it costs the player $11. Is this a good deal for the player or not?

Three wheels on a slot machine work independently, which means that the probability of a certain outcome is found by multiplying the individual probabilities on each wheel. Suppose that each wheel on a slot machine has 13 symbols as follows: 1 bar, 2 bells, 2 cherries, 2 lemons, 3 plums, and 3 oranges. Find the probability in Problems 20–27.

20. What is the probability of a jackpot (three bars)?

21. What is the probability of a cherry on the first wheel?

22. What is the probability of cherries on the first two wheels?

23. What is the probability of cherries on the first two wheels and a bar on the third wheel?

24. What is the probability of three cherries?

25. What is the probability of three oranges?

26. What is the probability of three plums?

27. What is the probability of three bells?

CALCULATOR PROBLEMS

28. A company held a contest, and the following information was included in the fine print:

Prize	Number of prizes available	Approximate probability of winning
$1,000	13	.000005
100	52	.00002
10	520	.0002
1	28,900	.010989
Total	29,485	.011111

Read this information carefully, and calculate the expectation for this contest.

29. A company held a bingo contest for which the following chances of winning were given:

Playing one card, your chances of winning are at least:

	One card 1 time	One card 7 times	One card 13 times
$25 prize	1 in 21,252	1 in 3,036	1 in 1,630
$3 prize	1 in 2,125	1 in 304	1 in 163
$1 prize	1 in 886	1 in 127	1 in 68
Any prize	1 in 609	1 in 87	1 in 47

What is the expectation for playing one card 13 times?

30. Calculate the expectation for the *Reader's Digest* Sweepstakes described below.

You have six chances to win

$100.00 A MONTH FOR LIFE:

Or, if you prefer, $24,000.00 cash or $2,000.00 a month for a year.
Your six Sweepstakes Numbers will give you six chances to win it --
plus any of 2,815 other cash prizes.

IMPORTANT! Mail this Official Sweepstakes Record Card
not later than October 20, 1972...to have six chances to win any prize.

----------Tear off here—You may keep top portion for your records if you wish—Mail Official Record Card today!------------

OFFICIAL PRIZE LIST

1 Prize of
$20,000.00 CASH

1 Prize of
$10,000.00 CASH

3 Prizes of
$5,000.00 CASH

10 Prizes of
$1,000.00 CASH

2,800 Prizes of
$25.00 CASH

© 1972, The Reader's Digest Association, Inc.

How the Sweepstakes Works:

1. All prizes will be awarded. Any prize set aside for a winning number that is not returned will be awarded in a drawing conducted by The Reuben H. Donnelley Corporation from all eligible entries submitted.

2. The approximate numerical odds of winning each prize are: Grand Prize—one in 6,851,000; $20,000—one in 15,970,000; $10,000—one in 15,970,000; $5,000—one in 5,323,000; $1,000—one in 1,597,000; $25—one in 5,704.

3. Names, cities and states of residence of major prize winners will be published in Reader's Digest; all winners will be notified by mail; a printed list of winners will be sent upon request if you enclose a self-addressed, stamped envelope.

4. No purchase is required.

5. This Sweepstakes will begin when the first entry is received and the final closing date is December 26, 1972. (Postmarks determine dates.) Closing date for the Grand Prize is October 20, 1972.

6. Winners must verify that they are not employees of Reader's Digest or The Reuben H. Donnelley Corporation or members of their immediate families.

7. This is a national Sweepstakes.

31. Using Problems 20–27, compute the expectation for playing a slot machine that gives the following payoffs:

First one cherry	2 coins
First two cherries	5 coins
First two cherries and a bar	10 coins
Three cherries	10 coins
Three oranges	14 coins
Three plums	14 coins
Three bells	20 coins
Three bars (jackpot)	50 coins

32. **St. Petersburg paradox.** Suppose you toss a coin and will win $1 if it comes up heads. If it comes up tails, you toss again. This time you will receive $2 if it comes up heads. If it comes up tails, toss again. This time you will receive $4 if it is heads. You continue in this fashion until you finally toss a head.

Would you pay $100 for the privilege of playing this game? What is the mathematical expectation for this game?

Gambler's Financial Ruin

In all casinos, the payoffs are adjusted so that the "house" has the advantage. This means that your expectation is less than the amount you must pay to play the game. (There is one exception; that is at certain times in the game of blackjack. A discussion of this is beyond the scope of this course, but the classic book on this topic is *Beat the Dealer* by Edward O. Throp.) Thus, if you continue to play long enough, regardless of your original bankroll, you will meet your financial ruin. These *times to ruin* are based on the expectation for the game and assume that a $1 bet is made each minute—with no time off.

"How did you do with your new blackjack system?"

Time to Gambler's Ruin

Game	Expectation per game ($1 bet)	Time required to lose $100 by betting $1 per minute		
		Days	Hours	Minutes
Roulette (with 0 and 00)				
1. Black or Red bet	$.9474	1	7	41
2. Odd or Even bet	.9474	1	7	41
3. Single number bet	.9211	0	21	7
4. Two number bet	.8947	0	15	49
5. Three number bet	.8684	0	12	39
6. Four number bet	.8421	0	10	33
7. Five number bet	.7895	0	7	55
8. Six number bet	.7895	0	7	55
9. Twelve number bet	.6316	0	4	31
Dice				
1. Eleven (or Three)	.8333	0	9	59
2. Twelve (or Two)	.8333	0	9	59
3. Craps	.7778	0	7	30
4. Seven	.6667	0	5	0
5. Any doubles	.6667	0	5	0
6. Field	.4444	0	3	0
7. Under Seven (or Over)	.4167	0	2	51
Slot machine (our hypothetical machine Problem 31)	.9203	0	20	55
Keno				
1. One Spot	.7500	0	6	40
2. Two Spot	.7515	0	6	42
3. Three Spot	.7412	0	6	26
4. Four Spot	.7502	0	6	40
5. Five Spot	.7458	0	6	33
6. Six Spot	.7492	0	6	38
7. Seven Spot	.7477	0	6	36
8. Eight Spot	.7345	0	6	16
9. Nine Spot	.7453	0	6	32
10. Ten Spot	.7498	0	6	39
11. Eleven Spot	.7445	0	6	31
12. Twelve Spot	.7448	0	6	31
13. Thirteen Spot	.7490	0	6	38
14. Fourteen Spot	.7468	0	6	34
15. Fifteen Spot	.7444	0	6	31

11.5 Review Problems

1. A die is rolled. What is the probability that it comes up:
   a. Three? **b.** Three *or* four? **c.** Three *and* four?
   Section 11.2

2. A card is selected from an ordinary deck of cards. What is the probability that it is:
   a. A diamond? **b.** A diamond *and* a two?
   c. A diamond *or* a two?
   Section 11.2

3. Define probability.
   Section 11.2

4. According to the Internal Revenue Service, the odds against a corporate tax return being examined are 15 to 1. What is the probability that a corporate tax return will be audited?
   Section 11.3

5. Two dice are tossed. What is the probability that the sum is 6 or at least one of the two faces is odd?
   Section 11.3

6. A jar contains three orange and two purple balls, and two are chosen at random. What is the probability of obtaining two orange balls if we know that the first drawn was orange and we draw the second ball:
   a. After replacing the first ball?
   b. Without replacing the first ball?
   Section 11.2

7. Two dice are thrown. What is the probability of at least one being a six? What is the probability of both being a six?
   Section 11.2

8. If $P(A) = \frac{5}{11}$, what is $P(\overline{A})$?
   Section 11.3

9. If Ferdinand the Frog has twice as much chance of winning the jumping contest as the other two champion frogs:
   a. What is the probability that Ferdinand will lose?
   b. What are the odds in favor of Ferdinand's winning?
   Section 11.3

10. A lottery offers a prize of a color TV (value $500), and 800 tickets are sold. What is the expectation if you buy three tickets? If the tickets cost $1 each, should you buy them? How much should you be willing to pay for the three tickets?
    Section 11.4

Statistics

12.1 Frequency Distributions and Graphs

SITUATION   "Nine out of ten dentists recommend Trident for their patients who chew gum." "Crest has been shown to be . . . ." "You can clearly see that Bufferin is the most effective . . . ." "Pennzoil is better suited . . . ." "Sylvania was preferred by . . . ."

"How can anyone analyze the claims of the commercials we see and hear on a daily basis?" asked Betty. "I even subscribe to *Consumer Reports*, but so many of the claims seem to be unreasonable. I hate to buy items by trial and error, but I also hate to believe all the claims in advertisements."

The first step in dealing with and understanding statistics is to be able to think critically and understand some of the ways that data can be organized. In this section, Betty will learn how to make a frequency distribution and then how to organize the data into a bar graph or a line graph.

THERE ARE THREE KINDS OF LIES ... LIES, DAMN LIES, AND STATISTICS!

Undoubtedly, you have some idea about what is meant by the term *statistics*. For example, we hear about:

1. The latest statistics on the cost of living
2. Statistics on population growth
3. The Gallup Poll's use of statistics to predict election outcomes
4. The Nielsen ratings, which indicate that one show has 30% more viewers than another
5. Baseball statistics

We could go on and on, but you can see from these examples that there are two main uses for the word **statistics**. First, we use the term to mean a mass of data, including charts and tables. Second, the word refers to *statistical methods*, which are techniques used in the collection of numerical facts, called *data*; the analysis and interpretation of these data; and, finally, the presentation of these data. In this chapter we'll introduce some of these statistical methods.

In the last chapter, we performed experiments and gathered data. For example, we rolled a pair of dice 50 times and the outcomes in one trial were:

3	2	6	5	3	8	8	7	10	9	7	5	12	9	6	11	8
11	11	8	7	7	7	10	11	6	4	8	8	7	6	4	10	
7	9	7	9	6	6	9	4	4	6	3	4	10	6	9	6	11

We can organize the above data in a convenient way by using a **frequency distribution**, as shown in Table 12.1.

TABLE 12.1 Frequency Distribution for an Experiment of Rolling a Pair of Dice 50 Times

Outcome	Tally	Frequency
2	|	1
3	|||	3
4	⊥⊦⊦	5
5	||	2
6	⊥⊦⊦ ||||	9
7	⊥⊦⊦ |||	8
8	⊥⊦⊦ |	6
9	⊥⊦⊦ |	6
10	||||	4
11	⊥⊦⊦	5
12	|	1

EXAMPLE 1

Make a frequency distribution for the information in Table 8.1, page 293.

Solution

Make three columns: First, list the sales tax categories; next, tally; finally, count the tallies to list the frequency of each:

Sales tax	Tally	Frequency
0%	⊥⊦⊦	5
1%		0
2%	|	1
3%	⊥⊦⊦ ⊥⊦⊦ ⊥⊦⊦ |	16
$3\frac{3}{4}$%	|	1
4%	⊥⊦⊦ ⊥⊦⊦ ||||	14
$4\frac{1}{8}$%	|	1
$4\frac{1}{2}$%	|	1
5%	⊥⊦⊦ |||	8
6%	|||	3
$7\frac{1}{2}$%	|	1
Total		51

■

To account for certain irregularities (such as $4\frac{1}{8}\%$ or $3\frac{3}{4}\%$) the categories can sometimes be grouped. One possibility for the information in Example 1 is shown by the horizontal lines. When the data are grouped as indicated, the frequency distribution looks a little more uniform:

Sales tax	Tally	Frequency
0–1%	⊬Ħ	5
1⁺–2%	I	1
2⁺–3%	⊬Ħ ⊬Ħ ⊬Ħ I	16
3⁺–4%	⊬Ħ ⊬Ħ ⊬Ħ	15
4⁺–5%	⊬Ħ ⊬Ħ	10
5⁺–6%	III	3
Over 6%	I	1

A disadvantage of a frequency distribution is that it lacks visual appeal. Data classified into groups are often represented by means of **bar graphs**. To construct a bar graph, draw a horizontal axis and a vertical axis, as shown in Figure 12.1. We have labeled the vertical axis "Frequency," and the horizontal axis represents the outcomes of the experiment of rolling a pair of dice. The data are from Table 12.1.

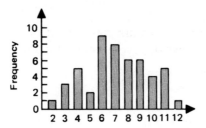

FIGURE 12.1   Outcomes of experiment of rolling a pair of dice

EXAMPLE 2

Construct a bar graph for the data given in Example 1. Use the grouped categories as shown in the table just following the example.

Solution

To construct a bar graph, draw and label horizontal and vertical axes, as shown in Figure 12.2a. It is helpful to use graph paper, although it isn't entirely necessary.

Next, draw in marks indicating the frequency, as shown in Figure 12.2b. Finally, complete the bars and shade them as shown in Figure 12.2c. ∎

FIGURE 12.2 Bar graph for sales tax data

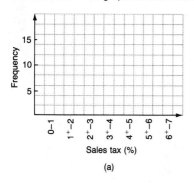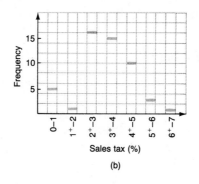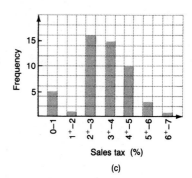

(a)

(b)

(c)

Another type of graph is the **line graph**, which uses dots that are marked on a grid and then connected in order.

EXAMPLE 3

Draw a line graph for the data given in Example 1. Use the grouped categories.

Solution

The line graph uses points instead of bars to designate the locations of the frequencies. These points are then connected by lines as shown in Figure 12.3. To plot the points, use the frequency distribution on page 426. First, find the category (0–1%, for example); then, plot a point showing the frequency (5 in this example). This step is shown in Figure 12.3a. The last step is to connect the dots, as shown in Figure 12.3b. ■

FIGURE 12.3 Line graph for sales tax data

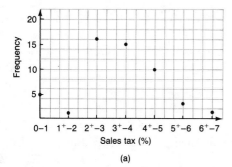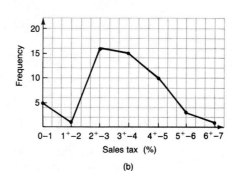

(a)

(b)

In addition to organizing data and constructing graphs, you should be able to read graphs.

EXAMPLE 4

The monthly rainfall for Honolulu and New Orleans is shown in Figure 12.4.

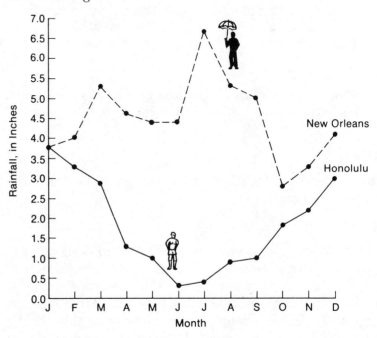

FIGURE 12.4 Monthly normal rainfall for two U.S. cities

a. During which month is there the most rain in New Orleans? In Honolulu?
b. During which month is there the least rain in New Orleans? In Honolulu?

Solution

a. It rains most in New Orleans in July and in Honolulu in January.
b. The least rain falls in New Orleans in October and in Honolulu in June. ■

Problem Set 12.1

1. The heights of 30 students are as follows (rounded to the nearest inch):

```
66  68  64  70  67  67  68  64  65  66
64  70  72  71  69  64  63  70  71  63
68  67  67  65  69  65  67  66  69  69
```

Give the frequency distribution, and draw a bar graph to represent these data.

2. The wages of the employees of a small accounting firm are as follows:

```
$10,000  $15,000  $15,000  $15,000
$20,000  $20,000  $25,000  $40,000
$60,000   $8,000   $8,000   $8,000
 $8,000   $8,000   $6,000   $4,000
```

Give the frequency distribution, and draw a bar graph to represent these data.

3. The following line graph shows the expenses for two sales-people of the Leadwell Pencil Company for each month of last year.

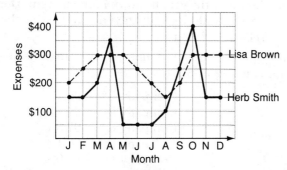

a. During which month did Herb incur the most expense?
b. During which month did Lisa incur the least expense?

4. The purchasing power of the dollar is shown below, to the nearest cent (where 1967 is considered the base year; that is, $1967 = \$1.00$):

```
1960  $1.13      1962  $1.10      1964  $1.07
1966  $1.03      1968  $0.96      1970  $0.86
1972  $0.80      1974  $0.68      1976  $0.59
1978  $0.49      1980  $0.38
```

Show these figures by using a line graph.

5. The amount of time it takes for three leading pain relievers to reach your bloodstream is as follows:

```
Brand A    480 seconds
Brand B    490 seconds
Brand C    500 seconds
```

a. Make a bar graph using the scale shown in Figure a.
b. Make a bar graph using the scale shown in Figure b.

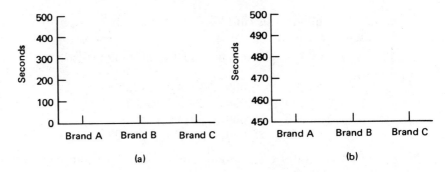

(a)   (b)

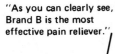

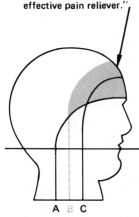

"As you can clearly see, Brand B is the most effective pain reliever."

A B C

Brands of Pain Relievers

c. If you were an advertiser working on a promotion campaign for Brand C, which graph would seem to give your product a more distinct advantage?
d. Consider the graph shown in the margin. Does it tell you anything at all about the effectiveness of the three pain relievers? Discuss.

6. The percentage of claims paid by five leading insurance companies is as follows:

Company A   96.2%
Company B   94.1%
Company C   97.6%
Company D   96.1%
Company E   95.9%

a. Make a bar graph using the scale shown in Figure a.
b. Make a bar graph using the scale shown in Figure b.

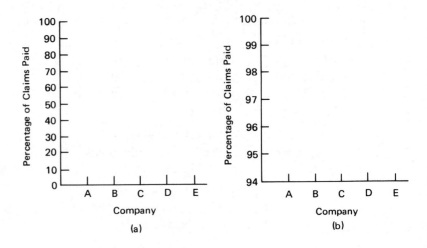

(a)   (b)

   c. If you were an advertiser working on a promotion campaign for Company C, which graph would seem to give your client a more distinct advantage?

7. Comment on the following graph and advertising statement:

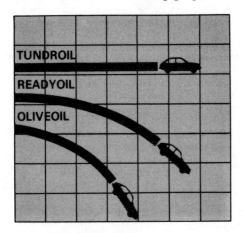

To prove that Tundroil is better suited for smaller, hotter, higher-revving engines we tested Tundroil against Readyoil and Oliveoil. As the graph above plainly shows, only Tundroil didn't break down.

8. Use the graphs shown here to answer the following questions:
   a. How many deaths occurred in 1975?
   b. What was the death rate per 100 million miles of travel in 1975?
   c. How many miles were traveled in 1976?
   d. Which year had the fewest number of deaths, and which year had the lowest death rate?

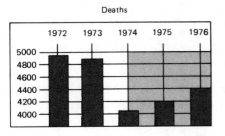

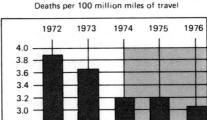

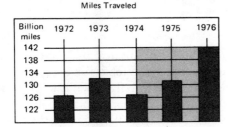

Shaded area reflects the speed limit change in 1974 to 55 mph.

9. Following is a list of holidays in 1984. Find the frequency distribution for the number of holidays celebrated on each day of the week, and draw a bar graph to represent the data.

January

1 (Sun)	New Year's Day
8 (Sun)	Battle of New Orleans (Louisiana)
16 (Mon)	Arbor Day (Florida)
19 (Thur)	Robert E. Lee's Birthday

February

12 (Sun)	Lincoln's Birthday
14 (Tue)	Admission Day (Arizona)
20 (Mon)	Washington's Birthday

March

2 (Fri)	Texas Independence Day (Texas) Town Meeting Day (Vermont)
25 (Sun)	Maryland Day (Maryland)
26 (Mon)	Kuhio Day (Hawaii) Seward's Day (Alaska)

April

2 (Mon)	Pascua Florida Day (Florida)
13 (Fri)	Jefferson's Birthday (Missouri)
19 (Thur)	Patriot's Day (Maine & Massachusetts)
20 (Fri)	Good Friday
21 (Sat)	San Jacinto Day (Texas)

22 (Sun)	Arbor Day (Nebraska)
26 (Thur)	Fast Day (New Hampshire) Confederate Memorial Day (Alabama & Mississippi)
30 (Mon)	Arbor Day (Utah)

May

4 (Fri)	Rhode Island Independence (Rhode Island)
10 (Thur)	Confederate Memorial Day (North Carolina & South Carolina)
20 (Sun)	Mecklenburg Day (North Carolina)
28 (Mon)	Memorial Day

June

3 (Sun)	Confederate Memorial Day (Kentucky & Louisiana)
4 (Mon)	Jefferson Davis's Birthday (Alabama & Mississippi)
14 (Thur)	Flag Day (Pennsylvania)
20 (Wed)	West Virginia Day (West Virginia)

July

4 (Wed)	Independence Day
24 (Tue)	Pioneer Day (Utah)

August

6 (Mon)	Colorado Day (Colorado)
13 (Mon)	Victory Day (Rhode Island)
16 (Thur)	Bennington Battle Day (Vermont)
17 (Fri)	Admission Day (Hawaii)
27 (Mon)	Lyndon Johnson's Birthday (Texas)
30 (Thur)	Huey B. Long's Birthday (Louisiana)

September

3 (Mon)	Labor Day
9 (Sun)	Admission Day (California)
12 (Wed)	Defenders Day (Maryland)

October

8 (Mon)	Columbus Day
31 (Wed)	Nevada Day (Nevada)

November

1 (Thur)	All Saints Day (Louisiana)
6 (Tue)	Election Day
11 (Sun)	Veterans Day
22 (Thur)	Thanksgiving

December

10 (Mon)	Wyoming Day (Wyoming)
19 (Wed)	Hanukkah
25 (Tue)	Christmas

10. The home-run champions for the American and National Leagues are listed in the table for every year from 1901 to 1980. Divide the data into five even categories, and construct a line graph for the number of players in each category. Do the National League and the American League separately, but put the graphs on the same coordinate system.

National League

Year	Player	No. of home runs	Year	Player	No. of home runs	Year	Player	No. of home runs	Year	Player	No. of home runs
1901	Crawford	16	1921	Kelly	23	1939	Mize	28	1959	Mathews	46
1902	Leach	6	1922	Hornsby	42	1940	Mize	43	1960	Banks	41
1903	Sheckard	9	1923	Williams	41	1941	Camilli	34	1961	Cepeda	46
1904	Lumley	9	1924	Fournier	27	1942	Ott	30	1962	Mays	49
1905	Odwell	9	1925	Hornsby	39	1943	Nicholson	29	1963	Aaron	44
1906	Jordan	12	1926	Wilson	21	1944	Nicholson	33		McCovey	
1907	Brain	10	1927	Wilson	30	1945	Holmes	28	1964	Mays	47
1908	Jordan	12		Williams		1946	Kiner	23	1965	Mays	52
1909	Murray	7	1928	Wilson	31	1947	Kiner	51	1966	Aaron	44
1910	Beck	10		Bottomley			Mize		1967	Aaron	39
	Schulte		1929	Klein	43	1948	Kiner	40	1968	McCovey	36
1911	Schulte	21	1930	Wilson	56		Mize		1969	McCovey	45
1912	Zimmerman	14	1931	Klein	31	1949	Kiner	54	1970	Bench	45
1913	Cravath	19	1932	Klein	38	1950	Kiner	47	1971	Stargell	48
1914	Cravath	19		Ott		1951	Kiner	42	1972	Bench	40
1915	Cravath	24	1933	Klein	28	1952	Kiner	37	1973	Stargell	44
1916	Robertson	12	1934	Ott	35		Sauer		1974	Schmidt	36
	Williams			Collins		1953	Mathews	47	1975	Schmidt	38
1917	Robertson	12	1935	Berger	34	1954	Kluszewski	49	1976	Schmidt	38
	Cravath		1936	Ott	33	1955	Mays	51	1977	Foster	52
1918	Cravath	8	1937	Ott	31	1956	Snider	43	1978	Foster	40
1919	Cravath	12		Medwick		1957	Aaron	44	1979	Kingman	48
1920	Williams	15	1938	Ott	36	1958	Banks	47	1980	Schmidt	48

American League

Year	Player	No. of home runs	Year	Player	No. of home runs	Year	Player	No. of home runs	Year	Player	No. of home runs
1901	Lajoie	13	1921	Ruth	59	1941	Williams	37	1962	Killebrew	48
1902	Seybold	16	1922	Williams	39	1942	Williams	36	1963	Killebrew	45
1903	Freeman	13	1923	Ruth	41	1943	York	36	1964	Killebrew	49
1904	Davis	10	1924	Ruth	46	1944	Etten	22	1965	Conigliaro	32
1905	Davis	8	1925	Meusel	33	1945	Stephens	24	1966	Robinson	49
1906	Davis	12	1926	Ruth	47	1946	Greenberg	44	1967	Yastrzemski	44
1907	Davis	8	1927	Ruth	60	1947	Williams	32		Killebrew	
1908	Crawford	7	1928	Ruth	54	1948	DiMaggio	39	1968	Howard	44
1909	Cobb	9	1929	Ruth	46	1949	Williams	43	1969	Killebrew	49
1910	Stahl	10	1930	Ruth	49	1950	Rosen	37	1970	Howard	44
1911	Baker	9	1931	Gehrig	46	1951	Zernial	33	1971	Melton	33
1912	Baker	10		Ruth		1952	Doby	32	1972	Allen	37
1913	Baker	12	1932	Foxx	58	1953	Rosen	43	1973	Jackson	32
1914	Baker	8	1933	Foxx	48	1954	Doby	32	1974	Allen	32
	Crawford		1934	Gehrig	49	1955	Mantle	37	1975	Jackson	36
1915	Roth	7	1935	Foxx	36	1956	Mantle	52		Scott	
1916	Pipp	12		Greenberg		1957	Sievers	42	1976	Nettles	32
1917	Pipp	9	1936	Gehrig	49	1958	Mantle	42	1977	Rice	39
1918	Ruth	11	1937	DiMaggio	46	1959	Colavito	42	1978	Rice	46
	Walker		1938	Greenberg	58		Killebrew		1979	Thomas	45
1919	Ruth	29	1939	Foxx	35	1960	Mantle	40	1980	Jackson	41
1920	Ruth	54	1940	Greenberg	41	1961	Maris	61		Oglivie	

11. Refer to the table of random numbers on page 384 (Table 11.10).
    a. Select 100 random digits. Make a frequency distribution for the digits 0, 1, 2, . . . , 8, 9. Draw a line graph.
    b. Repeat part a for a different 100 random digits. Draw a line graph.
    c. Combine the frequencies for the digits from parts a and b, and draw a line graph.

12. The typical percentage of family income allotted to various items in the budget is shown in the figure. Use this figure to answer the following questions.

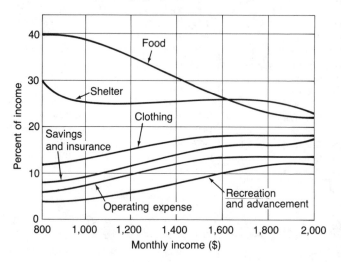

    a. If a family's monthly income increases from $1,000 to $1,200, which items remain about the same in percentage?
    b. If a family's income increases from $800 to $1,400, what is the percentage of change in the amount spent for food?
    c. Does a family with a monthly income of $800 spend more for food than a family with a monthly income of $1,200? How much more or less?

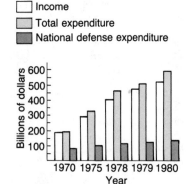

☐ Income
▨ Total expenditure
▰ National defense expenditure

Government income and expenditures for selected years

13. Answer the following questions by referring to the accompanying figure.
    a. In which year was the ratio of national defense expenditures to total expenditures the greatest?
    b. In which year was the ratio of national defense expenditures to income the greatest?
    c. In which year was the total expenditures over income the greatest?

14. Read the SAAB advertisement carefully. There is a fallacy in the statistics given about the car. See if you can find it.

There are two cars built in Sweden. Before you buy theirs, drive ours.

When people who know cars think about Swedish cars, they think of them as being strong and durable. And conquering some of the toughest driving conditions in the world.

But, unfortunately, when most people think about buying a Swedish car, the one they think about usually isn't ours. (Even though ours doesn't cost any more.)

Ours is the SAAB 99E. It's strong and durable. But it's also a lot different from their car.

Our car has Front-Wheel Drive for better traction, stability and handling.

It has a 1.85 liter, fuel-injected, 4-cylinder, overhead cam engine as standard in every car. 4-speed transmission is standard too. Or you can get a 3-speed automatic (optional).

Our car has four-wheel disc brakes and a dual-diagonal braking system so you stop straight and fast every time.

It has a wide stance. (About 55 inches.) So it rides and handles like a sports car.

Outside, our car is smaller than a lot of "small" cars. 172″ overall length, 57″ overall width.

Inside, our car has bucket seats up front and a full five feet across in the back so you can easily accommodate five adults.

It also has more headroom than a Rolls Royce and more room from the brake pedal to the back seat than a Mercedes 280. And it has factory air conditioning as an option.

There are a lot of other things that make our car different from their car. Like roll cage construction and a special "hot seat" for cold winter days.

So before you buy their car, stop by your nearest SAAB dealer and drive our car. The SAAB 99E. We think you'll buy it instead of theirs. **SAAB 99E**

Phone 800-243-6000 toll-free for the name and location of the SAAB dealer nearest you. In Connecticut, call 1-800-942-0655.

15. Collect examples of good statistical graphs and examples of misleading graphs. Use some of the leading newspapers and national magazines.

16. Read *How to Lie with Statistics*, by Darrell Huff (New York: Norton, 1954), and make a presentation to the class. Illustrate your presentation with graphs and examples from the book.

12.2 Descriptive Statistics

SITUATION   "We need privacy and a consistent wind," said Wilbur. "Did you write to the Weather Bureau to find a suitable location?"

"Well," replied Orville, "I received this list of possible locations and Kitty Hawk, North Carolina, looks just like what we want. Look at this . . . ."

However, Orville and Wilbur spent many days waiting in frustration after they arrived in Kitty Hawk because the winds weren't suitable. The Weather Bureau's information gave the averages, but the Wright brothers didn't realize that an acceptable average can be produced by unacceptable extremes. In this section, we'll discuss not only averages, but also measures of dispersion which would have helped Wilbur and Orville find a more suitable location for their experiments.

© 1961 United Feature Syndicate, Inc.

In the last section, we organized data into a frequency distribution and then discussed their representation as a graph. However, there are some properties of data that can help us to interpret masses in information.

Violet is using one of these properties in the cartoon. Do you suppose her dad necessarily bowled better on Monday nights than on Thursday nights? Don't be too hasty to say "yes" unless you first look at the scores making up these averages.

	Monday night	*Thursday night*
Game 1:	175	180
Game 2:	150	130
Game 3:	160	161
Game 4:	180	185
Game 5:	160	163
Game 6:	183	185
Game 7:	287	186
Totals:	1,295	1,190

To find the averages used by Violet, we divide the total scores by the number of games:

Monday night	Thursday night
$\dfrac{1{,}295}{7} = 185$	$\dfrac{1{,}190}{7} = 170$

However, if we consider the games separately, we see that Violet's dad did better on Thursday in five out of seven games. Are there any other properties of the bowling scores that would tell us this fact?

The average used by Violet is only one kind of statistical measure that can be used. It is the measure that most of us think of when we hear someone use the word *average*. It is called the **mean**. Other statistical measures, called **averages** or **measures of central tendency**, are given by the following definitions:

---

1. **Mean:** The number found by adding the data and then dividing by the number of data. The mean is usually denoted by $\bar{x}$.
2. **Median:** The middle number when the numbers are arranged in order of size. If there are two middle numbers (in the case of an even number of data), the median is the mean of these two middle numbers.
3. **Mode:** The value that occurs most frequently. If there is no number that occurs more than once, there is no mode. It is possible to have more than one mode.

Measures of Central Tendency: Mean Median Mode

Consider these other measures of central tendency for Violet's dad's bowling scores.

a. The median:

Monday night		Thursday night
150		130
160		161
160	The middle	163
175 ←	number is the →	180
180	median.	185
183		185
287		186

b. The mode:

Monday night	Thursday night
150	130
160	161
160	163
175	180
180	185
183	185
287	186

The mode is the number that occurs most frequently.

When we compare the three measures of the bowling scores, we find the following:

	Monday night	Thursday night
Mean:	185	170
Median:	175	180
Mode:	160	185

We are no longer convinced that Violet's dad did better on Monday nights than on Thursday nights.

EXAMPLE 1

Find the mean, median, and mode for the following sets of numbers:

a. 3, 5, 5, 8, 9   **b.** 4, 10, 9, 8, 9, 4, 5
c. 6, 5, 4, 7, 1, 9

Solution

a. *Mean:* $\dfrac{\text{Sum of the terms}}{\text{Number of terms}} = \dfrac{3 + 5 + 5 + 8 + 9}{5}$

$$= \frac{30}{5}$$

$$= 6$$

Median: Arrange in order: 3, 5, 5, 8, 9.
The middle term, 5, is the median.

Mode: The term that occurs most frequently is the mode, which is 5.

b. *Mean:* $\dfrac{4 + 10 + 9 + 8 + 9 + 4 + 5}{7} = \dfrac{49}{7}$

$$= 7$$

Median: 4, 4, 5, 8, 9, 9, 10
The median is 8.
Mode: This set of data is **bimodal**, with modes 4 and 9.

c. *Mean:* $\dfrac{6 + 5 + 4 + 7 + 1 + 9}{6} = \dfrac{32}{6}$

≈ 5.33

Median: 1, 4, $\underbrace{5, 6}$, 7, 9

$\dfrac{5 + 6}{2} = \dfrac{11}{2} = 5.5$

Mode: There is no mode. ∎

In the last section you worked with frequency distributions. Example 2 shows you how to find the measures of central tendency when you are presented with a frequency distribution.

EXAMPLE 2

Table 12.2 shows the number of days one must wait for a marriage license in the various states in the United States. Find the average wait for a marriage license using this information.

Solution

Mean: To find the mean, we could, of course, add all 50 individual numbers. But, instead, notice that

0 occurs 15 times, so write 0×15
1 occurs 3 times,   so write 1×3
2 occurs 2 times,   so write 2×2
3 occurs 21 times, so write 3×21
$\vdots$ $\vdots$
7 occurs 1 time,   so write 7×1

TABLE 12.2

Days' wait for a marriage license	Frequency (number of states)
0	15
1	3
2	2
3	21
4	1
5	7
6	0
7	1
Total	50

Thus, the mean is

$$\bar{x} = \frac{0\times15 + 1\times3 + 2\times2 + 3\times21 + 4\times1 + 5\times7 + 6\times0 + 7\times1}{50}$$

$$= \frac{0 + 3 + 4 + 63 + 4 + 35 + 0 + 7}{50}$$

$$= \frac{116}{50} = 2.32$$

Median: Since there are 50 values, the mean of the 25th and 26th largest values is the median. From Table 12.2, we see that the 25th term is 3 and the 26th term is 3, so the median is

$$\frac{3 + 3}{2} = \frac{6}{2} = 3$$

Mode: The mode is the value that occurs most frequently, which is 3.   ∎

The measures we've been discussing can help us interpret information, but they do not give the whole story. For example, consider these sets of data:

First example	Second example		First example	Second example
8	2	*Mean:*	9	9
9	9	*Median:*	9	9
9	9	*Mode:*	9	9
9	12			
10	13			

Notice that the two examples have the same mean, median, and mode, but the second set of data is more spread out than the first set. There are three **measures of dispersion** that we'll consider: the **range**, the **variance**, and the **standard deviation**.

The simplest measure of dispersion is the **range**.

Range

> The **range** of a set of data is the difference between the largest and the smallest numbers in the set.

EXAMPLE 3

Find the ranges for the data given above.

a. $10 - 8 = 2$ **b.** $13 - 2 = 11$

Even though the averages are the same, you can see from the ranges that the second example is much more dispersed.   ∎

Notice that the range is determined only by the largest and the smallest numbers in the set; it does not give us any information about the other numbers. It thus seems reasonable to invent another measure of dispersion that takes into account all the numbers in the data.

The **variance**, denoted by **var**, is a measure of dispersion which is found as follows:

1. Determine the mean of the numbers.
2. Subtract each number from the mean.
3. Square each of these differences.
4. Find the sum of the squares of these differences.
5. Divide this sum by 1 less than the number of pieces of data.

Variance

EXAMPLE 4

Find the variance for each set of data:

a. 8, 9, 9, 9, 10   **b.** 2, 9, 9, 12, 13

Solution

Remember that the mean, median, and mode for both these examples is the same (it is 9). The range for Example a is 2 and for Example b is 11. Now we want to find the second measure of dispersion, called *variance*. The procedure is lengthy, so make sure you follow each step carefully.

Step 1. Find the mean.

a. $\bar{x} = 9$ **b.** $\bar{x} = 9$

Step 2. Subtract each number from the mean.

a.

Data	Difference from the mean
8	1
9	0
9	0
9	0
10	−1

b.

Data	Difference from the mean
2	7
9	0
9	0
12	−3
13	−4

Step 3. Square each of these differences. Notice that some of the differences in Step 2 are positive and others are negative. Remember, we wish to find a measure of total dispersion. But if we add all these differences, we won't obtain the total variability. Indeed, if we simply add the differences for either example, the sum is zero. But we don't wish to say there is no dispersion. To resolve this difficulty with positive and negative differences, we square each difference, so the result will always be nonnegative.

a.

Data	Difference from the mean	Square of the difference
8	1	1
9	0	0
9	0	0
9	0	0
10	−1	1

b.

Data	Difference from the mean	Square of the difference
2	7	49
9	0	0
9	0	0
12	−3	9
13	−4	16

Step 4. Find the sum of these differences.

a. $1 + 0 + 0 + 0 + 1 = 2$　　**b.** $49 + 0 + 0 + 9 + 16 = 74$

Step 5. Divide this sum by 1 less than the number of terms. In each of these examples there are five pieces of data, so to find the variance, divide by 4:

a. $\text{var} = \dfrac{2}{4} = .5$　　　　　　**b.** $\text{var} = \dfrac{74}{4} = 18.5$　　■

The larger the variance, the more dispersion there is in the original data.

A third measure of dispersion, and by far the most commonly used, is called the **standard deviation**. This is found by taking the square root of the variance.

Standard Deviation

> The **standard deviation**, denoted by σ, is the square root of the variance. That is,
>
> $$\sigma = \sqrt{\text{var}}$$

The most efficient way of finding square roots is to use a

key marked ☑ on your calculator. However, if you don't have a calculator, or if your calculator doesn't have this key, you can use Table 12.3.

TABLE 12.3 Squares and Square Roots

n	n^2	$\sqrt{n}$	$\sqrt{10n}$	n	n^2	$\sqrt{n}$	$\sqrt{10n}$
1	1	1.000	3.162	51	2601	7.141	22.583
2	4	1.414	4.472	52	2704	7.211	22.804
3	9	1.732	5.477	53	2809	7.280	23.022
4	16	2.000	6.325	54	2916	7.348	23.238
5	25	2.236	7.071	55	3025	7.416	23.452
6	36	2.449	7.746	56	3136	7.483	23.664
7	49	2.646	8.367	57	3249	7.550	23.875
8	64	2.828	8.944	58	3364	7.616	24.083
9	81	3.000	9.487	59	3481	7.681	24.290
10	100	3.162	10.000	60	3600	7.746	24.495
11	121	3.317	10.488	61	3721	7.810	24.698
12	144	3.464	10.954	62	3844	7.874	24.900
13	169	3.606	11.402	63	3969	7.937	25.100
14	196	3.742	11.832	64	4096	8.000	25.298
15	225	3.873	12.247	65	4225	8.062	25.495
16	256	4.000	12.649	66	4356	8.124	25.690
17	289	4.123	13.038	67	4489	8.185	25.884
18	324	4.243	13.416	68	4624	8.246	26.077
19	361	4.359	13.784	69	4761	8.307	26.268
20	400	4.472	14.142	70	4900	8.367	26.458
21	441	4.583	14.491	71	5041	8.426	26.646
22	484	4.690	14.832	72	5184	8.485	26.833
23	529	4.796	15.166	73	5329	8.544	27.019
24	576	4.899	15.492	74	5476	8.602	27.203
25	625	5.000	15.811	75	5625	8.660	27.386
26	676	5.099	16.125	76	5776	8.718	27.568
27	729	5.196	16.432	77	5929	8.775	27.749
28	784	5.292	16.733	78	6084	8.832	27.928
29	841	5.385	17.029	79	6241	8.888	28.107
30	900	5.477	17.321	80	6400	8.944	28.284
31	961	5.568	17.607	81	6561	9.000	28.460
32	1024	5.657	17.889	82	6724	9.055	28.636
33	1089	5.745	18.166	83	6889	9.110	28.810
34	1156	5.831	18.439	84	7056	9.165	28.983
35	1225	5.916	18.708	85	7225	9.220	29.155
36	1296	6.000	18.974	86	7396	9.274	29.326
37	1369	6.083	19.235	87	7569	9.327	29.496
38	1444	6.164	19.494	88	7744	9.381	29.665
39	1521	6.245	19.748	89	7921	9.434	29.833
40	1600	6.325	20.000	90	8100	9.487	30.000
41	1681	6.403	20.248	91	8281	9.539	30.166
42	1764	6.481	20.494	92	8464	9.592	30.332
43	1849	6.557	20.736	93	8649	9.644	30.496
44	1936	6.633	20.976	94	8836	9.695	30.659
45	2025	6.708	21.213	95	9025	9.747	30.822
46	2116	6.782	21.448	96	9216	9.798	30.984
47	2209	6.856	21.679	97	9409	9.849	31.145
48	2304	6.928	21.909	98	9604	9.899	31.305
49	2401	7.000	22.136	99	9801	9.950	31.464
50	2500	7.071	22.361	100	10000	10.000	31.623

EXAMPLE 5

Find the standard deviation for the data:

a. 8, 9, 9, 9, 10   **b.** 2, 9, 9, 12, 13

Solution

We did most of the work in finding the standard deviation in Example 4. Now, all we need to do is find the square root of the answers to Example 4.

a. $\sqrt{.5} \approx .71$ **b.** $\sqrt{18.5} \approx 4.30$ ■

Problem Set 12.2

Find the mean, median, mode, range, variance, and standard deviation for each set of values in Problems 1–12.

1. 1, 2, 3, 4, 5

2. 17, 18, 19, 20, 21

3. 103, 104, 105, 106, 107

4. 765, 766, 767, 768, 769

5. 4, 7, 10, 7, 5, 2, 7

6. 15, 13, 10, 7, 6, 9, 10

7. 3, 5, 8, 13, 21

8. 1, 4, 9, 16, 25

9. 79, 90, 95, 95, 96

10. 70, 81, 95, 79, 85

11. 1, 2, 3, 3, 3, 4, 5

12. 0, 1, 1, 2, 3, 4, 16

13. Compare Problems 1–4. What do you notice about the mean and standard deviation?

14. By looking at Problems 1–4 and discovering a pattern, find the mean and standard deviation of the numbers 217,849, 217,850, 217,851, 217,852, and 217,853.

15. Find the mean, median, and mode of the following salaries of employees of the Moe D. Lawn Landscaping Company:

Salary	Frequency
$ 5,000	4
8,000	3
10,000	2
15,000	1

16. G. Thumb, the leading salesperson for the Moe D. Lawn Landscaping Company, turned in the summary of sales for the week of October 23–28 given in the margin. Find the mean, median, and mode.

Date	Number of clients contacted by G. Thumb
Oct. 23	12
Oct. 24	9
Oct. 25	10
Oct. 26	16
Oct. 27	10
Oct. 28	21

17. Find the mean, median, and mode of the following scores:

Test score	Frequency
90	1
80	3
70	10
60	5
50	2

18. A class obtained the following scores on a test:

Score	Frequency
90	1
80	6
70	10
60	4
50	3
40	1

Find the mean, median, mode, and range for the class.

19. A class obtained the test scores listed in the margin. Find the mean, median, mode, and range for the class.

Score	Frequency
90	2
80	4
70	9
60	5
50	3
40	1
30	2
0	4

20. Suppose a variance is zero. What can you say about the data?

21. A professor gives five exams. Two students' scores have the same mean, although one student seemed to do better on all tests except one. Give an example of such scores.

22. A professor gives six exams. Two students' scores have the same mean, although one student's scores have a small standard deviation and the other student's scores have a large standard deviation. Give an example of such scores.

23. The salaries for the executives of a certain company are shown below.

Position	Salary
President	$80,000
1st VP	30,000
2nd VP	30,000
Supervising Manager	24,000
Accounting Manager	20,000
Personnel Manager	20,000

Find the mean, median, and mode. Which measure seems to best describe the average executive salary for the company?

24. The number of miles driven on each of five tires was 17,000, 19,000, 19,000, 20,000, and 21,000 miles. Find the mean, range, and standard deviation for these mileages.

25. What is the average length of the words in the first two paragraphs of the Situation at the beginning of this section? Find the mean, median, and mode. Find the range and standard deviation of the word size in this problem.

26. Repeat Problem 25 for the last paragraph of the Situation at the beginning of this section.

27. Roll a single die until all six numbers occur at least once. Repeat the experiment 20 times. Find the mean, median, mode, and range of the number of tosses.

28. Roll a pair of dice until all eleven numbers occur at least once. Repeat the experiment 20 times. Find the mean, median, mode, and range of the number of tosses.

CALCULATOR PROBLEMS

Find the variance and standard deviation for the data indicated in Problems 29–34.

29.	Problem 17	30.	Problem 18	31.	Problem 19
32.	Problem 15	33.	Problem 16	34.	Problem 23

35. If you roll a die 36 times, the expected number of times for rolling each of the numbers is summarized in the table. A graph of these data is shown in the figure. Find the mean, variance, and standard deviation for this model.

Expected frequency from rolling a pair of dice 36 times

Outcome	Expected frequency
2	1
3	2
4	3
5	4
6	5
7	6
8	5
9	4
10	3
11	2
12	1

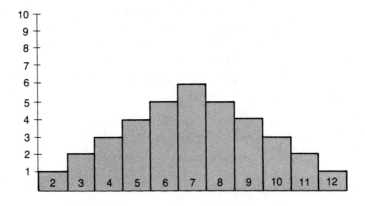

36. **Continuation of Problem 35.** Roll a pair of dice 36 times. Construct a table and a graph similar to the ones shown for Problem 35. Find the mean, variance, and standard deviation.

37. **Continuation of Problems 35 and 36.** Compare the results of Problems 35 and 36. If this is a class problem, you might wish to pool the entire class's results for Problem 36 before making the comparison.

38. Prepare a report or exhibit showing how statistics are used in baseball.

39. Prepare a report or exhibit showing how statistics are used in educational testing.

40. Prepare a report or exhibit showing how statistics are used in psychology.

41. Prepare a report or exhibit showing how statistics are used in business. Use a daily report of transactions on the New York Stock Exchange. What inferences can you make from the information reported?

42. Investigate the work of Quetelet, Galton, Pearson, Fisher, and Nightingale. Prepare a report or an exhibit of their work in statistics.

12.3 The Normal Curve

SITUATION "I signed up for Hunter in History 17," said Ben.

"Why did you sign up for him?" asked Ted. "Don't you know he has given only three As in the last fourteen years?!"

"That's just a rumor. Hunter grades on a curve," Ben replied.

"Don't give me that 'on the curve' stuff," continued Ted, "I'll bet you don't even know what that means. Anyway, my sister-in-law had him last year and said it was so bad that . . . ."

In this section, Ted and Ben will learn what it means to grade on a curve. Many everyday examples can be represented by bell-shaped, or normal, curves. Ted and Ben will see how to make certain predictions based on this normal distribution.

The cartoon suggests that there are more children with above-normal intelligence than with normal intelligence. But what do we mean by *normal* or *normal intelligence*?

Suppose we survey the results of 20 children's scores on an IQ test. The scores (rounded to the nearest 5 points) are: 115, 90, 100, 95, 105, 105, 95, 105, 105, 95, 125, 120, 110, 100, 100, 90, 110, 100, 115, and 80. The mean is 103, the standard deviation is 10.93, and the frequency is shown in Figure 12.5a.

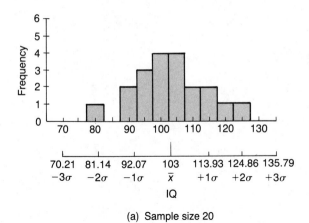

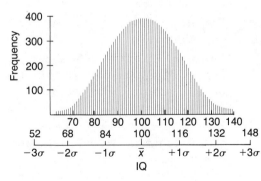

(a) Sample size 20

(b) Sample size 10,000

FIGURE 12.5   Frequencies of IQs

If we consider 10,000 IQ scores instead of only 20, we might obtain the frequency distribution shown in Figure 12.5b. As we can see, the frequency distribution approximates a curve. If we connect the end points of the bars in Figure 12.5b, we obtain a curve that is very close to something called a *normal frequency curve*, or simply a **normal curve**, as shown in Figure 12.6.

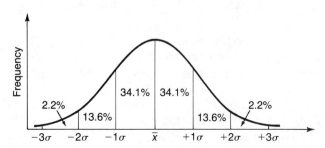

FIGURE 12.6   A normal curve

The distribution represented by the normal curve is very common and occurs in a wide variety of circumstances. For example, the heights of horses, the nodes on a jellyfish, baseball averages, the size of peas, the density of stars, the life span of light bulbs, chest size, and IQs all are normally distributed. That is, if we obtain the frequency distribution of a large number of measurements (as with the IQ example), the corresponding graph tends to look like a normal, or **bell-shaped**, curve.

The interesting and useful property of the normal curve is that roughly 68% of all values lie within one standard deviation above and below the mean. About 96% lie within two standard deviations, and virtually all (99.8%) values lie within three standard deviations of the mean. These percentages are the same regardless of the particular mean or standard deviation.

EXAMPLE 1

Use the percentages given in Figure 12.6 to predict the distribution of IQ scores of 1,000 people if you know that the mean is 100 and the standard deviation is 15.

Solution

First find the breaking points around the mean:

$$\bar{x} = 100; \quad \sigma = 15$$

$$\bar{x} + \sigma = 115 \quad \text{and} \quad \bar{x} - \sigma = 85$$
$$\bar{x} + 2\sigma = 130 \quad \text{and} \quad \bar{x} - 2\sigma = 70$$
$$\bar{x} + 3\sigma = 145 \quad \text{and} \quad \bar{x} - 3\sigma = 55$$

From Figure 12.6, about 34.1% of the people are within 1 standard deviation above and below the mean:

$$.341 \times 1,000 = 341$$

About 13.6% are between 1 and 2 standard deviations:

$$.136 \times 1,000 = 136$$

Finally, about 2.2% are between 2 and 3 standard deviations from the mean:

$$.022 \times 1,000 = 22$$

The remainder, if any, are divided above and below 3 standard deviations. The distribution is shown in the table:

Range	Percentage	Expected number
Below 55	.1	1
55–69	2.2	22
70–84	13.6	136
85–99	34.1	341
100–114	34.1	341
115–129	13.6	136
130–144	2.2	22
Above 144	.1	1
Totals	100.0	1,000

EXAMPLE 2

Professor Hunter claims to grade "on a curve." This means that he believes that the scores on a particular test are normally distributed. If 200 students take an exam with a mean of 73 and a standard deviation of 9, what are the grades Hunter would give?

Solution

First, draw a normal curve with mean 73 and standard deviation 9, as shown in Figure 12.7. The range of 73 to 82

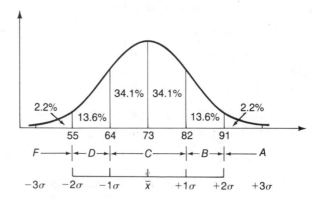

FIGURE 12.7

will contain about 34% of the class, the range of 82 to 91 will contain about 14% of the class, and about 2% of the class will score higher than 91. The lower section of the curve is figured the same way, so Hunter would grade this test according to the following table:

Grade on final	Letter grade	Number receiving grade	Percentage of class
Above 91	A	4	2%
83–91	B	28	14%
65–82	C	136	68%
55–64	D	28	14%
Below 55	F	4	2%

■

EXAMPLE 3

The Eureka Light Bulb Company tested a new line of light bulbs and found them to be normally distributed, with a

mean life of 98 hours and a standard deviation of 13.
a. What percentage of bulbs will last fewer than 72 hours?
b. What is the probability that a bulb selected at random will last more than 111 hours?

Solution

Draw a normal curve with mean 98 and standard deviation 13, as shown in Figure 12.8.

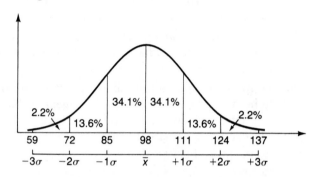

FIGURE 12.8

a. About 2% (2.2%) will last under 72 hours.
b. We see that about 16% (15.8%) of the bulbs last longer than 111 hours, so

$$P(\text{Bulb will last longer than 111 hours}) \approx .16 \quad \blacksquare$$

Problem Set 12.3

In Problems 1–5 suppose that people's heights (in centimeters) are normally distributed, with a mean of 170 and a standard deviation of 5. We take a sample of 50 people.

1. How many would you expect to be between 165 and 175 cm tall?

2. How many would you expect to be taller than 160 cm?

3. How many would you expect to be taller than 175 cm?

4. If a person is selected at random, what is the probability that he or she is taller than 165 cm?

5. What is the variance for this sample?

In Problems 6–10 suppose that for a certain exam, a teacher grades on a curve. It is known that the mean is 50 and the standard deviation is 5. There are 45 students in the class.

6. How many students should receive a C?

7. How many students should receive an A?

8. What score would be necessary to obtain an A?

9. If an exam paper is selected at random, what is the probability that it will be a failing paper?

10. What is the variance for this exam?

11. Suppose the breaking strength of a rope (in pounds) is normally distributed, with a mean of 100 pounds and a standard deviation of 16. What is the probability that a certain rope will break with a force of 132 pounds or less?

12. The diameter of an electric cable is normally distributed, with a mean of .9 inch and standard deviation of .01. What is the probability that the diameter will exceed .91 inch?

13. Suppose the annual rainfall in Ferndale, California, is known to be normally distributed, with a mean of 35.5 inches and a standard deviation of 2.5. About 2.2% of the time, the rainfall will exceed how many inches?

14. In Problem 13, what is the probability that the rainfall will exceed 30.5 inches in Ferndale?

15. The diameter of a pipe is normally distributed, with a mean of .4 inch and a variation of .0004. What is the probability that the diameter will exceed .44 inch?

16. The breaking strength (in pounds) of a certain new synthetic is normally distributed, with a mean of 165 and a variance of 9. The material is considered defective if the breaking strength is under 159 pounds. What is the probability that a sample chosen at random will be defective?

17. Suppose the neck size of men is normally distributed, with a mean of 15.5 inches and a standard deviation of .5. A shirt manufacturer is going to introduce a new line of shirts. How many of each of the following sizes should be included in a batch of 1,000 shirts?

    a. 14   **b.** 14.5   **c.** 15   **d.** 15.5   **e.** 16
    f. 16.5   **g.** 17

18. A package of Toys Galore Cereal is marked "Net Wt. 12 oz." The actual weight is normally distributed, with a mean of 12 oz and a variance of .04.

    a. What percentage of the packages will weigh under 12 oz?
    b. What weight will be exceeded by 2.2% of the packages?

19. Instant Dinner comes in packages that are normally distributed, with a standard deviation of .3 oz. If 2.2% of the dinners weigh more than 13.5 oz, what is the mean weight?

20. Select something that you think might be normally distributed (for example, the ring size of students at your college). Next, select a sample of 100 people, and make the appropriate measurements (in this example, ring size). Calculate the mean and standard deviation. Are your data normally distributed? Make a presentation of your findings to the class.

12.4 Sampling

> SITUATION   Suppose John hands you a coin to flip and wants to bet on the outcome. Now, John has tried this sort of thing before and you suspect that the coin is "rigged." You decide to test this hypothesis by taking a sample. You flip the coin twice, and it is heads both times. You say, "Aha, I knew it was rigged."
>
> John replies, "Don't be silly. Any coin can come up heads twice in a row."
>
> In this section you'll see how to test a hypothesis and you'll take a brief look at sampling, polls, and statistical predictions.

The first sections of this chapter dealt with the accumulation of data, measures of central tendency, and dispersion. However, a more important part of statistics is its ability to help us make predictions about a population based on a sample from that population. A **sample** is a small group of items chosen to represent a larger group. The larger group is called a **population**. Sampling necessarily involves some error, and therefore statistics is also concerned with estimating the error involved in predictions based on samples.

In national polls, everyone can't be interviewed; instead, predictions are based on the responses of a small sample. The inferences drawn from a poll can, of course, be wrong. In 1936 the *Literary Digest* predicted that Alfred Landon would defeat Franklin D. Roosevelt—who was subsequently reelected by a landslide. (The magazine ceased publication the following year.) In 1948 the *Chicago Daily Tribune* drew an incorrect conclusion from its polls and declared with a headline that Dewey was elected President. As recently as 1976, the *Milwaukee Sentinel* printed the erroneous headline shown here, which was based on the result of its polls.

In an attempt to minimize error in their predictions, statisticians follow very careful procedures:

Step 1. Propose some hypothesis about a population.
Step 2. Gather a sample from the population.
Step 3. Accept or reject the hypothesis. (It is also important to be able to estimate the error involved with making this decision.)

Suppose you want to decide whether a certain coin is a "fair" coin. You decide to test the hypothesis "This is a fair coin," and you test it by flipping the coin 100 times. This provides your *sample*. Suppose the result is

Heads: 55   Tails: 45

Do you accept or reject the hypothesis "This is a fair coin"? The expected number of heads is 50, but certainly a fair coin might well produce the results above.

As you can readily see, there are two types of error possible:

Type I: Rejection of the hypothesis when it is true.
Type II: Acceptance of the hypothesis when it is false.

How, then, can we proceed to minimize the possibility of making either error?

Let's carry our example further and repeat the experiment of flipping the coin 100 times:

Trial number	Number of heads
1	55
2	52
3	54
4	57
⋮	⋮

If the coin is fair and we repeat the experiment a large number of times, the distribution should be normal, with a mean of 50 and a standard deviation of 5, as shown in Figure 12.9.

Suppose you are willing to accept the coin as fair only if the number of heads falls between 45 and 55 ($\pm 1\sigma$). You

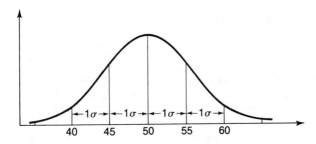

FIGURE 12.9

know you will be correct 68% of the time. But your friend says "Wait, you're rejecting a lot of fair coins! If you accept coins between 40 and 60 ($\pm 2\sigma$), you'll be correct 96% of the time."

You say "But suppose the coin really favors heads, so that the mean is 60 with a standard deviation of 5. I'd be accepting all the coins in the shaded region of Figure 12.10."

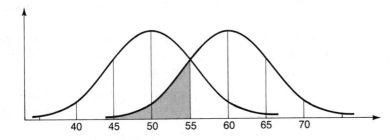

FIGURE 12.10

As you can see, a decrease in Type I error increases the Type II error, and vice versa. Deciding which type of error to minimize depends on the stakes involved as well as on some statistical calculations that are beyond the scope of this course. Consider a company that produces two types of valves. The first type is used in jet aircraft, and the failure of this valve might cause many deaths. A sample of the valves is taken, and the company must accept or reject the entire shipment. Under these circumstances, they would rather reject many good valves than accept a bad one.

On the other hand, the second valve is used in toy airplanes; the failure of this valve would cause the crash of the model. In this case, the company wouldn't want to reject too many good valves, so they would minimize the probability of rejecting the good valves.

Poll-taker to boss: "Our latest opinion poll showed that 90% of the people aren't interested in the opinions of others."

Problem Set 12.4

1. You are interested in knowing the number and ages of children (0–18 years) in a part (or all) of your community. You will need to sample 50 families and find out the number of children in each family and the age of each child. It is important that you select the 50 families at random.

    Step 1. Determine the geographical boundaries of the area with which you are concerned.

    Step 2. Consider various methods for selecting the families at random. For example, could you: (a) Select the first 50 homes at which someone is at home when you call? (b) Select 50 numbers from the phone book?

    Step 3. Consider different ways of asking the question. Can the way the family is approached affect the response?

    Step 4. Gather your data.

    Step 5. Organize your data. Construct a frequency distribution for the children, with integral values from 0 to 18.

    Step 6. Find out the number of families actually living in the area you've selected. If you can't do this, assume that the area has 1,000 families.

    a. What is the average number of children per family?

    b. What percentage of the children are in the first grade (age 6)?

    c. If all the children of ages 12–15 are in junior high, how many are in junior high for the geographical area you are considering?

    d. See if you can actually find out the answers to parts b and c, and compare these answers with your projections.

    e. What other inferences can you make from your data?

2. Five identical containers (shoe boxes, paper cups, etc.) must be prepared for this problem, with contents as follows:

Box	Contents (marbles, beads, colored pieces of paper, or poker chips)
#1	15 red, 15 white
#2	30 red, 0 white
#3	25 red, 5 white
#4	20 red, 10 white
#5	10 red, 20 white

Select one of the boxes at random so that you don't know its contents.

Step 1. Shake the box.
Step 2. Select one marker, note the result, and return it to the box.
Step 3. Repeat the first two steps 20 times with the same box.

a. What do you think is inside the box you've sampled?
b. Could you have guessed the contents by repeating the experiment five times? Ten times? Do you think you should have more than 20 observations? Discuss.

12.5 Review Problems

1. Make a frequency table and draw a bar graph for the following data: 7, 8, 8, 7, 8, 9, 6, 9, 8, 6, 8, 9, 10, 7, 7.

   Section 12.1

2. Consider the data in the table, which show the U.S. government's expenditures for social welfare.

   Section 12.1

   Social Welfare Expenditures

Year	Expenditure (in billions of dollars)
1975	289
1976	331
1977	361
1978	393

   a. *Viewpoint 1: Social welfare expenditures rose only 32 billion dollars from 1977 to 1978.* Draw a graph showing very little increase in the expenditures.
   b. *Viewpoint 2: Social welfare expenditures rose $32,000,000,000 from 1977 to 1978.* Draw a graph showing a large growth in expenditures.

3. Consider the following data: 23, 24, 25, 26, 27. Find:

   Section 12.2

   a. Mean **b.** Median **c.** Mode **d.** Range
   e. Variance **f.** Standard deviation

4. A small grocery store stocked several sizes of Copycat Cola last year. The sales figures were as shown in the table. Find the mean, median, and mode for these data. If the store manager decides to cut back the variety and stock only one size, which measure of central tendency will be most useful in making this decision?

   Section 12.2

Size	Number of cases sold
6 oz	5
10 oz	10
12 oz	35
16 oz	50

Section 12.2   **5.** A student's scores in a certain math class are 72, 73, 74, 85, and 91. Find:
                a. Mean   **b.** Median   **c.** Mode   **d.** Range
                e. Variance   **f.** Standard deviation

Section 12.2   **6.** The earnings of the employees of a small real estate company for the past year were (in thousands of dollars) 12, 18, 15, 9, 11, 22, 18, 17, and 10. Find the mean, median, and mode. Which measure of central tendency is most representative of the employee's earnings?

Section 12.3   **7.** The heights of 10,000 women students at Eastern College are known to be normally distributed, with a mean of 67 inches and a standard deviation of 3.
       a. How many women would you expect to be between 64 and 70 inches tall?
       b. How many women would you expect to be shorter than 5'1"?

Section 12.3   **8.** Determine the upper and lower scores for the middle 68% of a normally distributed test in which

$$\bar{x} = 73 \qquad \text{and} \qquad \sigma = 8$$

Section 12.3   **9. a.** Two instructors gave the same Math 1A test in their classes. Both classes had the same mean, but one class had a standard deviation twice as large as that of the other class. Which class do you think would be easier to teach, and why?
       b. A student received a score of 91 on a history test for which the mean was 86 and the standard deviation was 5. She also received a score of 79 on a mathematics test for which the mean was 75 and the standard deviation was 2. On which test did she have the better score in comparison to her classmates?

Section 12.4   **10.** How would you obtain a sample of one-syllable words used in this text? How would you apply those results to the whole book?

Graphs

CHAPTER 13

13.1 Ordered Pairs and the Cartesian Coordinate System

SITUATION   Jack had not been to San Francisco before and he was looking forward to seeing Fisherman's Wharf, North Beach, Nob Hill, Chinatown, and the Civic Center. However, the first thing he wanted to do was to look up his old friend, Charlie, who lives on Main Street in San Francisco. He bought a map and looked up Charlie's street in the index. It was listed as (7, B). What does (7, B) mean?

In mathematics, such a representation is called an *ordered pair* and has many useful applications. The first one we'll consider in this section is called a *Cartesian coordinate system* and sets the stage for the remainder of this chapter.

Have you ever drawn a picture by connecting the dots or found a city or street on a map? If you have, then you've used the ideas we'll be discussing in this section. Suppose we

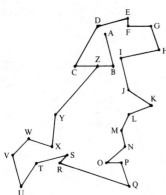

FIGURE 13.1   Map of San Francisco

wish to find Fisherman's Wharf in San Francisco. We look in the index for the map in Figure 13.1 and find Fisherman's Wharf listed as (6, D). Can you locate section (6, D) in Figure 13.1?

Have you ever been given the coordinates for a street and still not been able to find it easily? For example, suppose we want to find Main Street. Without the index the task is next to impossible, but even with the index listing Main at (7, B) the task of finding the street is not an easy one. How could we improve the map?

Let's create a smaller grid, as shown in Figure 13.2. Now Main Street is located at (9, G) and is easier to find. However, a smaller grid means that a lot of letters are needed for the vertical scale, and we might need more letters than we have. So let's use a notation that will allow us to represent points on the map with pairs of numbers. A pair of numbers written as

$$(2, 3)$$

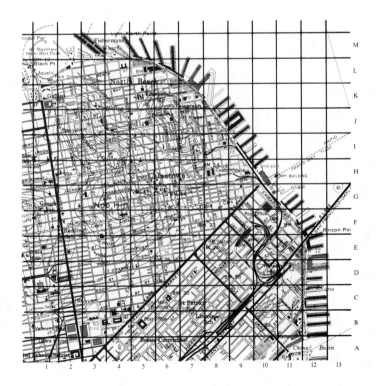

FIGURE 13.2 Map of
San Francisco with refined grid

is called an **ordered pair** to remind you that the order in which the numbers 2 and 3 are listed is important. In this example, 2 is the **first component** and 3 is the **second component**.

First component ⌐
$$(x, y) \quad \text{is an ordered pair}$$
└── Second component

For our map, suppose we relabel the vertical scale with numbers and change both scales so that we label the lines instead of the spaces, as shown in Figure 13.3. (By the way, most technical maps number lines instead of spaces.) Now we can fix the location of any street on the map quite precisely. Notice that if we use an ordered pair of numbers (instead of

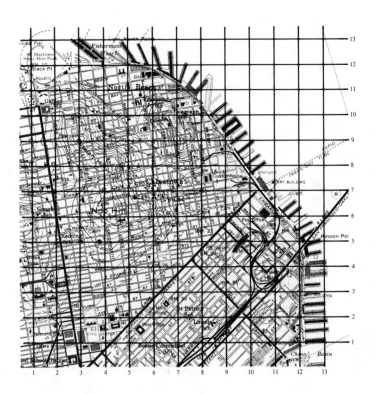

FIGURE 13.3 Map of
San Francisco with coordinate grid

a number and a letter), it is important to know which component of the ordered pair represents the horizontal distance and which component represents the vertical distance. Now, if you use Figure 13.3, you'll see that the coordinates of Main Street are about (9.5, 6.3), or we can even say that Main Street runs from about (9.5, 6.3) to (11.2, 4.5). Notice that, by using ordered pairs and numbering the lines instead of the spaces, we have refined our grid. We've refined it even more with decimal components. When using pairs of numbers instead of numbers and letters, we must remember that the first component is on the horizontal axis and the second component is on the vertical axis.

EXAMPLE 1

Find the major landmarks located in Figure 13.3 by the following coordinates:

		Answers:
a.	(3.5, 1.5)	Civic Center
b.	(3.5, 12.5)	Fisherman's Wharf
c.	(5, 11)	North Beach
d.	(4, 6)	Nob Hill
e.	(6, 7.3)	Chinatown ■

There are many ways to use ordered pairs to find particular locations. For example, a teacher may make out a seating chart like the one shown in Figure 13.4.

	1	2	3	4	5
5	Otis Morehouse	Wayne Savick	Harvey Dunker	Richard Giles	Shannon Smith
4	Vicki Switzer	Harold Peterson	Bob Anderson	Josephine Lee	Milt Hoehn
3	Jeff Atz	Carol Olmstead	Todd Humann	Jim Kintzi	Sharon Boschen
2	Terry Shell	Ralph Earnest	Steve Switzer	Clint Stevenson	
1	Rosamond Foley	Amy Olmstead	Missy Smith	Niels Sovndal	Eva Mikalson

Front

FIGURE 13.4 Seating chart

In the grade book, the teacher records

Anderson $(3, 4)$ Remember, first component is horizontal direction,
Atz $(1, 3)$ second component is vertical direction.
$\vdots$

Anderson's seat is represented in column 3, row 4. Can you think of some other ways in which ordered pairs could be used to locate a position?

The idea of using an ordered pair to locate a certain position is associated with particular terminology. **Axes** are two perpendicular real number lines, such as those in Figure 13.5a. The point of intersection of the axes is called the **origin**. Remember that in Chapter 2 we associated direction to the right or up with positive numbers. The arrows in Figure 13.5a are pointing in the positive direction. These perpendicular lines are usually drawn so that one is horizontal and the other is vertical. The horizontal axis is called the **x-axis**, and the vertical axis is called the **y-axis**. Notice that these axes divide the plane into four parts, which are called **quadrants** and are arbitrarily labeled as shown in Figure 13.5b.

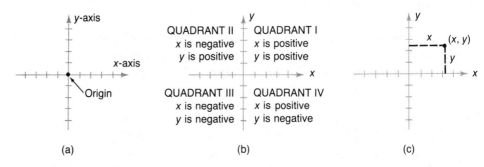

(a)   (b)   (c)

FIGURE 13.5   Cartesian coordinate system

We can now label points in the plane by using ordered pairs. The first component of the pair gives the horizontal distance and the second component gives the vertical distance, as shown in Figure 13.5c. If x and y are components of the point, then this representation of the point (x, y) is called the **rectangular** or **Cartesian coordinates** of the point.

EXAMPLE 2

Plot (graph) the given points.

a. $(5, 2)$ **b.** $(3, 5)$ **c.** $(-2, 1)$
d. $(0, 5)$ **e.** $(-6, -4)$ **f.** $(3, -2)$
g. $(\frac{1}{2}, 0)$ **h.** $(0, 0)$ **i.** $(5, 0)$

Solution

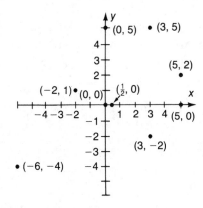

Problem Set 13.1

Find the landmarks given by the coordinates in Problems 1–9. Use the map in Figure 13.3.

1. $(7, 10)$ **2.** $(1, 2)$ **3.** $(11, 6)$
4. $(12, 0.5)$ **5.** $(11, 7.3)$ **6.** $(1.5, 11.3)$
7. $(6.5, 2.5)$ **8.** $(6.5, 4.5)$ **9.** $(6.9, 9.8)$

Using the seating chart in Figure 13.4, name the occupant of the seat given by the coordinates in Problems 10–15.

10. $(5, 1)$ **11.** $(4, 1)$ **12.** $(1, 2)$
13. $(4, 2)$ **14.** $(2, 3)$ **15.** $(3, 1)$

Find the points given in Problems 16–20. Use the indicated scale.

16. Scale: 1 square on your paper = 1 unit
     a. $(1, 2)$ **b.** $(3, -3)$ **c.** $(-4, 3)$
     d. $(-6, -5)$ **e.** $(0, 3)$

17. Scale: 1 square on your paper = 1 unit
     a. $(-1, 4)$ **b.** $(6, 2)$ **c.** $(1, -5)$
     d. $(3, 0)$ **e.** $(-1, -2)$

18. Scale: 1 square on your paper = 10 units
   a. $(10, 25)$ **b.** $(-5, 15)$ **c.** $(0, 0)$
   d. $(-50, -35)$ **e.** $(-30, -40)$

19. Scale: 1 square on your paper = 50 units
   a. $(100, -225)$ **b.** $(-50, -75)$ **c.** $(0, 175)$
   d. $(-200, 125)$ **e.** $(50, 300)$

20. Scale, x-axis: 1 square on your paper = 1 unit
   Scale, y-axis: 1 square on your paper = 10 units
   a. $(4, 40)$ **b.** $(-3, 0)$ **c.** $(2, -80)$
   d. $(-5, -100)$ **e.** $(0, 0)$

21. Plot the following coordinates on graph paper, and connect each point with the preceding one: $(-2, 2)$, $(-2, 9)$, $(-7, 2)$, $(-2, 2)$.   Start again: $(-6, -6)$, $(-6, -8)$, $(10, -8)$, $(10, -6)$, $(6, -6)$, $(6, 0)$, $(10, 0)$, $(10, 2)$, $(6, 2)$, $(6, 16)$, $(-2, 14)$, $(-11, 2)$, $(-9, -1)$, $(-2, -1)$, $(-2, -6)$, $(-6, -6)$.

22. Plot the following coordinates on graph paper, and connect each point with the preceding one: $(9, 0)$, $(7, -\frac{1}{2})$, $(6, -2)$, $(8, -5)$, $(5, -8)$, $(-7, -10)$, $(-1, -5)$, $(0, -2)$, $(-2, -1)$, $(-6, -5)$, $(-5, -3)$, $(-6, -2)$, $(-5, -1)$, $(-6, 0)$, $(-5, 1)$, $(-6, 2)$, $(-5, 3)$, $(-6, 4)$, $(-5, 5)$, $(-6, 7)$, $(-2, 3)$, $(0, 4)$, $(-1, 10)$, $(-7, 15)$, $(3, 10)$, $(6, 5)$, $(4, 3)$, $(5, 2)$, $(7, 2)$, $(9, 0)$.   Finally, plot a point at $(7, 1)$.

23. Suppose we place coordinate axes on the connect-the-dots figure shown in the margin. Let A be $(2, 9)$ and Z be $(1, 5)$. Write directions similar to those of Problems 21 and 22 about how to sketch this figure.

24. Connect the points $(0, 2)$, $(1, 5)$, $(2, 8)$, $(-1, -1)$, $(-2, -4)$.   What do you observe?

25. **a.** How many points lie on a line?
   b. How many points do you need to plot in order to determine a line?

26. Plot five points with 2 as the first component. Connect the plotted points.

27. Plot five points with 3 as the second component. Connect the plotted points.

28. Draw a picture similar to the ones in Problems 21 and 22, and then describe the picture using ordered pairs.

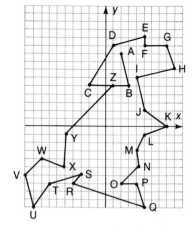

13.2 Lines

SITUATION   Linda is on a business trip and needs to rent a car for the day. The car rental agency has the following options on the car she wants to rent:

>   *Option A:* $40 plus 50¢ per mile
>   *Option B:* Flat $60 with unlimited mileage

Which car should she rent? The answer, of course, depends on the number of miles she intends to drive. In this section, Linda will learn about certain relationships called *lines*, and how to use the ideas of a Cartesian coordinate system to help graph linear relationships.

Many situations involve two variables or unknowns related in some specific fashion. The situation described above relates the cost of a car rental, c, to the number of miles driven in the following way:

Option A:

COST = BASIC CHARGE + MILEAGE CHARGE

┌─ 50¢ per mile

COST =   40   + .5(NUMBER OF MILES DRIVEN)

$$c = 40 + .5m$$

Option B: COST = FLAT FEE

$$c = 60$$

Suppose we represent these relationships on a **graph** as described by a Cartesian coordinate system. We will find ordered pairs (m, c) that make these equations true.

Option A: Let $m = \mathbf{10}$: $c = 40 + .5m$
$$c = 40 + .5(\mathbf{10})$$
$$= 40 + 5$$
$$= 45$$

If Linda drives 10 miles ($m = 10$), the cost of the rental is $45. Plot the point $(10, 45)$ as shown in Figure 13.6.

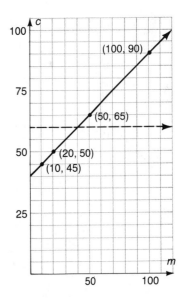

FIGURE 13.6   Car rental problem

Let $m = \mathbf{20}$: $c = 40 + .5m$
$c = 40 + .5(\mathbf{20})$
$= 40 + 10$
$= 50$

If she drives 20 miles, the cost of the rental is $50. Plot (20, 50).

Let $m = \mathbf{50}$: $c = 40 + .5m$
$c = 40 + .5(\mathbf{50})$
$= 40 + 25$
$= 65$

Plot (50, 65).

Let $m = \mathbf{100}$: $c = 40 + .5m$
$c = 40 + .5(\mathbf{100})$
$= 40 + 50$
$= 90$

Plot (100, 90).

From the graph, we can see that these points all lie on the same line. Draw a line through these points, as shown in Figure 13.6.

Option B: $c = 60$

The second component of the ordered pair (m, c) is always 60 in this case, regardless of the number of miles, m. This is shown as a dashed line in Figure 13.6.

EXAMPLE 1

Use Figure 13.6 to estimate the mileage for which both rates are the same.

Solution

The solution is the point of intersection. Estimate the coordinates, as shown in Figure 13.7—it looks like (40, 60). This means that if Linda expects to drive more than 40 miles, she should take the fixed rate.   ■

In order to graph a line, find two ordered pairs that lie on the line. Then find a third point as a check. Draw the line passing through these three points. If the three points don't lie on a straight line, then you have made an error.

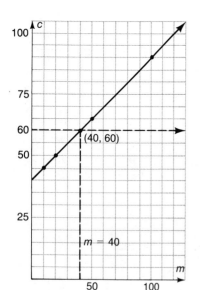

FIGURE 13.7 Comparing car rental rates

EXAMPLE 2

Graph: $y = 2x + 2$

Solution

It is generally easier to pick x and find y.

If $x = 0$: $y = 2 \times 0 + 2$
$\qquad\qquad = 0 + 2$
$\qquad\qquad = 2$ Plot $(0, 2)$ as shown in Figure 13.8.

If $x = 1$: $y = 2 \times 1 + 2$
$\qquad\qquad = 2 + 2$
$\qquad\qquad = 4$ Plot $(1, 4)$.

If $x = 2$: $y = 2 \times 2 + 2$
$\qquad\qquad = 4 + 2$
$\qquad\qquad = 6$ Plot $(2, 6)$. This is the check point. Two points determine a line—this point checks your work.

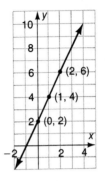

FIGURE 13.8   Graph of $y = 2x + 2$

Draw the line passing through the three plotted points. ∎

Problem Set 13.2

Graph the lines in Problems 1–21.

1. $y = x + 3$
2. $y = x + 1$
3. $y = 2x + 1$
4. $y = 3x - 2$
5. $y = x - 1$
6. $y = 2x + 3$
7. $x + y = 0$
8. $x - y = 0$
9. $2x + y = 3$
10. $3x + y = 10$
11. $3x + 4y = 8$
12. $x + 2y = 4$
13. $x = 5$
14. $x = -4$
15. $x = -1$
16. $y = -2$
17. $y = 6$
18. $y = 4$
19. $x + y + 100 = 0$
20. $5x - 3y = 45$
21. $y = 50x$

22. Suppose a car rental agency gave you the following choices:

    Option A: $30 per day plus 40¢ per mile
    Option B: Flat $50 per day

    Write equations for options A and B.

23. Graph the equations you found in Problem 22.

24. Use the graph in Problem 23 to estimate the mileage for which both rates are the same.

25. A paint sprayer rents for $4 per hour or $24 per day. Write equations for these options, assuming that you need the sprayer for one day or less.

26. Graph the equations you found in Problem 25.

27. Use the graph in Problem 26 to estimate the number of hours for which both rates are the same.

28. If an amount of money A results from investing a sum P at a simple interest rate of 7% for 10 years,

$$A = P + .7P$$
$$A = 1.7P$$

Graph this equation. Let $P = 0$, $P = 5{,}000$, and $P = 10{,}000$ be your choices for the first component.

29. An independent distributor bought a vending machine for $2,000 with a probable scrap value of $100 at the end of its expected 10 year life. The value V at the end of n years is given by

$$V = 2{,}000 - 190n$$

Graph this equation. Let $n = 0$, $n = 5$, and $n = 10$ be your choices for the first component.

CALCULATOR PROBLEM

30. Internal Revenue Service (IRS) regulations allow taxpayers a $1,000 deduction for each dependent and a 15% standard deduction on the amount of income remaining. For a person with income I and four dependents (including herself), the amount to be taxed can be found as follows:

$$\text{INCOME} - \left(\begin{matrix} \text{DEPENDENT} \\ \text{DEDUCTION} \end{matrix} \right) - \left(\begin{matrix} \text{STANDARD} \\ \text{DEDUCTION} \end{matrix} \right)$$

$$= \quad I \quad - \quad 4(1{,}000) \quad - \quad .15I$$
$$= .85I - 4{,}000$$

If the tax is 25% of the final amount, then the relationship between income and tax can be expressed as

$$T = .25(.85I - 4{,}000)$$
$$T = .2125I - 1{,}000$$

Graph this equation. Choose $I = 10{,}000$, $I = 12{,}000$, and $I = 14{,}000$ as your choices for the first component.

13.3 Parabolas

SITUATION   "Well, Billy, can I make it?" asked Evel. "If you can accelerate to the proper speed, and if the wind is not blowing too much, I think you can," answered Billy. "It will be one huge money maker, but I want some assurance that it can be done!" retorted Evel. In 1974 Evel Knievel attempted a skycycle ride across the Snake River.

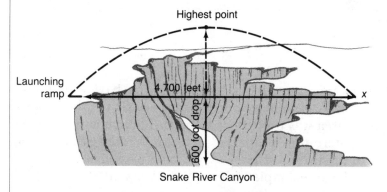

In this section, Evel will learn about a mathematical curve called a *parabola* and the fact that a projectile propelled upward will follow the path of a parabola. Although the mathematics for Evel's ride is rather involved, the mathematics necessary for answering Evel's question is developed in this section (see Problem 24).

When cannons were first introduced in the 13th century, their primary use was to demoralize the enemy. It was much later that they were employed for strategic purposes. In fact, cannons existed nearly three centuries before enough was known about the behavior of a projectile to use them with accuracy. The cannonball does not travel in a straight line, it was discovered, because of an unseen force called *gravity*. The path described by a projectile is called a **parabola**. Any projectile—a ball, an arrow, a bullet, a rock from a slingshot, and even water from the nozzle of a hose or sprinkler—will travel a parabolic path. Consider Figure 13.9 (page 472). It is a scale drawing (a graph) of the path of a cannonball fired in a particular way.

Note that the parabolic curve has a maximum height and is symmetric about a vertical line through that height. In other words, the ascent and descent paths are symmetric.

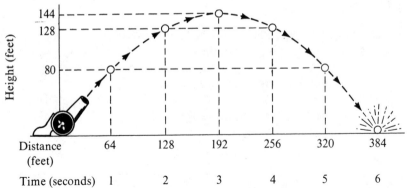

FIGURE 13.9

Perhaps this curve looks familiar. If you have ever used a flashlight to illuminate your way in the dark, the shape of the lighted area was probably parabolic. This smooth curve has many interesting properties and applications. Every equation of the form $y = ax^2 + bx + c$ has a parabola as its graph. Let's take an example with $a = 1$, $b = 0$, and $c = 0$.

EXAMPLE 1

Graph $y = x^2$.

Solution

We will choose x-values and find some corresponding y-values.

Let $x = \mathbf{0}$: $y = \mathbf{0}^2$
                         $= 0$ Plot $(0, 0)$.
Let $x = \mathbf{1}$: $y = \mathbf{1}^2$
                         $= 1$ Plot $(1, 1)$.
Let $x = \mathbf{-1}$: $y = (\mathbf{-1})^2$
                         $= 1$ Plot $(-1, 1)$.

Notice that the points do not fall in a line. If we find two more points, we can see the shape of the graph.

Let $x = \mathbf{2}$: $y = \mathbf{2}^2$
                         $= 4$ Plot $(2, 4)$.
Let $x = \mathbf{-2}$: $y = (\mathbf{-2})^2$
                         $= 4$ Plot $(-2, 4)$.

Connect the points to form a smooth curve, as shown in Figure 13.11.   ∎

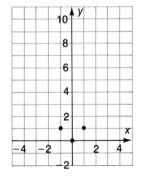

FIGURE 13.10

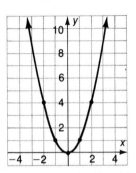

FIGURE 13.11   Graph of $y = x^2$

The curve shown in Figure 13.11 is called a **parabola** and is said to **open up**. The lowest point, $(0, 0)$ in Example 1, is called the **vertex**. In this book, we'll only consider parabolas that open up or down. Example 2 shows a parabola that **opens down**. Notice that the number a in $y = ax^2$ is negative.

EXAMPLE 2

Sketch $y = -\frac{1}{2}x^2$.

Solution

$$
\begin{aligned}
\text{Let } x = \mathbf{0}: \quad & y = 0 & & \text{Plot } (0, 0). \\
x = \mathbf{1}: \quad & y = -\tfrac{1}{2}(\mathbf{1})^2 & & \\
& = -\tfrac{1}{2} & & \text{Plot } (1, -\tfrac{1}{2}). \\
x = \mathbf{-1}: \quad & y = -\tfrac{1}{2}(\mathbf{-1})^2 & & \\
& = -\tfrac{1}{2} & & \text{Plot } (-1, -\tfrac{1}{2}). \\
x = \mathbf{2}: \quad & y = -\tfrac{1}{2}(\mathbf{2})^2 & & \\
& = -2 & & \text{Plot } (2, -2). \\
x = \mathbf{4}: \quad & y = -\tfrac{1}{2}(\mathbf{4})^2 & & \\
& = -8 & & \text{Plot } (4, -8).
\end{aligned}
$$

Connect these points to form a smooth curve, as shown in Figure 13.12. ■

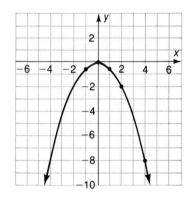

FIGURE 13.12 Graph of $y = -\frac{1}{2}x^2$

Notice from Example 2 that you can use symmetry for points on one side of the vertex.

It is possible that the vertex is not at $(0, 0)$, as shown by Example 3.

EXAMPLE 3

Graph $y = x^2 - 6x + 4$.

Solution

We use the same procedure to find points on the graph by choosing x-values and finding the corresponding y-values.

$$
\begin{aligned}
\text{Let } x = \mathbf{0}: \quad & y = \mathbf{0}^2 - 6 \cdot \mathbf{0} + 4 & & \\
& = 4 & & \text{Plot } (0, 4). \\
x = \mathbf{1}: \quad & y = \mathbf{1}^2 - 6 \cdot \mathbf{1} + 4 & & \\
& = 1 - 6 + 4 & & \\
& = -1 & & \text{Plot } (1, -1).
\end{aligned}
$$

$$x = 2: \quad y = 2^2 - 6 \cdot 2 + 4$$
$$= 4 - 12 + 4$$
$$= -4 \qquad \text{Plot } (2, -4).$$
$$x = 3: \quad y = 3^2 - 6 \cdot 3 + 4$$
$$= 9 - 18 + 4$$
$$= -5 \qquad \text{Plot } (3, -5).$$
$$x = 4: \quad y = 4^2 - 6 \cdot 4 + 4$$
$$= 16 - 24 + 4$$
$$= -4 \qquad \text{Plot } (4, -4).$$
$$x = 5: \quad y = 5^2 - 6 \cdot 5 + 4$$
$$= 25 - 30 + 4$$
$$= -1 \qquad \text{Plot } (5, -1).$$
$$x = 6: \quad y = 6^2 - 6 \cdot 6 + 4$$
$$= 36 - 36 + 4$$
$$= 4 \qquad \text{Plot } (6, 4).$$

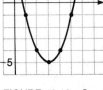

FIGURE 13.13 Graph of $y = x^2 - 6x + 4$

Connect these points into a smooth curve as shown in Figure 13.13. From the graph, you can see that the vertex is $(3, -5)$. ■

Problem Set 13.3

Sketch the graph of each equation in Problems 1–21.

1. $y = 3x^2$
2. $y = 2x^2$
3. $y = -x^2$
4. $y = -2x^2$
5. $y = -5x^2$
6. $y = 5x^2$
7. $y = \frac{1}{3}x^2$
8. $y = -\frac{1}{3}x^2$
9. $y = \frac{1}{10}x^2$
10. $y = -\frac{1}{10}x^2$
11. $y = x^2 - 4$
12. $y = 9 - x^2$
13. $y = -3x^2 + 4$
14. $y = 2x^2 - 3$
15. $y = x^2 - 2x + 1$
16. $y = x^2 + 4x + 4$
17. $y = -2x^2 + 4x - 2$
18. $y = \frac{1}{4}x^2 - \frac{1}{2}x + \frac{1}{4}$
19. $y = \frac{1}{2}x^2 + x + \frac{1}{2}$
20. $y = x^2 - 2x + 3$
21. $y = 3x^2 + 12x + 14$

22. In 1635 Galileo discovered that the distance, d, an object falls can be described in terms of the length of time, t, that it falls. Consider the *B.C.* cartoon shown here.

The equation for a falling object is

$$d = 16t^2$$

if we disregard air resistance and measure d in feet and t in seconds. Use this equation to find the depth of the well in the cartoon.

23. **a.** Graph $d = 16t^2$.
    b. Does the parabola in part a open up or down?
    c. If we relate the graph in part a with the falling object of Problem 22, can t be negative? Draw the graph of $d = 16t^2$ for nonnegative values of t.

CALCULATOR PROBLEM

24. The path of Evel Knievel's skycycle described in the situation at the beginning of this section is

$$y = -0.0005x^2 + 2.39x$$

assuming that the ramp is at the origin and x is the horizontal distance traveled. Graph this relationship. Using your graph, answer Evel's question: Will he make it? Assume that the actual distance the skycycle must travel is 4,700 ft.

13.4 Exponential and Logarithmic Graphs

SITUATION Marsha and Tony have owned a piece of property within the city limits of their small town for a number of years. Now, a developer is threatening to build a large housing project on the land next to their property, but to do so the developer is seeking a zoning change from the City Planning Commission. The hearing is next week, and Tony figures that the population projections presented by the developer are vague and unfounded. What Tony needs is a mathematical method for projecting population. He also needs to do it so that it can be understood by the Planning Commission. In this section, Marsha and Tony will learn how to draw a graph to make population projections, and they will also see how to draw logarithmic graphs.

In the first three sections of this chapter we saw how equations of lines and parabolas could be graphed to form a visual representation for an algebraic representation of variables. In this section, we'll extend that idea in order to draw the graphs of two less elementary equations.

An **exponential equation** is one in which a variable appears as an exponent. Consider the equation

$$y = 2^x$$

This equation represents a doubling process. The graph of such an equation is called an **exponential curve**, and is illustrated by Example 1.

EXAMPLE 1

Sketch the graph of $y = 2^x$ for nonnegative values of x.

Solution

Choose x-values and find corresponding y-values. These values form ordered pairs (x, y). Plot enough ordered pairs so that you can see the general shape of the curve, and then connect the points with a smooth curve.

$$\text{If } x = 0: \quad y = 2^0$$
$$= 1 \qquad \text{Plot } (0, 1).$$
$$x = 1: \quad y = 2^1$$
$$= 2 \qquad \text{Plot } (1, 2).$$
$$x = 2: \quad y = 2^2$$
$$= 4 \qquad \text{Plot } (2, 4).$$
$$x = 3: \quad y = 2^3$$
$$= 8 \qquad \text{Plot } (3, 8).$$
$$x = 4: \quad y = 2^4$$
$$= 16 \qquad \text{Plot } (4, 16).$$

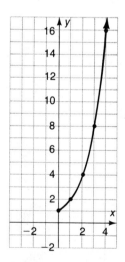

FIGURE 13.14   Graph of $y = 2^x$

Connect the points with a smooth curve, as shown in Figure 13.14.   ■

Population growth is described by an exponential equation. The population, P, at some future date can be predicted if you know the present population, P_0, and the growth rate, r. The predicted population in t years is given by the equation

$$P = P_0(2.72)^{rt}$$

EXAMPLE 2

Tony calls his local Chamber of Commerce and finds that the growth rate of his town is now .5%. Also, according to the 1980 census, the population is 2,500. Draw a graph showing the population between the years 1980 and 2000. (A calculator is necessary for this example.)

Solution

$P_0 = 2,500$ and $r = .005$ Remember, to change a percent to a decimal, move the decimal point two places to the left.

The equation to graph is

$$P = 2,500(2.72)^{.005t}$$

If $t = 0$, then: $P = 2,500(2.72)^0$ This is the 1980 base year,
                        $= 2,500$ and corresponds to the **present time** (even if it is now 1983). Thus, if $t = 5$, then the population is for 1985. If $t = 10$, the population is for 1990.

If $t = 10$, then: $P = 2,500(2.72)^{.005(10)}$
                        $= 2,500(2.72)^{.05}$

Press: | 2.72 | | Yˣ | | .05 | | = | | × | | 2500 | | = |
Display: 2628.2608   This means that the predicted 1990 population is 2,628. Plot the point (10, 2,628).

If $t = 20$, then: $P = 2,500(2.72)^{.005(20)}$
                        $= 2,500(2.72)^{.1}$

Press: | 2.72 | | Yˣ | | .1 | | = | | × | | 2500 | | = |
Display: 2763.1019   This means that the predicted population in the year 2000 is 2,763. Plot (20, 2,763).

When we talked about graphs representing data in Section 12.1, we noted the effect that the chosen scale can have on the visual impact and interpretation of the data. This example illustrates the same point. Figure 13.15a shows the data as represented by Tony, who wants to show little predicted growth; and Figure 13.15b shows the same data as represented by the developer, who wants to show considerable growth.

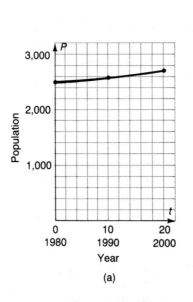

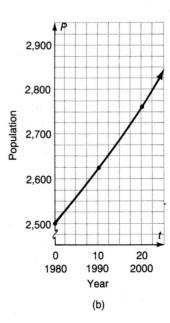

FIGURE 13.15 Graph of $P = 2,500(2.72)^{.005t}$ shown using two different scales

(a)

(b) ∎

In Chapter 3 we defined a logarithm. The graph of a logarithm is similar to the graph of an exponential, as shown by Example 3. For this section, we'll need another notation for logarithm. Remember,

$$y = \log x$$

means that y is the logarithm to the base 10 of x. Now write

$$y = \log_b x$$

to mean that y is the logarithm to the base b of x.

EXAMPLE 3

Graph $y = \log_2 x$.

Solution

This equation means y is the logarithm to the base 2 of x. A table of values is shown on page 116, but part of it is repeated here for convenience:

x	1	2	4	8	16
$y = \log_2 x$	0	1	2	3	4

Remember to plot (x, y), as shown in Figure 13.16.

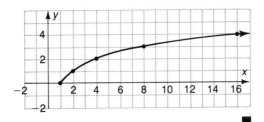

FIGURE 13.16 Graph of $y = \log_2 x$

The graphs of $y = 2^x$ and $y = \log_2 x$ are shown together in Figure 13.17 for comparison. If you were to place a mirror on the line between the two curves, you would see that these curves are mirror images. This is true for any logarithmic and exponential curves with the same base.

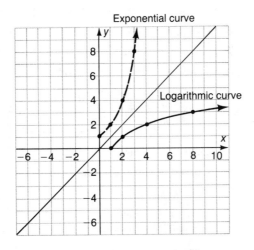

FIGURE 13.17 Graphs of $y = 2^x$ and $y = \log_2 x$

We conclude with one additional logarithmic curve, the familiar common log.

EXAMPLE 4

Graph $y = \log x$.

Solution

For $y = \log x$, use Table 3.1 on page 120 or use a calculator.

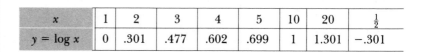

x	1	2	3	4	5	10	20	$\frac{1}{2}$
$y = \log x$	0	.301	.477	.602	.699	1	1.301	−.301

$$\begin{aligned} \log 20 &= \log(2 \times 10) \\ &= 1.301 \\ \log .5 &= \log(5 \times 10^{-1}) \\ &= .699 - 1 \\ &= -.301 \end{aligned}$$

The graph is shown in Figure 13.18.   ∎

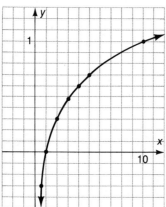

FIGURE 13.18   Graph of $y = \log x$

Problem Set 13.4

Sketch the graph of the equations in Problems 1–9 for nonnegative values of x.

1. $y = 3^x$ **2.** $y = 4^x$ **3.** $y = 5^x$
4. $y = 6^x$ **5.** $y = 7^x$ **6.** $y = 10^x$
7. $y = (\frac{1}{2})^x$ **8.** $y = (\frac{1}{3})^x$ **9.** $y = (\frac{1}{10})^x$

10. Consider the following table for $y = \log_3 x$:

x	1	3	9	27	81
$y = \log_3 x$	0	1	2	3	4

Graph $y = \log_3 x$.

11. Consider the following table for $y = \log_4 x$:

x	1	4	16	64	256
$y = \log_4 x$	0	1	2	3	4

Graph $y = \log_4 x$.

12. Consider the following table for $y = \log_5 x$:

x	1	5	25	125	625
$y = \log_5 x$	0	1	2	3	4

Graph $y = \log_5 x$.

CALCULATOR PROBLEMS

13. Change the growth rate in Example 2 to 1%, and graph the population curve.

14. Change the growth rate in Example 2 to 3.5%, and graph the population curve.

15. Change the growth rate in Example 2 to 1.5%, and graph the population curve.

16. Predict the population of your city or state in the year 2000.

13.5 Review Problems

1. a. Plot five points so that the first and second components are the same. Draw the line passing through the plotted points.

    b. Plot five points so that the first component is 5. Draw the line passing through the plotted points.

Section 13.1

2. Graph $y = 3x - 2$.

Section 13.2

3. Graph $y = 25$.

Section 13.2

4. A rototiller rents for $8 per day or $40 per week. Steve wants to rent the tiller for one week or less.

    a. Write equations for these options.

    b. Graph the equations.

    c. Use the graph to find the number of days for which both rates are the same.

Section 13.2

5. Graph $y = 4x^2$.

Section 13.3

6. Graph $y = 5 - x^2$.

Section 13.3

7. Graph $y = x^2 + 6x + 11$.

Section 13.3

8. Graph $y = (\frac{1}{4})^x$ for nonnegative x.

Section 13.4

9. Graph $y = 2(\log x)$ for x greater than or equal to one. Let $x = 0$, $x = 10$, and $x = 100$ to help you find points for graphing this curve.

Section 13.4

CALCULATOR PROBLEM

10. Graph $P = P_0(2.72)^{rt}$ for $P_0 = 53{,}000$ and $r = 1.3\%$. Let $t = 0$, $t = 10$, and $t = 20$ to help you find points for graphing this curve.

Computers

CHAPTER 14

14.1 Introduction to Computers

SITUATION   "I'll bet Dave is the best Space Invader player in the county," said Turner. "I heard he even bought his own computer so that he could practice at home."

"Well," retorted Jimmy, "I don't know how he could justify spending $2,000 for a computer to play a game. What is this world coming to?"

"Oh, Dave does a lot more with his computer than play games," said Turner. "He uses it to analyze the stock market, write letters, keep his records, and he is even tied in to a service to receive his news directly through his computer."

"I'll tell you," Jimmy continued, "You'll never see me dependent on computers. I've got plenty to spend my money on, and it won't be a computer. I think we should just forget about computers. They are an invasion of my privacy anyway!"

In this section, Jimmy and Turner will be introduced to computers. Jimmy's negative feelings toward computers may be caused by the fact that he knows nothing about them. Jimmy will learn that computers are already so much a part of our lives that none of us could do without them. Both Jimmy and Turner will learn about the five main components of a computer in this section, and then in the later sections of this chapter they will learn how to talk to a computer.

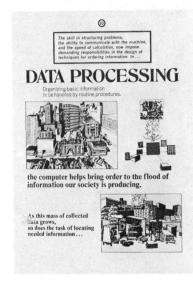

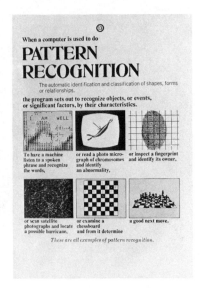

Computers are used for a variety of purposes. **Data processing** involves large numbers of data on which relatively few calculations need to be made. For example, a credit card company could use the system to keep track of the balances of its customers' accounts. **Information retrieval** involves locating and displaying material from a description of its content. For example, a police department may want information about a suspect, or a realtor may want to find listings that meet the criteria of a client. **Pattern recognition** is the identification and classification of shapes, forms, or relationships. For example, a chess player will examine the chessboard carefully to determine a good next move. Finally, a computer can be used for **simulation** of a real situation. For example, a prototype of a new spacecraft can be tested under various circumstances to determine its limitations.

The latest, and perhaps most significant, development in computers is the appearance of **personal computers**. The Apple, TRS-80, and Atari 8800 are all personal desktop computers that are designed for individual use.

Radio Shack TRS-80 line of microcomputers

The machine, or computer itself, is referred to as **hardware**. A basic microcomputer might sell for under $1,000, but many added "extras," called **peripherals**, can increase the price by several thousand dollars.

Contrary to some of the advertising you might have seen, these personal computers take a certain amount of know-how to operate. You can purchase packages or programs that instruct the computer to perform certain operations or tasks. These are referred to as **software**. A Space Invaders game is an example of a software package. There are also software packages to keep track of a family budget, calculate your income tax, play games, or analyze stock market trends. In any case, the owner of a computer must be willing to put in many hours learning to use the computer. This chapter will give you an introduction to using a computer.

A computer carries out tasks by accepting information, performing mathematical and logical operations with the information, and then supplying the results of the operations as new information. Since there are two ways of dealing with numbers and quantities (counting and measuring), historically there were two types of computers: (1) **digital** computers (counting) and (2) **analog** computers (measuring). The fundamental difference between the two types of computers is that digital computers work with discrete numbers, whereas analog computers work with continuous information. The turnstile counter at a fair represents a digital device, since the indicator shows only discrete numbers from 00000 to 99999. On the other hand, the odometers (mileage indicator) on many cars are analog devices. They show continuous change and may stop at any place. Table 14.1 lists some familiar analog and digital de-

TABLE 14.1   Analog and Digital Devices

Analog	Digital
Odometer	Turnstile
Thermometer	Telephone dial
Burning candle	Calendar
Spring balance	Adding machine
Watch with hands	Digital watch

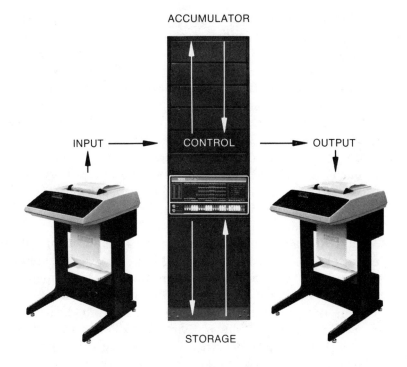

ACCUMULATOR

INPUT ⟶ CONTROL ⟶ OUTPUT

STORAGE

FIGURE 14.1 Relationships among the five main components of a digital computer

vices. Today, when we say *computer*, we mean *digital computer*, so the rest of this chapter refers only to digital computers.

Every computing system is made up of five main components: **input**, **storage** or memory, **accumulator** or arithmetic unit, **control**, and **output**. (See Figure 14.1.) **Input** refers to the procedure of entering data or instructions into a computer. The most common method used by most students is on a **CRT keyboard** (see Figure 14.2a). You type in your

FIGURE 14.2 Methods of inputting information into a computer

(a) CRT terminal

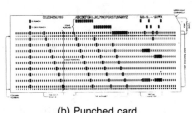

(b) Punched card

(c) Magnetic tape

instructions or messages manually, just as on a typewriter, and you see it on a video screen. The computer then responds on the video screen. This response is called **output**. Another common output device is a **line printer**, which has the advantage of outputting a great deal of data at a rate that is much faster than can be read by an operator (Figure 14.1).

Another method of inputting data is with the familiar **punched card** (see Figure 14.2b). Punched cards are prepared on a machine called a **keypunch**, which has a keyboard similar to a typewriter and can produce new cards or duplicate old ones. Many computers read **mark-sense cards**, which are like punched cards but do not need to have the holes punched in the cards.

Magnetic tape is another way to store or enter information into a computer (see Figure 14.2c). This is a very fast method of inputting information into a computer. Many home computers use ordinary cassette tapes for this purpose.

Some computers have unconventional input devices that allow them to "see," "hear," or "speak," but we will not discuss these devices here because they are still considered experimental. (See page 494.)

The **storage** unit stores information until it is needed for some purpose. Storage might be visualized as a long corridor in a hotel. Each door is numbered, and behind each door is a limited number of "slots" to hold information. Each compartment (door) is capable of holding one piece of information or one of the instructions in a program.

The number associated with each of these slots is called the *location* or **address**. The information inside the box is called a **word**. During the process of solving a problem, the storage unit contains:

ADDRESS 1	INFORMATION
ADDRESS 2	INFORMATION
ADDRESS 3	IS
ADDRESS 4	STORED
ADDRESS 5	IN
ADDRESS 6	EACH
ADDRESS 7	NUMBERED
ADDRESS 8	LOCATION

1. A set of instructions that specify the sequence of operations required to complete the problem
2. The input data
3. The intermediate results for as long as they are required in the development of the problem
4. The final result before it is sent to the output unit

In practice, each number, letter of the alphabet, punctuation mark or other symbol, and special computer instruction is **coded** using only two symbols (represented as 0 and 1). This is called a **binary coding**, and it is used because it assumes only two states (on–off, hole–no hole, right–left, and so on). Figure 14.3 illustrates a binary coding device. One possible coding is shown below. (Can you figure out the coding?)

0 = 000 000	A = 100 000
1 = 000 001	B = 100 001
2 = 000 010	C = 100 010
3 = 000 011	⋮
⋮	⋮
	S = 110 011
	T = 110 100
	⋮
	⋮
	SPACE = 111 111

FIGURE 14.3 Light bulbs serve as a good example of two-state devices. The 1 is symbolized by "on," and the 0 is symbolized by "off." The number coded in this figure is 011001. Can you tell what this represents? .

To store the information "3 cats" in a computer, we would need 6 locations or addresses:

Address numbers	*Information*
1000	000011
1001	111111
1010	100010
1011	100000
1100	110100
1101	110011

Notice that an address number has nothing to do with the contents of that address. The early computers stored these individual digits (0 or 1), called **bits**, with switches, vacuum tubes, or cathode ray tubes. In the "3 cats" example, each word contains 6 bits. In practice, however, a word may contain 36 or more bits, which allows the use of much larger numbers. That is, a location might look like the following:

Address 00101 | 000111110000000110110110000000001101 |

Modern memory devices fall into one of two categories— moving-surface devices or electronic devices. The moving-surface devices are adapted from audiocassette technology and are called *magnetic disks*. They provide information storage in localized areas of a thin magnetic film that is coated onto a nonmagnetic supporting surface. Information is stored in the form of tiny magnetized spots in the magnetic film. The magnetic film and the read/write head move in relation to each other to bring a storage site into position for writing or reading of information. These electronic devices utilize the latest computer chip technology. These chips have greatly reduced the cost of computers.

The **accumulator** is the part of the computer in which addition, subtraction, multiplication, or other operations are carried on. It is also called the *processor* or *arithmetic unit*. The accumulator keeps a running tally of whatever is put into it. This is accomplished with logic circuits, which enable it to do these operations. We do not have to be concerned with how these operations occur; we will merely assume that the computer "knows how" to carry them out.

One problem that frequently arises in computer applications is that of number scale. Assume that a computer is capable of working with eight-place accuracy, and you wish to write a very large number—say a googol. We have discussed the use of scientific notation to represent such large numbers. Computers use a form of scientific notation called **floating-point notation**. In floating-point notation, the power of 10 is expressed as E + exponent. Thus,

$$650 = 6.5 \times 10^2 = 6.5E + 2$$
$$650,000 = 6.5 \times 10^5 = 6.5E + 5$$
$$.000065 = 6.5 \times 10^{-5} = 6.5E - 5$$

Therefore, $4.7E + 17$ means $470,000,000,000,000,000$, and a googol can be written as

$$1 \times 10^{100} = 1.0E + 100$$

Numbers such as 650, -489.5678, and 470 are said to be in **fixed-point form**.

EXAMPLE 1

Express the given numbers in floating-point notation.

a. 745   **b.** .00573

Solution

a. $745 = 7.45 \times 10^2 = 7.45E + 2$
b. $.00573 = 5.73 \times 10^{-3} = 5.73E - 3$ ∎

EXAMPLE 2

Express the given numbers in fixed-point notation.

a. $1.23E + 6$ **b.** $6.239E - 7$

Solution

a. $1.23E + 6 = 1.23 \times 10^6 = 1,230,000$
b. $6.239E - 7 = 6.239 \times 10^{-7} = .0000006239$ ∎

The **control** unit of the computer directs the flow of data to and from the various components of the system. It obtains instructions, one at a time, from the storage unit and causes the arithmetic unit to carry out the operations.

When data and instructions are read into the computer, control will usually direct them to storage. Both the data and the instructions can be placed anywhere in storage, although they are usually put apart from each other and kept in some predetermined sequence.

In solving a problem, the computer is directed to start at the first instruction of the program stored at a given location. Once the first instruction has been executed, the control circuits usually get the next instruction from the next higher storage position.

Problem Set 14.1

1. Explain the difference between an analog and a digital computer.

2. Name the five main components of a computer, and briefly explain the function of each component.

3. Name three methods for inputting information into a computer.

4. Explain what we mean by "address," "word," and "bit" when we speak of computers.

5. What is floating-point notation?

Express the numbers in Problems 6–10 in floating-point notation.

6.	**a.**	546	**b.**	1,700.82
7.	**a.**	.0000537	**b.**	3.14159
8.	**a.**	23,000	**b.**	45.2
9.	**a.**	378,596	**b.**	500.2
10.	**a.**	.000035	**b.**	45,000,000,000

Express the numbers in Problems 11–16 in fixed-point notation.

11.	**a.**	4.56E + 3	**b.**	1.79×10^{-4}
12.	**a.**	5.800E − 7	**b.**	3.98×10^4
13.	**a.**	3.05E + 4	**b.**	7.056E + 2
14.	**a.**	7.9E + 14	**b.**	6.02E − 5
15.	**a.**	5.9326E + 6	**b.**	3.217E − 6
16.	**a.**	5.72E + 7	**b.**	5.87E − 11

17. **a.** Think of five additional examples of analog devices to add to Table 14.1.
    b. Think of five additional examples of digital devices.

18. Give examples of:
    a. Data processing   **b.** Information retrieval
    c. Pattern recognition   **d.** Simulation

19. It has been said that "computers influence our lives increasingly every year, and the trend will continue." Do you see this as a benefit or a detriment to mankind? Explain your reasons.

20. What do you mean by "thinking" and "reasoning"? Try to formulate these ideas as clearly as possible, and then discuss the following question: "Can computers think or reason?"

Express your present opinions—not any knowledge of computers or computer hardware that you may have.

21. The following are some basic definitions for "thinking." *According to each of the given definitions,* answer the question "Can computers think?"
   a. To remember
   b. To subject to the process of logical thought
   c. To form a mental picture of
   d. To perceive or recognize
   e. To have feeling or consideration for
   f. To create or devise

22. In his article "Toward an Intelligence beyond Man's" (*Time,* February 20, 1978), Robert Jastrow claims that by the 1990s computer intelligence will match that of the human brain. He quotes Dartmouth president John Kemeny, a leading mathematician and computer scientist, as saying that he "sees the ultimate relation between man and computer as a symbiotic union of two living species, each completely dependent on the other for survival." He calls the computer a new form of life. Do you agree or disagree with the hypothesis that a computer could be "a new form of life"?

23. Several classics from literature, as well as contemporary movies and books, deal with attitudes toward computers and automated machinery. For example, HAL in *2001: A Space Odyssey* tries to take command of the spacecraft, and PROTEUS IV in *Demon Seed* even attempts to procreate a living child with a human "bride." Books such as *Frankenstein, Brave New World*, and *1984* precede computers but offer warnings about future supremacy of technology over humans. In light of these works and the recent developments in computer science, what moral responsibility do you think scientists must take for their creations?
   References: Clarke, Arthur C., *2001: A Space Odyssey* (New York: New American Library, 1968).

   Huxley, Aldous, *Brave New World* (New York: Harper & Row, 1932).

   Orwell, George, *1984* (New York: Harcourt Brace Jovanovich, 1949).

   Shelley, Mary, *Frankenstein* (New York: Macmillan, 1970).

SCIENTIST SAYS COMPUTERS THAT TALK A REALITY

PASADENA (UPI)—Talking computers are a reality, according to a scientist at the California Institute of Technology.

Computers can now be taught to speak, including such niceties as stressing important words and pausing for punctuation, said Dr. John R. Pierce, a communications expert and former executive director of Bell Laboratories communications sciences division.

"Computers that imitate human speech with a remarkable degree of accuracy already are a reality." Pierce said Monday in a lecture at Cal Tech.

Noriko Umeda, a Japanese linguist, has succeeded in synthesizing human speech from a computer equipped with a pronouncing dictionary in its memory system, Pierce said.

Some computers are now equipped to give voice responses in some circumstances, by triggering a tape recorded message. But teaching the computer itself to talk leads to greater accuracy—because the computer will no longer repeat a mistake made by a human announcer—and "economy, rapidity and flexibility," Pierce said.

"Tape recorded answers have a very limited range and only give a few responses. A computer would have a large body of information stored in it alphabetically. The computer would be able to put these words together in an intelligible form which could answer questions." Pierce said.

Although talking computers are still imperfect, the idea is now workable, he said.

Talking computers can give instructions to workmen, he said, and could be linked to a telephone to provide advice services, such as counseling housewives on what dishes could be prepared from leftovers.

A computer that can see gets its first real job.

General Motors Research Laboratories
Warren, Michigan 48090

A "seeing" computer developed by the General Motors Research Laboratories has recently become the first of its kind to go to work on a U.S. automotive production line. The employer: GM's Delco Electronics Division.

COMPUTERS TO 'HEAR' COMMANDS

By Edward O'Brien
Newhouse News Service

WASHINGTON—Pentagon research sometimes wanders into fields far removed from the battlefield. Now Defense officials predict confidently that in two or three years they will have developed computers that take instruction from a human voice.

"The use of natural spoken English as an input language to computers will revolutionize the effectiveness and utility of computer systems," Stephen Lukasik, director of defense Advanced Research Projects, has reported to Congress.

The study is at the midpoint and has produced computer systems that understand continuous speech with a vocabulary of about 100 words. This is much more sophisticated than anything done by government or private industry.

The next 900 words will come more easily because of progress already made in teaching the computers to break speech into separate words, find individual sounds within words, and use clues provided by the vocabulary, grammar, and subject matter.

"Speech is the most natural means for a person to express himself, and we will be able to integrate the computer much more closely to our daily workings if we can develop means for it to understand natural speech," Lukasik said.

"Fast typists (feeding information into computers) can normally type 60 to 70 words per minute, while non-typists may be able to achieve only six to 10 words.

"Almost everybody, however, talks in excess of 100 words per minute, and no special manual skill or equipment is required . . . with voice access, we can expect computers to be as easy to use as telephones are today."

Voice control of computers, however, may not be the ultimate.

14.2 BASIC Programming

SITUATION "Bah Humbug!" exclaimed Jerry. "I'll never
believe in thinking computers. A computer is nothing more
than hardware along with some software to make it operate."

"How do you know?" asked Louise. "Do you know anything
about computers or computer programming?"

"No, but this poem by David H. H. Diamond says it well."

The computer is a funny beast
He's not aggressive in the least.
He has no malice, fad or whim,
His moods are neither gay nor grim.
He cannot walk (though he can run),
He never makes a joke or pun.
He treats both friend and foe the same,
Incurring neither praise nor blame.
He does just what you tell him to,
And thus he puts the blame on you.

"Look at you," said Louise. "You read me a poem that calls
a computer 'he'! Why not 'she'? If it is hardware only, why
not 'it'? I don't think that computers can think either, but
before I can reach a conclusion I need to know more about
computers. Just the other day I saw an article about
computer poetry and I understand that computers have also
composed original music and have even proved new
mathematical theorems."

In this section, Jerry and Louise will find out what we
mean by programming and take a look at the four steps used
when writing a program.

To get a computer to operate, a complete set of step-by-
step instructions must be provided. These sets of instructions
are known as **programs**, and the process of setting them up is
known as **programming**.

Each step or instruction for the solution of a problem must
be provided for the computer in detail. For example, if a
teacher asks a pupil to add 7 and 4, the student will generally
say 11; the programmer, however, not only must ask the
computer to add 7 and 4 but also must ask it to indicate the
answer.

To write a program, we should carry out the following
five steps:

1. Analyze the problem and break it down into its basic parts.
2. Prepare a **flowchart** or write out a description of a process for carrying out the task.
3. Put the flowchart or description into a language that the machine can "understand."
4. Test the program to see if it does what it is supposed to do.
5. **Debug** the program, if necessary; this means that you remove any deficiencies in the program.

A **flowchart** is an outline that lists the steps to be performed in sequence. The directions are usually listed within squares, circles, or other geometric figures, as shown in Figure 14.4. The direction, or flow, of the process is shown by arrows.

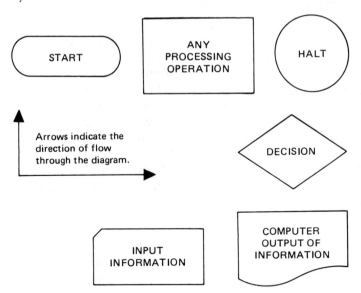

FIGURE 14.4   Common flowchart symbols

The flowchart provides a way of checking the overall analysis of a process or problem, and it also provides a way for describing the process to someone else once the flowchart is completed. Using the flowchart as a guide, the details of the program are easier to write.

Flowcharts can direct us in other types of activities besides writing computer programs. For example, the General In-

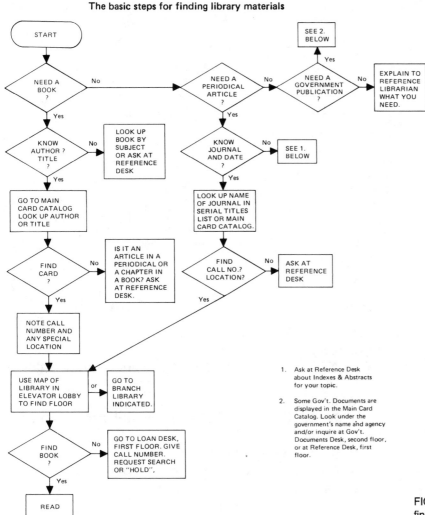

The basic steps for finding library materials

FIGURE 14.5 The basic steps for finding library materials

formation Leaflet of The General Library of The University of California, Davis, uses the flowchart shown in Figure 14.5 to explain how to find library materials. Note that no detail should be overlooked in a flowchart.

EXAMPLE 1

Follow the directions given by the flowchart on page 498.

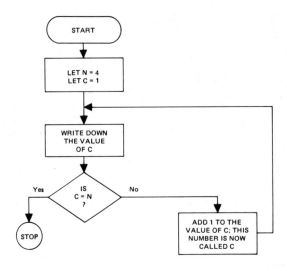

Solution

START: N = 4, C = 1

Write down the value of C: **1**

Is C = N? No (C = 1 and N = 4), so add 1 to the value of
    C: C = 1 + 1 = 2

    Write down the value of C: **2**

Is C = N? No (C = 2 and N = 4), so add 1 to the value of
    C: C = 2 + 1 = 3

    Write down the value of C: **3**

Is C = N? No (C = 3 and N = 4), so add 1 to the value of
    C: C = 3 + 1 = 4

    Write down the value of C: **4**

Is C = N? Yes, so STOP. The output answer is: 1
                                                 2
                                                 3
                                                 4   ■

EXAMPLE 2

Draw a flowchart that will direct someone to write down the
square numbers up to 50.

Solution

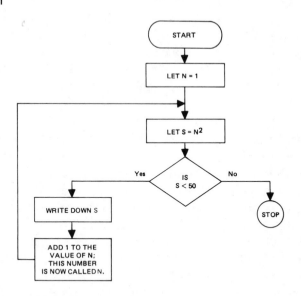

After preparing a flowchart, the next step is to write the program. Computers use many different languages. Some widely used languages are BASIC, PASCAL, FORTRAN, ALGOL, and COBOL. In order to communicate with a computer, you need to learn one of these languages. We'll concentrate on a very popular conversational language

FRENCH ENGLISH RUSSIAN GERMAN CHINESE FORTRAN

called **BASIC**. BASIC consists of short, easy-to-learn commands that are very similar to normal English. We will talk to the computer on a CRT keyboard, and the computer will respond on that same device.

A program is made up of a sequence of statements called **commands**. Different commands instruct the computer to carry out different processes. We'll discuss some of these commands in the BASIC language in this and the following sections. In order to tell a computer the order of commands, **line numbers** are used. That is, each command in a program is preceded by a counting number (called its *line number*) followed by a space. When running a program, the computer will execute the commands in the order of ascending line numbers, regardless of the order in which they are typed on the CRT. For this reason, we usually use the line numbers 10, 20, 30 (rather than 1, 2, 3) so that we can insert new statements to debug or modify a program at some later time.

You are now ready for a simple program. Every computer has some sort of a log-in procedure. These procedures vary from machine to machine and facility to facility, and they must be kept relatively secret so that not just anyone can use the computer. You will need to find out the log-in procedure for the computer or terminal you will be using. All the examples in this book will assume that you have correctly followed the log-in procedure for your particular computer.

Suppose you wish to have the computer type the phrase "I love you." In order to do this, you will use a command called **PRINT**.

This is a counting number; it is the **line number** of the command.

```
10   PRINT  "I LOVE YOU."
```

Whatever is enclosed in quotation marks will be printed by the computer.

This is a **command**; in this case, it is a PRINT command.

```
20   END
```

The end of the program is designated by **END** (yes, even a program as short as this example).

This program, along with another example, is shown below as it would look on the CRT.

Program A	*Program B*
10 PRINT "I LOVE YOU."	10 PRINT "I AM ROBBIE THE ROBOT."
20 END	20 PRINT "I LOVE YOU."
	30 PRINT "GOODBYE."
	40 END

Now that we've written a program, the next step is to run the program. The **RUN** command causes the computer to begin the program. When you tell the computer to RUN, followed by a line feed (return key), it will begin at the lowest line number and will execute in consecutive order all commands in the program. A RUN command is called a **system command** and is not preceded by a line number. A system command is a command that is not stored in the memory, but is executed immediately. Other system commands will be considered later.

In the course of the program, the computer will probably output data or information; it may ask you questions or do some calculations. When it is finished, it will type

READY

to signify that it has finished the program.

EXAMPLE 3

Run Programs A and B.

Solution

a. Type RUN and return (which we'll denote by RUN↵), and the computer will respond with

I LOVE YOU.
READY

b. RUN↵

I AM ROBBIE THE ROBOT.
I LOVE YOU.
GOODBYE
READY

Notice that each new line number generates a new line of output. ∎

When you need to use a computer to do arithmetic operations, you will use * for multiplication, / for division, and ↑ for exponentiation, as shown in Table 14.2.

TABLE 14.2 Mathematical Symbols in BASIC

Symbol		Math notation	BASIC
↑	Exponentiation	3^4	3↑4
*	Multiplication	3×4	3*4
/	Division	$3 \div 4$ or $\frac{3}{4}$	3/4
+	Addition	$3 + 4$	3+4
−	Subtraction	$3 - 4$	3-4

EXAMPLE 4

Write a program to calculate $5 \times 3 - 6$.

Solution

```
10  PRINT 5*3 - 6
20  END
```

Notice that no quotation marks are used when using the PRINT command to do calculations. ∎

EXAMPLE 5

Write a program to calculate the area of a circle with a radius equal to 6.3.

Solution

Recall that the formula for the area of a circle is $A = \pi r^2$, where π is approximately 3.1416.

```
10  PRINT 3.1416*6.3↑2
20  END
```
∎

Debugging means checking and revising your program to make sure it does what it is supposed to do. The programs in this section will be easy to debug, but if your program is more complex and involves several decisions and options, it will be much more difficult to check.

There are two types of mistakes. The first is a logic error, which is an error in your thinking. A flowchart may help you overcome this type of error if the problem is complicated. The second type of error is a coding error, which is a clerical error.

In the next section, we will introduce the BASIC commands you will use when writing a program. We will also write some simple programs.

Problem Set 14.2

1. Discuss the steps to follow when writing a program.
2. What is a flowchart?
3. Draw the symbol used in a flowchart to denote:
    a. Start **b.** Stop **c.** A decision **d.** Input data
    e. Output data
4. Tell what each of the following BASIC operation symbols mean:
    a. − **b.** * **c.** / **d.** + **e.** ↑
5. In your own words, describe what is meant by debugging a program.
6. Consider the following program:

```
10  PRINT "GOOD MORNING."
20  PRINT "CAN YOU DO THIS PROBLEM CORRECTLY?"
30  END
```

Without using a computer, explain what the computer will type after you give a RUN command.

7. Suppose you have just written the program shown in Problem 6. Then you type

```
25  PRINT 2 + 3
35  END
```

Without using a computer, explain what the computer will type after you give a RUN command.

In Problems 8–15 write each BASIC expression in ordinary algebraic notation.

8. **a.** `4*X + 3` **b.** `(5/4)*Y + 14↑2`
9. **a.** `35*X↑2 - 13*X + 2` **b.** `6*X - 7`
10. **a.** `5*X↑2 - 3*X + 4` **b.** `(X - 5)*(2*X + 4)↑2`
11. **a.** `6.29↑14 - 7` **b.** `17*X↑3 - 13*X↑2 + (15/2)`
12. `4*(X↑2 + 5)*(3*X↑3 - 3)↑2`
13. `(16.34/12.5)*(42.1↑2 - 64)`
14. `47*X↑3 + (13/2)*X↑2 + (15/4)*X - 17`
15. `36*X↑3 + (5/2)*X + 135/2`

In Problems 16–23 write each expression in BASIC language notation.

16. **a.** $\frac{2}{3}x^2$ **b.** $3x^2 - 17$
17. **a.** $5x^3 - 6x^2 + 11$ **b.** $14x^3 + 12x^2 + 3$
18. **a.** $12(x^2 + 4)$ **b.** $\dfrac{15x + 7}{2}$

19. **a.** $(5 - x)(x + 3)^2$ **b.** $6(x + 3)(2x - 7)^2$

20. $(x + 1)(2x - 3)(x^2 + 4)$ **21.** $(2x - 3)(3x^2 + 1)$

22. $\frac{1}{4}x^2 - \frac{1}{2}x + 12$ **23.** $\frac{2}{3}x^2 + \frac{1}{3}x - 17$

Write a simple BASIC program to carry out the tasks given in each of Problems 24–29.

24. Say "I'm a happy computer."

25. Say "I will demonstrate my computational skill."

26. Calculate $\dfrac{53 + 87}{2}$.   **27.** Calculate $(1 + .08)^{12}$.

28. Calculate $5,000(1 + .06)^9$.   **29.** Calculate $\dfrac{23.5^2 - 5(61.1)}{2}$.

30. Follow the directions shown in the flowchart. Is the answer predicted in the flowchart correct?

Flowchart for Problem 30

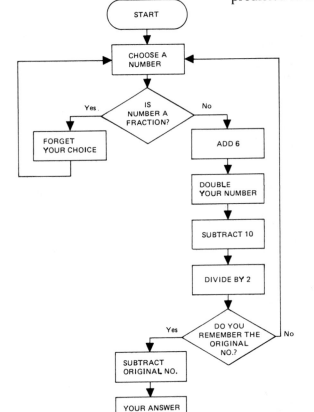

Flowchart for Problem 31

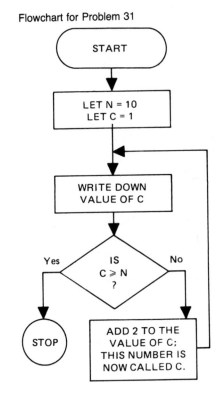

31. Follow the directions shown in the flowchart above.

32. Write a flowchart to describe how to add two numbers on an algebraic-logic calculator.

Computer Poetry

The four computer-written poems at the right were produced by Margaret Masterman and Robin McKinnon Wood using human–machine interaction at the Cambridge Language Research Unit. The program is written in the TRAC language.

"Only once in every generation is there a computer that can write poetry like this."

©DATAMATION

```
1 POEM eons deep in the
        ice
        I paint all time in
        a whorl
        bang the sludge
        has cracked

2 POEM eons deep in the
        ice
        I see gelled time
        in a whorl
        pffftt the sludge
        has cracked

3 POEM all green in the
        leaves
        I smell dark
        pools in the trees
        crash the moon
        has fled

4 POEM all white in the
        buds
        I flash snow
        peaks in the
        spring
        bang the sun has
        fogged
```

This poem is a translation of a computer-written German poem. It was done at Stuttgart's Technical College from a program written by Professor Max Bense. Professor Bense has been trying to discover if it is possible to formulate problems in esthetics mathematically.

```
THE POEM

The joyful dreams rain down
The heart kisses the blade of grass
The green diverts the tender lover
Far away is a melancholy vastness
The foxes are sleeping peacefully
The dream caresses the lights
Dreamy sleep wins an earth
Grace freezes where this glow dallies
Magically the languishing shepherd dances.
```

33. Write a flowchart that will tell someone how to add two numbers.

34. Prepare a flowchart to describe the process of taking attendance by calling roll. You must take into account the possibilities of both tardy and absent persons.

35. It has been said that anyone who does not know how to program a computer is functionally illiterate. Do you agree or disagree? Give some arguments to support your answer. Project this question into the future; do you think it will ever be true?

14.3 Communicating with a Computer

SITUATION   Future career opportunities in the rapidly growing world of computers seem to be practically limitless. These opportunities may be direct, as in the case of those who manufacture and operate computers, or indirect, as in the case of businesspeople, scientists, and others who use computer systems.

Increasingly great numbers of skilled personnel will be needed by the computer industry itself:

Designers and manufacturers of systems
Engineers and scientists for research and development
Sales personnel skilled in marketing methods
Systems analysts to analyze and meet special
    requirements of customers
Programmers who prepare programs to meet customers'
    needs
Computer operators to run systems
Personnel for clerical and data preparation jobs
Managers of computer operations
Management interpreters of computer systems, needs, and
    opportunities
Specialists in areas such as business, science, education,
    and government
Interdisciplinarians—those who can understand and meet
    the needs of persons from varied professions united on
    mutual projects

More than a thousand colleges and universities in the United States and Canada, according to a recent survey, now offer courses in the computer sciences and data processing. Computer usage is being taught in many high schools and even in some grammar schools. Many independent training schools exist for high school and college graduates.

The use of remote terminals that connect to a central computer system, sometimes thousands of miles away, is becoming commonplace. Industry experts say it's only a matter of time and cost reduction before the use of household terminals, for a variety of purposes ranging from information services to entertainment, becomes as ordinary as the use of the telephone.

Economists predict that by the end of the century, or earlier, the computer industry and directly associated industries will be the largest business in the United States.

In this section, we'll introduce some of the more common BASIC commands and write some programs. If you have access to a computer, it will be helpful to try to run some of these programs.

BASIC treats all numbers as decimal numbers; if the input number doesn't have a decimal, the computer assumes a decimal point after the last numeral. Also, if the answer to an arithmetic operation is $\frac{1}{3}$, the computer writes this as .3333333333. We also use variables in BASIC, which are formed by a single letter or by a letter followed by a digit.

SCR COMMAND*

To begin with a clean slate, we must first erase ("scratch") from its memory any program the computer is storing. Therefore, before we write a program, we would probably give the computer an SCR command. An SCR command is a system command, so it is executed immediately after pressing the return key.

LIST COMMAND

To find out what is in the computer's memory at any stage, we use a LIST command.

---

*BASIC has a number of variations, most of which are minor and depend on the computer being used. You should check with your instructor for possible variations in the system you are using.

LIST *n*   This command causes the computer to list line
              n only

LIST *n,m*   This command causes the computer to list
              lines *n* through *m*, inclusive.

LIST   This command causes the computer to list the
              entire program in its memory.

These LIST commands are system commands.

END COMMAND

The END command signifies that the program is complete.
If a program has a command with a higher line number,
then those commands will not be carried out. The program
will terminate upon reaching an END command.

PRINT COMMAND

As you saw in the last section, the PRINT command carries
out more than one function. The first is to perform arithmetical calculations on the computer. For example, suppose
we have erased any previous programs with a SCR command. Next, we type the following:

```
10  PRINT 7 + 4
50  END
```

Notice that the END command tells the computer that the
program is complete at this point. Now, if we give the
computer a RUN command, the computer will output the
answer.

```
SCR
READY
10  PRINT 7 + 4
50  END
RUN
11
READY
```

Several computations can be done with one program. Each
new line in the program will require that the answer be on a
new line. For example,

```
10  PRINT 55 + 11
20  PRINT 55*11
30  PRINT 55 - 11, 55/11
50  END
```

will cause the computer to output the answers for $55 + 11$ and $55*11$ on successive lines, but the answers for $55 - 11$ and $55/11$ will be recorded on the same line:

```
SCR
READY
10  PRINT 55 + 11
20  PRINT 55*11
30  PRINT 55 - 11, 55/11
50  END
RUN
 66
 605
 44      5
READY
```

A single PRINT statement can contain more than one expression, but the results will be printed on the same line (up to the permissible width on the page). If the output for one line is too long, the end of it is automatically printed on the next line. If a PRINT statement contains more than one expression, those expressions are separated by commas or semicolons.

The PRINT command can also be used for spaces in the output. For example, the command PRINT followed by no variables or numerals will simply generate a line feed (a blank line).

```
SCR
READY
10  PRINT 1 + 2
20  PRINT 2 + 3
30  PRINT
40  PRINT
50  PRINT 4 + 5
60  END
RUN
 3
 5


 9
READY
```

If the answer that we are calling for is too large, the computer will automatically print it out in *floating-point form*.

```
SCR
READY
 10   PRINT 1*2
 20   PRINT 2*3
 30   PRINT 6*4
 40   PRINT 24*5
 50   PRINT 120*6
 60   PRINT 720*7
 70   PRINT 5040*8
 80   PRINT 40320*9
 90   PRINT 362880*10
100   END
RUN
  2
  6
 24
120
720
5040
40320
362880
3.628800E+6  ←— Notice that when the answers get too large, the
READY             computer will automatically switch to floating-
                  point form.
```

In the last section we also used the PRINT command to type messages. For example, suppose we would like the computer to say "Hello." This can be accomplished by giving a PRINT command followed by the word enclosed in quotation marks:

```
10   PRINT "HELLO"
20   END
```

A run of this program would simply cause the computer to say "HELLO."

```
RUN
HELLO
```

Let's look at this use of the PRINT command more carefully by studying the following program:

```
10   PRINT "4 + 7 =", 4 + 7
20   END
```

What is the difference between "4 + 7 =" and 4 + 7 in the program? The "4 + 7 =" causes the computer to type out

$$4 + 7 =$$

and does no arithmetic. The second part, 4 + 7, tells the computer to do the arithmetic, which causes the computer to type out 11.

```
RUN
4 + 7 =        11
```

LET COMMAND

We assign values to locations by using a LET command. For example, we might say

```
 10  LET A = 2
 20  LET B = 5
 30  LET P = 3.1416
 40  PRINT A,B
 50  PRINT P
 60  PRINT A + B
 70  PRINT
 80  PRINT P*B↑A
100  END
```

We usually write flowcharts only for more involved programs, but the flowchart for this program is shown in Figure 14.6. Notice that each instruction on the flowchart is translated into a separate line in the program.

INPUT COMMAND

Suppose we wish to compute the amount of money present after one year for various interest rates. The simple interest formula is

$$A = P(1 + rt)$$

where

P = Principal (amount invested)
r = Interest rate (written as a decimal)
t = Time (in years)
A = Amount present (after t years)

For purposes of this example, we will let P = \$10,000 and t = 1. We must have some way of telling the computer the

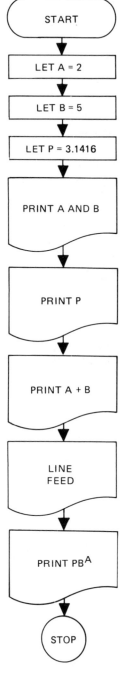

FIGURE 14.6

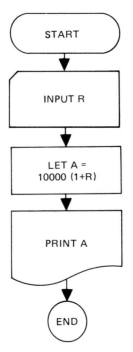

FIGURE 14.7 Flowchart for
$A = P(1 + rt)$, where $P = \$10,000$
and $t = 1$

value of r without rewriting the program each time. A BASIC command called INPUT will be used. It is used to supply data while the program is running. When the computer receives an INPUT command, it will type out a question mark, ?, and wait for the operator to input the data. When you are finished inputting data, depress the return key (⮠). Consider the following BASIC program (the flowchart is shown in Figure 14.7):

```
10   INPUT R
20   LET A = 10000*(1 + R)
30   PRINT A
40   END
```

When a RUN command is given, the computer will type

?

The operator must then type the value for r (three runs of the program are shown for interest rates 5%, $5\frac{3}{4}$%, and 7%):

```
RUN⮠
?    .05        Notice that interest rates are input as decimals.
  10500
READY           Can you tell which of these are operator typed and which
RUN⮠            are computer typed?
?    .0575
  10575
READY
RUN⮠
?    .07
  10700
READY
```

The programmer may find it helpful to remind the people using the program what is being asked for with a question mark, so the program can be modified as follows (notice that even the spaces must be included inside the quotation marks):

```
 5  PRINT "I WILL CALCULATE THE AMOUNT PRESENT FROM A $10,000 ";
 6  PRINT "INVESTMENT AFTER"
 7  PRINT "1 YEAR AT SIMPLE INTEREST. WHAT IS THE INTEREST RATE";
24  PRINT
25  PRINT "THE AMOUNT PRESENT IS $";
```

Combining these steps with the original program, we get:

```
 5  PRINT "I WILL CALCULATE THE AMOUNT PRESENT FROM A $10,000 ";
 6  PRINT "INVESTMENT AFTER"
 7  PRINT "1 YEAR AT SIMPLE INTEREST, WHAT IS THE INTEREST RATE";
10  INPUT R
20  LET A = 10000*(1 + R)
24  PRINT
25  PRINT "THE AMOUNT PRESENT IS $";
30  PRINT A
40  END
```

This second version of the program has several nonessentials from a programming standpoint, but it makes the computer much more conversational:

```
RUN
I WILL CALCULATE THE AMOUNT PRESENT FROM A $10,000 INVESTMENT AFTER
1 YEAR AT SIMPLE INTEREST, WHAT IS THE INTEREST RATE? .05

THE AMOUNT PRESENT IS $ 10500


READY
```

Notice the significance of the semicolon at the end of lines 5, 7, and 25. It means that there should be *no line feed* when going to the next line of the program.

It is possible to input several variables at once. Suppose we wish to input not only R but also P, as in the command

```
        INPUT P, R
```

We have the following variation:

```
10  INPUT P, R          Can you add PRINT commands
20  LET A = P*(1 + R)   to this program to make it more
30  PRINT A             conversational?
40  END
```

When we run this program, two variables must be input after the question mark:

```
10  INPUT P, R
20  LET A = P*(1 + R)
30  PRINT A
40  END
RUN
?  5000, .07
 5350
```

The first run of the program is calculating the simple interest formula for one year at 7% with $5,000 invested.

```
READY
RUN
?  30000, .075
 32250
```

Can you interpret what is being calculated in the second two runs of this program?

```
READY
RUN
?  30000, .09
 32700
READY
```

Problem Set 14.3

1. Give an example of each of the three uses for the PRINT command.

2. What is the significance of the semicolon at the end of a line beginning with a PRINT command?

3. Name one command that allows you to put information into the computer.

4. You have fed the following program into a computer:

```
10  PRINT 6 + 5
20  PRINT 6 - 5
30  PRINT 6*5, 6/3
40  END
```

What will the computer type if you give a RUN command?

5. You have fed the following program into a computer:

```
10  PRINT 4 + 5
20  PRINT "HELLO, ";
30  PRINT "I LIKE YOU."
40  PRINT "4 + 5 =";4 + 5
50  END
```

What will the computer type if you give a RUN command?

6. You have fed the following program into a computer:

```
10   PRINT 5 + 10
20   PRINT 5*10
30   PRINT 10/5, 5/10
40   END
```

What will the computer type if you give a RUN command?

7. You have fed the following program into a computer:

```
10   PRINT 7 + 12
20   PRINT 7 - 12, 7*12
30   END
```

What will the computer type if you give a RUN command?

8. Find the error in the following program:

```
10   PRINT "I WILL COMPUTE (2*X↑2)(X + 1)"
20   INPUT X
30   PRINT (2*X↑2)(X + 1)
40   END
```

9. Find the error in the following program:

```
10   INPUT X
20   PRINT (X↑2 - 3)(2*X + 1)
30   END
```

10. You have fed the following program into a computer:

```
10   LET A = 2
20   PRINT "WORKING"
30   LET B = 3
40   PRINT A + B
50   END
```

What will the computer type if you give a RUN command?

11. You have fed the following program into a computer:

```
10   "I WILL CALCULATE IQ."
20   "WHAT IS MENTAL AGE";
30   INPUT M
40   PRINT "WHAT IS CHRONOLOGICAL AGE";
50   INPUT C
60   PRINT
70   PRINT "IQ IS";
80   PRINT (M/C)*100
```

What will the computer type if the program is run for a mental age of 12 and a chronological age of 10?

12. What will the computer type if you run the following program for a 4 by 5 rectangle?

```
 10  PRINT "I WILL CALCULATE THE PERIMETER AND ";
 20  PRINT "AREA OF A RECTANGLE."
 30  PRINT "WHAT IS THE LENGTH";
 40  INPUT L
 50  PRINT "WHAT IS THE WIDTH";
 60  INPUT W
 70  PRINT "THE PERIMETER IS ";
 80  PRINT 2*(L + W);
 90  PRINT "AREA IS ";
100  PRINT L*W
110  END
```

13. Write a BASIC program to ask for a value for x, and then use this value to evaluate $5x^3 + 17x - 128$. Also, show the flowchart.

14. Write a BASIC program to ask for a value of x, and then use this value to evaluate $12.8x^2 + 14.76x + 6\frac{1}{3}$. Also, show the flowchart.

15. Suppose you are considering a job with a company as a sales representative. You are offered \$100 per week plus 5% commission on sales. Write a program that will compute your weekly salary if you input the total amount of sales. Be sure to include a flowchart.

16. A manufacturer of sticky widgets wishes to determine a monthly cost of manufacturing on an item that costs \$2.68 per item for materials and \$14 per item for labor. The company also has fixed expenses of \$6,000 per month (this includes advertising, taxes, plant facilities, and so on). Write a program that will compute monthly cost if you input the number of items manufactured. Be sure to include a flowchart.

17. The shortage of computer programmers has never been greater. In 1980 there were 150,000 people employed as programmers, and there were openings for at least 50,000 more. Check out the job opportunities for programmers in your area.

Problems 18–21 require that you have access to a computer.

18. Write a BASIC program to ask for a value of x, and then use this value to evaluate $24x^3 + 3x^2 - x + 6$ for $x = 1$, $x = 10$, and $x = 4.63$.

19. The Indians sold Manhattan Island to the Dutch in 1626 for goods worth about $24. (It has often been said that the Dutch took advantage of the Indians.) Suppose the Indians had put the $24 in a savings account at 7% interest. The formula for calculating the present amount, A, is

$$A = P(1 + i)^n$$

where P is the principal ($24 for this problem), i is the interest written as a decimal (.07 for this problem), and n is the number of years. For example, after 1 year:

$$A = 24(1 + .07)$$
$$= 25.62$$

After 2 years:

$$A = 24(1 + .07)^2$$
$$= 27.48$$

Find the value of $24 in 1976 if it had been deposited in 1626 at 7% interest. Write your answer in scientific notation and compare it with the present estimated value of $82.3 billion for Manhattan Island.

20. Think of any counting number. Add 9. Square. Subtract the square of the original number. Subtract 61. Multiply by 2. Add 24. Subtract 36 times the original number. Take the square root. What is the result?

a. Show algebraically why everyone who works the problem correctly gets the same answer.
b. Write a BASIC program to solve the puzzle. In BASIC, the square root of a number N is found by giving the command SQR(N).

21. Write a BASIC program for obtaining the pattern shown.

```
xxxxxx   xxxxxx   xxxxxx   xxxxxx   xxxxxx
xx   x   x        x    x   x        x
xxxxxx   xxxxxx   xxxxxx   x        xxxxxx
xx       x        x    x   x        x
xx       x        x    x   x        x
xx       xxxxxx   x    x   xxxxxx   xxxxxx
```

Computers and Music

Two Southwestern College professors have combined their talents and interests to utilize a computer for musical composing.

John Bibbo, mathematics and computer programming instructor at the college, has assisted Dr. Victor Saucedo, music instructor, in a computerized approach to musical composing.

Dr. Saucedo was requested to present his composition, "RAN. I.X. for Solo Clarinet and Tape," at the Western Regional Conference of the American Society of University Composers held at Fullerton State University in November 1977.

MATH PLUS MUSIC John Bibbo, left, and Dr. Victor Saucedo, instructors at Southwestern College, look at computer which they used to compose music. They were both recently honored by the American Society of University Composers for their innovative work.

This musical work was completed in April 1976. An algorithm was used to generate random numbers that define processes on the IBM 360/22 computer.

"The main difference between computers and other systems that use random selection," said Mr. Bibbo, "is the computer's ability to execute, repeat the execution, retain and/or modify the final outcome."

"This gives the composer an editing (revising) capability and control over the final 'random' product," said Dr. Saucedo. "It affords him control to alter or the ability to calculate the probability of a predetermined event taking place, or occurring within the composition."

"Classical composers such as Bach, Haydn and Mozart presumably employed unorthodox methods of composing analogous to those used by computers," stated Saucedo.

Bibbo and Saucedo have been working extensively with this concept of musical composition since 1973. Saucedo also presented a paper at the Conference, describing in detail the methods used by Bibbo and himself for defining parameters by means of a R.N.G. (random number generator).

Saucedo has since been invited to present two works—one for percussion and one for clarinet—at the National Conference of A.S.U.C. in Champaign, Illinois, in March.

14.4 Programming Repetitive Processes

SITUATION   "I would like to have an explanation of why I must receive my bank statement on the fifth instead of just before the first of each month," demanded Bill.

"I'm sorry, sir, there is nothing we can do. It is all done by computer."

As you learn about programming, keep in mind some of the deception that is carried out in the name of computers. People seem to think that "if a computer did it, then it must be correct." This is false, and the fact that a computer is involved has no effect on the validity of the results. Have you ever been told "The computer requires that . . ." or "The computer won't permit it"? What they really mean is that they don't want to write the program to permit it.

Up to now, most of the computer programs you've seen in this chapter could be done more easily by a calculator than by a computer. The power of the computer lies in its ability to go through large numbers of repetitive steps in a short time. In this section, Bill will learn about several new ways to program using the idea of a *loop*.

Repetition of arithmetic operations is a tedious task that can be assigned to computers. Suppose we wish to add the first 100 counting numbers:

$$1 + 2 + 3 + 4 + \cdots + 97 + 98 + 99 + 100$$

We could write a program similar to the ones of the previous section, but these methods would still be tedious. Instead, we will use a **loop**, which repeats a sequence of operations. The loop is eventually completed (we hope!), and the sequence of instructions proceeds to the stopping point.

The simplest command that sets up a loop is called the **GOTO** command. (GOTO is pronounced as two words: "go to.") This command transfers the control to a line number other than the next higher one. The flowchart in Figure 14.8 illustrates the GOTO command. This command can cause the computer to go into a loop, as shown by the program below:

```
10   LET N = 1
20   PRINT N
30   LET N = N + 1
40   GOTO 20
50   END
```

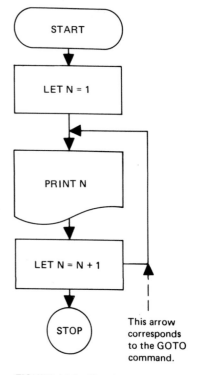

This arrow corresponds to the GOTO command.

FIGURE 14.8 Flowchart illustrating a GOTO command

If we type RUN, this program will continue forever typing out the successive counting numbers. Can you follow the steps of this program? The loop is set up using the GOTO statement. The problem is, however, how to "get out" of the loop. To do this, we introduce some additional commands.

The **READ** and **DATA** commands allow you to input data into the program itself rather than with the INPUT command we discussed in the last section. This command is illustrated in Example 1.

EXAMPLE 1

Consider the following program and determine the computer output if you give a RUN command:

```
10  PRINT "NUMBER","NUMBER SQUARED"
20  READ N ←————— The first time the computer "sees" N,
30  LET S = N↑2    it will read the first value under the
40  PRINT N,S       DATA statement (5 in this example).
50  GOTO 20
60  DATA 5,6,7,8,198,4.63
70  END
```

Transfers back to line 20. The second time through, it will read the next value in the DATA statement, namely 6.

Solution

The output will be:

```
RUN
NUMBER          NUMBER SQUARED
   5                  25
   6                  36
   7                  49
   8                  64
 198               39204
4.63              21.4369
     OUT OF DATA LINE 20
```

The computer continues to loop until all the data are processed. ■

A second, and perhaps more sophisticated, way of ending a loop is to use an **IF–THEN** or **IF–GOTO** command. These commands (whichever one your computer uses) give the computer its basic decision-making ability. Consider the following example:

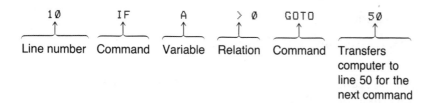

10	IF	A	> 0	GOTO	50
Line number	Command	Variable	Relation	Command	Transfers computer to line 50 for the next command

Some different relations that can be used are shown in Table 14.3.

TABLE 14.3 Relation Symbols That Can Be Used with IF–GOTO and IF–THEN Commands

Math symbol	BASIC symbol	Meaning
=	=	Equal to
>	>	Greater than
≥	>=	Greater than or equal to
<	<	Less than
≤	<=	Less than or equal to
≠	<>	Not equal to

On a flowchart, conditional transfers are symbolized by a diamond-shaped box, as shown in Figure 14.9.

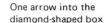

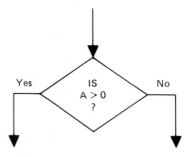

One arrow into the diamond-shaped box

Two arrows out of the box; one is labeled "Yes" and the other is labeled "No." The condition is written inside the box.

FIGURE 14.9

EXAMPLE 2

Consider the following program and determine the computer output if you give a RUN command:

```
10   PRINT "INPUT TWO NUMBERS, AND I WILL TELL YOU WHETHER THEIR ";
20   PRINT "PRODUCT IS "
30   PRINT "POSITIVE, NEGATIVE, OR ZERO."
40   PRINT "WHAT ARE THE NUMBERS";
50   INPUT A,B
60   PRINT
70   PRINT "THE PRODUCT IS ";
80   IF A*B = 0 GOTO 150
90   IF A*B > 0 GOTO 200
100  IF A*B < 0 GOTO 250
150  PRINT "ZERO."
160  GOTO 300
200  PRINT "POSITIVE."
210  GOTO 300
250  PRINT "NEGATIVE."
300  END
```

Solution

Here are some sample outputs for this program:

```
RUN
INPUT TWO NUMBERS, AND I WILL TELL YOU WHETHER THEIR PRODUCT IS
POSITIVE, NEGATIVE, OR ZERO.
WHAT ARE THE NUMBERS?   4,-5

THE PRODUCT IS NEGATIVE.

READY

RUN
INPUT TWO NUMBERS, AND I WILL TELL YOU WHETHER THEIR PRODUCT IS
POSITIVE, NEGATIVE, OR ZERO.
WHAT ARE THE NUMBERS?   -4,-5

THE PRODUCT IS POSITIVE.

READY

RUN
INPUT TWO NUMBERS, AND I WILL TELL YOU WHETHER THEIR PRODUCT IS
POSITIVE, NEGATIVE, OR ZERO.
WHAT ARE THE NUMBERS?   0,123

THE PRODUCT IS ZERO.

READY
```

The flowchart for this program is shown in Figure 14.10. ∎

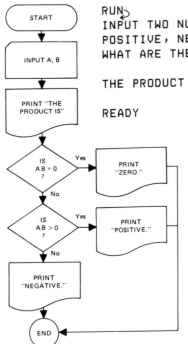

FIGURE 14.10 Flowchart for the example illustrating the IF–THEN or IF–GOTO command

EXAMPLE 3

We began this section by asking for a program to find the sum of the first 100 counting numbers. Write this BASIC program.

Solution

```
10   LET S = 1
20   LET T = 1
30   LET T = T + 1
40   LET S = S + T
50   IF T < 100 GOTO 30
60   PRINT S
70   END
RUN
5050
READY
```

∎

Problem Set 14.4

1. What is a loop?
2. What procedure can be used to terminate a loop?
3. Consider the following program and determine the computer output if you give a RUN command:

```
 10   PRINT "H";
 20   PRINT "I";
 30   PRINT " ";
 40   GOTO 70
 50   PRINT "END";
 60   GOTO 110
 70   PRINT "F";
 80   PRINT "R";
 90   PRINT "I";
100   GOTO 50
110   PRINT ".";
120   END
```

4. Consider the program shown in the margin and determine the computer output if you give a RUN command.

```
 10   PRINT "T";
 20   PRINT "H";
 30   GOTO 100
 40   PRINT "IS ";
 50   PRINT "C";
 60   PRINT "O";
 70   GOTO 120
 80   PRINT "ECT";
 90   GOTO 140
100   PRINT "IS ";
110   GOTO 40
120   PRINT "RR";
130   GOTO 80
140   PRINT "."
150   END
```

5. The program in Problem 3 is inefficient from a programmer's standpoint, since the same output can be generated by using only two commands. Refine the program in Problem 3 by writing a new program that is more efficient. By "refine" we mean write a better program that will generate the same output.

6. The program in Problem 4 is inefficient from a programmer's standpoint, since the same output can be generated by using only two commands. Refine the program in Problem 4 by writing a new program that is more efficient. By "refine" we mean write a better program that will generate the same output.

7. Suppose you are given the following program:

```
10   LET A = 7
20   IF A > 100 GOTO 80
30   IF A < 1 GOTO 60
40   PRINT "A IS BETWEEN 1 AND 100 INCLUSIVE."
50   GOTO 90
60   PRINT "A IS LESS THAN 1."
70   GOTO 90
80   PRINT "A IS GREATER THAN 100."
90   END
```

What will the computer type if you give a RUN command?

8. Repeat Problem 7 for $A = 107$.

9. Repeat Problem 7 for $A = -10$.

10. Write a program that will compare two numbers, A and B, and then print the larger number.

11. Consider the following program and determine the computer output if you give a RUN command:

```
10   PRINT "NUMBER","NUMBER CUBED"
20   READ N
30   LET S = N↑3
40   PRINT N,S
50   GOTO 20
60   DATA 1,2,4,3
70   END
```

12. Repeat Problem 11 where line 60 is changed to: DATA 0,5,4,2.

13. Consider the program shown in the margin and determine the computer output if you give a RUN command.

```
10   READ N
20   LET A = 2*N + 3
30   PRINT A
40   GOTO 20
50   DATA 1,5,2,4
60   END
```

14. Repeat Problem 13 where line 50 is changed to: DATA 3,6,8,5.

15. Consider the following program and determine the computer output if you give a RUN command:

```
10   LET N = 1
20   LET B = N↑2
30   PRINT B
40   IF N = 3, GOTO 70
50   LET N = N + 1
60   GOTO 20
70   END
```

16. Consider the following program and determine the computer output if you give a RUN command:

```
10   LET M = 1
20   LET C = 3*M - 2
30   PRINT C
40   IF M = 4 GOTO 70
50   LET M = M + 1
60   GOTO 20
70   END
```

17. The old song "100 Bottles of Beer on the Wall" is quite tedious because of the lyrics. Write a computer program that will generate an output:

```
100 BOTTLES OF BEER ON THE WALL.
 99 BOTTLES OF BEER ON THE WALL.
 98 BOTTLES OF BEER ON THE WALL.
                ⋮
  2 BOTTLES OF BEER ON THE WALL.
  1 BOTTLE OF BEER ON THE WALL.
    THAT'S ALL FOLKS!
```

18. Write a BASIC program that will print out the first one hundred numbers backwards:

$$100, 99, 98, 97, \ldots, 3, 2, 1$$

Problems 19 and 20 require a computer.

19. Write a BASIC program to find the solution to the equation $Ax + B = 0$. The flowchart is shown below.

20. Write a BASIC program to find the solution to the equation $Ax^2 + Bx + C = 0, A \neq 0$. The flowchart is shown here.

Flowchart for solving the equation
$Ax^2 + Bx + C = 0$ (Problem 20)

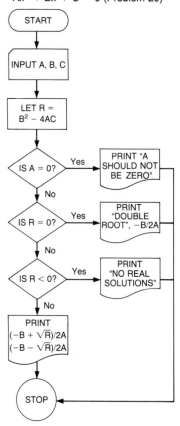

Flowchart for solving the equation
$Ax + B = 0$ (Problem 19)

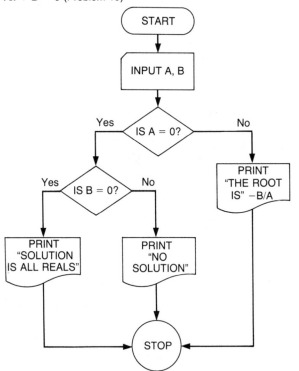

14.5 Review Problems

Section 14.1   **1.** Name the five main components of a computer, and briefly explain the function of each.

Section 14.1   **2.** What is a floating-point number? Why are floating-point numbers sometimes needed in working with computers?

Section 14.1   **3.** Express in floating-point notation:
    a. 576,000,000,000   **b.** .00000 2

Section 14.1   **4.** Express in fixed-point notation:
    a. 3.8E + 8   **b.** 5.74E − 8

Section 14.2   **5.** Briefly explain what we mean when we speak of computer "programming."

Section 14.2   **6.** **a.** Write 5*(X + 2)↑2 in ordinary mathematical notation.
    b. Write $3(x + 4)(2x − 7)^2$ in BASIC notation.

Section 14.3   **7.** Write a BASIC language program to ask the value of x, and then compute $18x^2 + 10$. Include a flowchart.

Section 14.4   **8.** Suppose you are given the following BASIC program:

```
10   PRINT "TH";
20   PRINT "E";
30   PRINT " ";
40   PRINT "M";
50   PRINT "IS";
60   PRINT "S";
70   PRINT "ISS";
80   GOTO 100
90   PRINT "H";
100  PRINT "IPP";
110  GOTO 130
120  PRINT "Y";
130  PRINT "I ";
140  PRINT "I";
150  PRINT "S ";
160  PRINT "WET";
170  GOTO 190
180  PRINT "NOODLE";
190  PRINT ","
200  END
```

What will the computer type if you give a RUN command?

Section 14.4   **9.** The program given in Problem 8 can be shortened considerably. Refine it.

Section 14.3   **10.** Write a BASIC program to calculate the areas of various circles; the radii are the input values. The formula for the area of a circle is $A = \pi r^2$, but use 3.1416 to approximate π.

Computer Graphics

Computers can be programmed to create a wide variety of designs. They can draw charts and maps in both two and three dimensions. They can also simulate art.

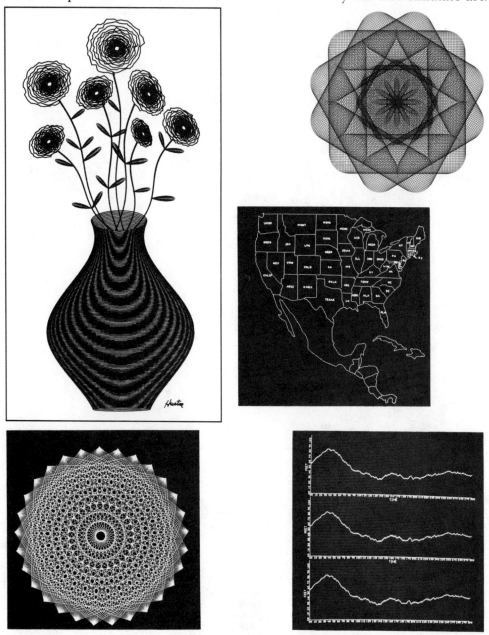

Answers to Selected Problems

Problem Set 1.1, Pages 7–8

There are, of course, no right or wrong answers to Problems 1–12. In the book from which these myths were obtained (*Mind Over Math*) there is a whole chapter justifying the fact that these are myths and the authors cite specific studies to justify their conclusions. The important aspect for you in answering these questions is that you begin thinking about mathematics and your own feelings about mathematics. Problems 15–18 are called Mind Bogglers. Don't worry if you can't answer *any* of these questions. They are presented here to remind you of the great variety of ways we use numbers. Also, remember that throughout the book, you may have to consult outside resources or do some library work to answer some of the Mind Bogglers.

15. a. The pendulum swings 120 times per minute.
b. It should be 5.6; as the *f*-numbers increase, each setting lets in only half as much light as the previous setting.
c. The person can read the letters on the test chart from 20 ft that a person with normal vision can read from 30 ft.
d. The first number is the percent of nitrogen, the second is the percent of phosphoric acid, and the third is the percent of potash in the fertilizer.
17. a. Increases   **b.** Clay   **c.** Alcohol content is 40%
d. Number of the Federal Reserve district where the money was issued

Problem Set 1.2, Pages 17–20

1. Answers vary, since there are many possible questions. For example:
"How many animals does the farmer have?"   *Answer:* $19 + 1 + 26 + 13 = 59$
"How many calves did the farmer have before the auction?"   *Answer:* $13 - 5 = 8$

3. Answers vary   **5.** Fourth diagonal   **7.** 12   **9.** Only dead people are buried in cemeteries.

11. 54 (each inning has 6 outs)   **13.** 2 down, 3 over; 10 paths (from Pascal's triangle)

15. 7 down, 4 over; 330 paths (from Pascal's triangle)

17. There are intermediate paths, so number the vertices; 37 paths

19. 11 (by numbering paths—if you took a path down to Leavenworth, I think you would be backtracking when you reach McAllister, so basically your choices are to come down Taylor or Jones Streets)

21. 20 miles (the fly flies for 1 hour at 20 mph)

23.

```
    Start      Move 1     Move 2     Move 3     Move 4     Move 5     Move 6     Move 7
     1                                                                            1
     2          2                    1          1                     2          2
     3          3       1    3 2 1   3 2        2 3     1 2 3     1   3          3
    ─────      ─────    ─────        ─────      ─────   ─────     ─────    ─────
    A B C      A B C    A B C        A B C      A B C   A B C     A B C    A B C
```

25. 5 rings; $(2 \times 2 \times 2 \times 2 \times 2) - 1 = 31$ moves   **27.** Answers vary

Problem Set 1.3, Pages 29–30

1. Common difference is 4; 57   **3.** Common difference is 6; 43   **5.** Not arithmetic

7. Common difference is 3; 131   **9.** Common difference is 252; 1056   **11.** Common ratio is 6; 2592

13. Not geometric   **15.** Not geometric   **17.** Common ratio is 10; 10,000

19. Common difference is 3; 17   **21.** Neither; the difference between terms increases by one; 22

23. Geometric; common ratio is 2; 96   **25.** Neither; subtract one, subtract two, . . . ; 85

27. Neither; number of fives between twos increases by one each time; 2

29. Neither; frontwards and then backwards letters of the alphabet; ɑ

31. Fibonacci numbers are numbers in the sequence 1, 1, 2, 3, 5, 8, 13, . . .

33. Write the sequence as 15×15, 25×25, 35×35, 45×45; the next term is $55 \times 55 = 3025$.

35. 7 (the numerals are arranged alphabetically)

37. Twelve (the number of letters in the preceding animal name)

Problem Set 1.4, Pages 33–36

1. Answers vary   **3.** It is the logic invented by Leibniz.   **5.** Deductive reasoning; explanations vary

7. Phoenix to Albuquerque is about 450 miles and Albuquerque to Oklahoma City is about 560 miles. If she gets up at 6:00 A.M., and he is in Phoenix, then about 3:00 P.M. he will be in Albuquerque if he is traveling by car. Another 12 hours and he'll be in Oklahoma and she'll probably be sleeping. Other modes of transportation would not fit the song as well, so he is probably using a car. Inductive reasoning

9. Answers vary; two persons, a woman and the speaker (don't know if the speaker is male, even though it is probable). Woman: (1) she works; (2) she has a door; (3) she can read; (4) she can talk; (5) she knows the speaker's name. Speaker: (1) left a note on her door; (2) is traveling east; (3) has left her before; (4) has informed her of intention of leaving at other times before this note; (5) speaker has (had) a phone. Deductive reasoning   **11.** The speaker. Deductive reasoning

13. It could not have taken place on Sunday ("Talkin' to him . . . last Sunday night"), Monday (or else would have said talked to him last night), or Saturday (or else would have said preacher is coming to dinner tomorrow). Possible days are Tuesday, Wednesday, Thursday, or Friday. Most probable day is Wednesday (or Thursday) because since Sunday night Billy Joe spent some time at the sawmill and on Choctaw Ridge with the speaker. This indicates that it probably wasn't Tuesday. The preacher would be likely to have dinner plans by Friday (or Thursday), so the most probable day is Wednesday. Inductive reasoning

15. The match   **17.** 12 (All months have 28 days.)

19. 1 hour (Take 1 now, 1 in half hour, and the last one in 1 hour.)

1.5 Review Problems, Pages 36–37

1. Answers vary   **2.** 2 blocks up, 4 over; 15 paths
3. $1 \times 1 = 1$; $11 \times 11 = 121$; $111 \times 111 = 12{,}321$; $111{,}111{,}111 \times 111{,}111{,}111 = 12{,}345{,}678{,}987{,}654{,}321$
4. a. Geometric   **b.** $\frac{1}{10}$ **c.** 1   **5. a.** Arithmetic   **b.** 10   **c.** 52
6. a. Neither   **b.** Add the two preceding terms   **c.** 80   **7. a.** Arithmetic   **b.** 11   **c.** 109
8. Answers vary   **9.** Smoke, from an electric locomotive?
10. It never will cover the first three rungs because as the tide rises the boat rises and thus the ladder will also rise.

Problem Set 2.1, Pages 48–49

1. a. 17   **b.** 12   **3. a.** 17   **b.** 7   **5. a.** 27   **b.** 19   **7.** 8   **9.** 8   **11.** 44
13. $7 \times 9 + 7 \times 4$ **15.** $7 \times 70 + 7 \times 3$ **17.** $6 \times 90 + 6 \times 7$ **19.** $6 \times 500 + 6 \times 30 + 6 \times 3$
21. $4 \times 700 + 4 \times 10 + 4 \times 5$ **23.** 900,950   **25.** $18,516   **27.** 261   **29.** 800   **31.** 1,080
33. 59   **35.** 27,000   **37.** 47,270   **39.** 2,323
41. a. 142,857   **b.** 285,714   **c.** 428,571   **d.** 571,428   **e.** 714,285   **f.** 857,142
g. The digits are the same.
43. a. 4,444,444,404   **b.** 5,555,555,505   **c.** 8,888,888,808   **d.** 9,999,999,909
45. a. 27   **b.** 63   **c.** 243   **d.** 432   **e.** 504   **47.** Answers vary

Problem Set 2.2, Pages 61–64

1. a. $1\frac{1}{2}$ **b.** $1\frac{1}{3}$ **c.** $1\frac{1}{4}$ **3. a.** $14\frac{1}{10}$ **b.** $16\frac{3}{10}$ **c.** $17\frac{7}{10}$ **5. a.** $2\frac{1}{16}$ **b.** $3\frac{6}{7}$ **c.** $4\frac{5}{21}$
7. a. $\frac{5}{2}$ **b.** $\frac{5}{3}$ **c.** $\frac{21}{4}$ **9. a.** $\frac{11}{3}$ **b.** $\frac{26}{5}$ **c.** $\frac{22}{5}$ **11. a.** $\frac{12}{5}$ **b.** $\frac{37}{10}$ **c.** $\frac{29}{10}$
13. a. $\frac{29}{15}$ **b.** $\frac{47}{17}$ **c.** $\frac{47}{12}$ **15. a.** .125   **b.** .375   **c.** .875   **17. a.** 2.5   **b.** $5.\overline{3}$ **c.** $6.\overline{6}$
19. a. $.\overline{5}$ **b.** $.\overline{7}$ **c.** $.\overline{285714}$ **21.** 2.3   **23.** 5.3   **25.** 12.82   **27.** $12.99   **29.** $15.00
31. 690   **33. a.** .333   **b.** .350   **c.** .417   **35.** $42.67   **37.** $7,333.33   **39.** $83.33
41. $127.14   **43. a.** $\boxed{3} \div \boxed{4} + \boxed{14} =$ **b.** $\boxed{2} \div \boxed{7} + \boxed{3} =$ **c.** $\boxed{3} \div \boxed{7} + \boxed{4} =$
45. a. $.\overline{46}$ **b.** $.3\overline{18}$ **c.** .26315789 (approx.)
47. Answers are not unique: $4 = 4 + (4 - 4) \div 4$; $5 = (4 \times 4 + 4) \div 4$; $6 = 4 + (4 + 4) \div 4$;
$7 = (44 \div 4) - 4$; $8 = 4 + 4 + 4 - 4$; $9 = 4 + 4 + (4 \div 4)$; $10 = (44 - 4) \div 4$

Problem Set 2.3, Pages 72–74

1. a. $\frac{1}{3}$ **b.** $\frac{2}{5}$ **c.** $\frac{1}{5}$ **d.** $\frac{2}{5}$ **e.** $\frac{3}{5}$ **3. a.** 2   **b.** 2   **c.** 46   **d.** 20   **e.** 7
5. a. $\frac{3}{5}$ **b.** $\frac{2}{3}$ **c.** $\frac{1}{8}$ **d.** $\frac{5}{3}$ or $1\frac{2}{3}$ **e.** $\frac{5}{3}$ or $1\frac{2}{3}$ **7. a.** $\frac{1}{24}$ **b.** $\frac{8}{15}$ **c.** $\frac{1}{8}$ **d.** $\frac{3}{10}$ **e.** $\frac{8}{9}$
9. a. $\frac{5}{2}$ or $2\frac{1}{2}$ **b.** $\frac{1}{10}$ **c.** $\frac{1}{12}$ **d.** 1   **e.** $\frac{4}{9}$
11. a. $\frac{9}{8}$ or $1\frac{1}{8}$ **b.** $\frac{2}{5}$ **c.** $\frac{5}{2}$ or $2\frac{1}{2}$ **d.** $\frac{5}{7}$ **e.** $\frac{7}{5}$ or $1\frac{2}{5}$ **13. a.** 4   **b.** 2   **c.** 6   **d.** $\frac{2}{15}$ **e.** $\frac{1}{8}$
15. a. $\frac{5}{6}$ **b.** $\frac{13}{6}$ or $2\frac{1}{6}$ **c.** Impossible   **d.** $\frac{9}{2}$ or $4\frac{1}{2}$ **e.** 20
17. a. $\frac{108}{25}$ or $4\frac{8}{25}$ **b.** $\frac{121}{6}$ or $20\frac{1}{6}$ **c.** $\frac{475}{48}$ or $9\frac{43}{48}$ **d.** $\frac{1}{2}$ **e.** 2
19. a. $\frac{1}{12}$ **b.** $\frac{1}{15}$ **c.** $\frac{1}{2}$ **21. a.** $\frac{19}{2}$ or $9\frac{1}{2}$ **b.** 1   **c.** $\frac{16}{9}$ or $1\frac{7}{9}$
23. a. $\frac{39}{50}$ **b.** $\frac{17}{20}$ **c.** $\frac{123}{500}$ **d.** $\frac{101}{200}$ **e.** $\frac{3}{200}$ **25. a.** $\frac{3}{8}$ **b.** $\frac{8}{9}$ **c.** $\frac{1}{3{,}000}$ **d.** $\frac{1}{12}$ **e.** $\frac{13}{400}$
27. $1\frac{1}{4}$ cups of sugar, 2 eggs, $1\frac{3}{4}$ cups of flour   **29.** $5,662.50   **31.** $10,000   **33.** 120 pounds

Problem Set 2.4, Pages 84–86

1. a. $\frac{3}{5}$ **b.** $\frac{8}{7}$ or $1\frac{1}{7}$ **c.** $\frac{8}{11}$ **3. a.** 1   **b.** $\frac{1}{2}$ **c.** $\frac{1}{4}$ **5. a.** 4   **b.** 9   **c.** 13
7. a. 8   **b.** 6   **c.** 10   **9. a.** 336   **b.** 630   **c.** 360   **11. a.** $\frac{7}{6}$ or $1\frac{1}{6}$ **b.** $\frac{7}{8}$ **c.** $\frac{1}{3}$
13. a. $\frac{1}{2}$ **b.** $\frac{35}{24}$ or $1\frac{11}{24}$ **c.** $\frac{7}{24}$ **15. a.** $\frac{19}{90}$ **b.** $\frac{7}{90}$ **c.** $\frac{1}{36}$ **17. a.** $7\frac{1}{4}$ **b.** $7\frac{1}{6}$ **c.** $8\frac{7}{8}$
19. a. $1\frac{5}{24}$ **b.** $3\frac{3}{4}$ **c.** $\frac{11}{16}$ **21. a.** $2\frac{23}{24}$ **b.** $1\frac{37}{70}$ **c.** $10\frac{31}{40}$ **23. a.** $\frac{8}{15}$ **b.** $\frac{19}{30}$ **c.** $\frac{12}{5}$ or $2\frac{2}{5}$
25. a. $\frac{19}{15}$ or $1\frac{4}{15}$ **b.** $\frac{43}{35}$ or $1\frac{8}{35}$ **c.** $\frac{9}{8}$ or $1\frac{1}{8}$ **27.** $7\frac{5}{12}$ lb   **29.** $18\frac{1}{16}$ in.   **31.** $\frac{3}{4}$ **33.** .875
35. .6767592593 (approx.)
37. a. 100 ft   **b.** 200 ft   **c.** 250 ft   **d.** 275 ft   **e.** 287.5 ft   **f.** 300 ft

2.5 Review Problems, Page 87

1. a. 22   **b.** 16   **2. a.** 768   **b.** 768   **3. a.** $16\frac{2}{7}$ **b.** 2.8   **4. a.** $\frac{14}{3}$ **b.** $.8\overline{3}$
5. a. $\frac{4}{5}$ **b.** $\frac{49}{110}$ **6. a.** $\frac{85}{6}$ or $14\frac{1}{6}$ **b.** $\frac{26}{15}$ or $1\frac{11}{15}$ **7. a.** $\frac{8}{7}$ or $1\frac{1}{7}$ **b.** $\frac{19}{24}$ **8. a.** $\frac{9}{20}$ **b.** $3\frac{11}{20}$
9. $\frac{1}{30}$ **10.** $\frac{16}{15}$ or $1\frac{1}{15}$

Problem Set 3.1, Pages 99–102

1. a. 14   **b.** −4   **c.** 4   **d.** −14   **e.** −1   **f.** 13
3. a. 2   **b.** −14   **c.** −2   **d.** 14   **e.** −18   **f.** 37
5. a. 4   **b.** 6   **c.** −14   **d.** 135   **e.** −19   **f.** −1
7. a. $9 + (−5) = 4$ **b.** $9 + 5 = 14$ **c.** $−9 + (−5) = −14$ **d.** $−9 + 5 = −4$
9. a. $−21 + (−7) = −28$ **b.** $−21 + 7 = −14$ **c.** $21 + 7 = 28$ **d.** $21 + (−7) = 14$
11. a. $8 + (−23) = −15$ **b.** $−8 + 23 = 15$ **c.** $−8 + (−23) = −31$ **d.** $8 + 23 = 31$
13. a. −54   **b.** −45   **c.** −94   **d.** −48   **15. a.** −20   **b.** −7   **c.** 11   **d.** −12
17. a. 0   **b.** −72   **c.** Impossible   **d.** 18   **19. a.** 30   **b.** −30   **c.** 40
21. a. −72   **b.** 35   **c.** 9   **23. a.** −1   **b.** −5   **c.** −5
25. a. Answers vary; 7 units to the right   **b.** +6   **c.** −8

27. a. 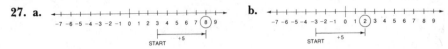   **b.**

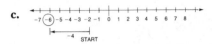   **c.**

29. 9° below 0, or −9°C   **31.** 8,773 ft   **33.** Down 2, or −2   **35. a.** −356   **b.** 462
37. a. −15,912   **b.** −5,847   **39.** Answers vary; $0 \div 5 = 0$ and $5 \div 0$ is impossible

Problem Set 3.2, Pages 112–115

1. a. One million   **b.** 10   **c.** 6   **d.** $10 \times 10 \times 10 \times 10 \times 10 \times 10$
3. a. 3.2×10^3 **b.** 2.5×10^4 **c.** 1.8×10^7 **d.** 6.4×10^2
5. a. 5.624×10^3 **b.** 2.379×10 **c.** 2.4006×10 **d.** 8.17×10^{-4}

7. a. 8.61×10^{-5} **b.** 2.49×10^8 **c.** 10^2 **d.** 1.15×10 **9.** 5.9×10^7 **11.** 4.184×10^7
13. a. .0021 **b.** .00000 046 **c.** .205 **d.** .03013 **15. a.** 49 **b.** 25 **c.** 64 **d.** 216
17. 5,000,000,000 **19.** 333,000 **21.** 10^6 **23.** 10^{12} **25. a.** 3 **b.** 1 **c.** 0 **d.** Impossible
27. a. 15 **b.** $\frac{5}{6}$ **c.** $\frac{10}{12}$ **d.** $\frac{3}{7}$ **29. a.** 35 **b.** 45 **c.** 98 **d.** 78
31. a. Rational; 5.0 **b.** Irrational; $2 < \sqrt{5} < 3$ **c.** Rational; 5.0 **d.** Rational; .25
33. a. Irrational; $3 < \sqrt{10} < 4$ **b.** Irrational; $3 < \sqrt{15} < 4$ **c.** Rational; 4.0
d. Irrational; $4 < \sqrt{17} < 5$
35. a. Rational; 49.0 **b.** Irrational; $49 < \sqrt{2,402} < 50$ **c.** Rational; 4.3
d. Irrational; $3 < \sqrt{12.4} < 4$
37. a. 3.87 **b.** 4.12 **c.** 4.47 **39. a.** 11.09 **b.** 10.91 **c.** 12.25
41. a. 15.81 **b.** 29.58 **c.** 64.88
43. Look for a pattern; consider each column separately:

Units column will contain	100,000 zeros
Tens column will contain	$100,000 -$ 9 zeros
Hundreds column will contain	$100,000 -$ 99 zeros
Thousands column will contain	$100,000 -$ 999 zeros
Ten-thousands column will contain	$100,000 -$ 9,999 zeros
Hundred-thousands column will contain	$100,000 - 99,999$ zeros
Sum	$600,000 - 111,105$

Thus, the total number of zeros is 488,895.

Problem Set 3.3, Pages 123–124

1.

131,072	262,144	524,288	1,048,576
17	18	19	20

3. 1,024 **5.** 32,768 **7.** 262,144 **9.** 256
11. 65,536 **13.** 1,048,576 **15.** 5 **17.** 8 **19.** 1 **21.** 7 **23.** 0.301 (0.301029996)
25. 0.886 (0.88649073) **27.** 0.146 (0.13987909) **29.** 0.491 (0.49715089) **31.** 7.924 (7.92427929)
33. 5.146 (5.15533604) **35.** 64 **37. a.** 81 **b.** 243 **c.** 729 **d.** 2,187 **e.** 6,561
39. 59,049 **41.** The number whose log is 21
43. There are approximately 17 pennies per inch. The stack will be about 4,902 ft or 0.93 mi, so the stack is almost a mile high.

Problem Set 3.4, Pages 129–131

1. 8.85 **3.** 6.8 **5.** 6.6 **7.** 8.5 **9.** 3.2 **11.** 80 decibels **13.** 65 decibels
15. 107 decibels **17.** 0.55 mi or about 2,900 ft **19.** At sea level **21.** 3.86 mi or about 20,400 ft
23. 1.87 mi or about 9,900 ft **25.** pH = 3 **27.** pH = 6.2 **29.** pH = 2.9
31. Growth rate of .6% **33.** Growth rate of -1.3% (population declined)

3.5 Review Problems, Page 132

1. a. -5 **b.** -15 **c.** -18 **d.** -16
2. a. 3.4×10^{-3} **b.** 4.0003×10^6 **c.** 1.74×10^4 **d.** 5
3. a. 64 **b.** 81 **c.** .000579 **d.** 401,000 **4. a.** 10 **b.** Impossible **c.** 0 **d.** 87

5. a. Rational; 23   **b.** Irrational; $31 < \sqrt{1,000} < 32$ **c.** Rational; $.\overline{6}$ **d.** Rational; 2.6
6. a. 12   **b.** 4   **c.** 0.903 (0.90308999)   **d.** 0.544 (0.54406804)   **e.** 0.204 (0.21703626)
7. a. 4.633 (4.63346846)   **b.** 8.623 (8.62117628)   **c.** 9.114 (9.10720997)   **d.** -2.347 (-2.34678749)
8. 3.7   **9.** 140 decibels   **10.** 5,100 ft

Problem Set 4.1, Pages 140–142

1. 5, 6, 8, 12   **3.** $-1, 1, 5, 13$ **5.** $-9, -6, -3, 4$ **7.** 10, 0, 10   **9.** $-90, 10, -90$ **11.** 7
13. -12 **15.** -22 **17.** -7 **19.** 8   **21.** -3 **23.** 1   **25.** -2 **27.** 49   **29.** 0
31. 15   **33.** 3   **35.** Answers vary: $x; x + 6; 2x + 12; 2x + 4; x + 2; 2$
37. Answers vary: Think of a number. Add three. Double it. Subtract two. Subtract twice the original number. The answer is four.   **39.** Answers vary

41. Easy (EZ) for (4) any (NE) one (1) to (2) see (C) why (Y) you (U) are overweight $\left(\dfrac{R}{WEIGHT}\right)$

Problem Set 4.2, Pages 148–150

1. 15   **3.** 2   **5.** -3 **7.** -9 **9.** -1 **11.** -88 **13.** 5   **15.** 1   **17.** -22 **19.** -4
21. 4   **23.** 24   **25.** 32   **27.** 21   **29.** -60 **31.** -44 **33.** 39   **35.** 40   **37.** 0
39. 11   **41.** 12   **43.** -14 **45.** -14 **47.** -45
49. Can't multiply equals times equals; can only multiply both sides by the same nonzero constant.

Problem Set 4.3, Pages 153–154

1. 4   **3.** 8   **5.** 5   **7.** 5   **9.** -54 **11.** -52 **13.** -1 **15.** 12   **17.** -19 **19.** -13
21. -4 **23.** -12 **25.** 14   **27.** 41   **29.** 0   **31.** 8   **33.** 21   **35.** 70   **37.** -14
39. 17   **41.** 3   **43.** 15   **45.** -1 **47.** -25
49. __ __ I __ HOPE __ YOU __ ARE __ __ FINDING __ THIS __ EASY __ AND __ ARE __ LEARNING __ __ USEFUL __ MATHEMATICS.

51. $\left(\begin{matrix} BANANA \\ UNPEELED \end{matrix}\right) = \left(\begin{matrix} PEELED \\ BANANA \end{matrix}\right) + \dfrac{7}{8}$ *Also:* $PEEL = \dfrac{1}{8}\left(\begin{matrix} BANANA \\ UNPEELED \end{matrix}\right)$

$\left(\begin{matrix} PEELED \\ BANANA \end{matrix}\right) + PEEL = \left(\begin{matrix} PEELED \\ BANANA \end{matrix}\right) + \dfrac{7}{8}$ $8(PEEL) = \left(\begin{matrix} BANANA \\ UNPEELED \end{matrix}\right)$

$PEEL = \dfrac{7}{8}$ $\longrightarrow$ $8\left(\dfrac{7}{8}\right) = \left(\begin{matrix} BANANA \\ UNPEELED \end{matrix}\right)$

$7 = \left(\begin{matrix} BANANA \\ UNPEELED \end{matrix}\right)$

The banana (unpeeled) weighs 7 oz.

Problem Set 4.4, Pages 159–166

1. a. 22   **b.** 286   **c.** NUMBER OF GALLONS   **d.** 286   **e.** 13
3. a. 8   **b.** 78   **c.** 9.75   **d.** $12\left(\begin{matrix} PRICE\ PER \\ OUNCE \end{matrix}\right)$ **e.** $12G$ **f.** 8.25   **g.** 9.75   **h.** 8.25   **i.** 1.5

5. a. VALUE OF LOT **b.** $5V$ **c.** 112,200 **d.** 112,200 **e.** 18,700

7. $\left(\begin{array}{c}\text{MILES PER}\\\text{GALLON}\end{array}\right)\left(\begin{array}{c}\text{NUMBER OF}\\\text{GALLONS}\end{array}\right) = \left(\begin{array}{c}\text{DISTANCE}\\\text{TRAVELED}\end{array}\right)$

$24\left(\begin{array}{c}\text{NUMBER OF}\\\text{GALLONS}\end{array}\right) = 312$

Let G = NUMBER OF GALLONS.

$$24G = 312$$
$$G = 13$$

It has a 13 gallon tank.

9. $\left(\begin{array}{c}\text{NUMBER OF}\\\text{OUNCES}\end{array}\right)\left(\begin{array}{c}\text{PRICE PER}\\\text{OUNCE}\end{array}\right) = \left(\begin{array}{c}\text{TOTAL}\\\text{PRICE}\end{array}\right)$

Chicken of the Sea:

$7\left(\begin{array}{c}\text{PRICE PER}\\\text{OUNCE}\end{array}\right) = 75$

Let C = PRICE PER OUNCE for Chicken of the Sea.

$$7C = 75$$
$$C \approx 10.7$$

Star Kist:

$5\left(\begin{array}{c}\text{PRICE PER}\\\text{OUNCE}\end{array}\right) = 63$

Let S = PRICE PER OUNCE for Star Kist.

$$5S = 63$$
$$S = 12.6$$

SAVINGS $= \left(\begin{array}{c}\text{COST PER OUNCE OF}\\\text{MORE EXPENSIVE TUNA}\end{array}\right) - \left(\begin{array}{c}\text{COST PER OUNCE OF}\\\text{LESS EXPENSIVE TUNA}\end{array}\right)$

$= 12.6 - 10.7$

$= 1.9$

Chicken of the Sea is cheaper by 1.9¢ per ounce.

11. $\left(\begin{array}{c}\text{PRICE OF LESS}\\\text{EXPENSIVE CABINET}\end{array}\right) + \left(\begin{array}{c}\text{PRICE OF MORE}\\\text{EXPENSIVE CABINET}\end{array}\right) = \left(\begin{array}{c}\text{TOTAL}\\\text{PRICE}\end{array}\right)$

$\left(\begin{array}{c}\text{PRICE OF LESS}\\\text{EXPENSIVE CABINET}\end{array}\right) + 4\left(\begin{array}{c}\text{PRICE OF LESS}\\\text{EXPENSIVE CABINET}\end{array}\right) = 2{,}075$

Let P = PRICE OF LESS EXPENSIVE CABINET.

$$P + 4P = 2{,}075$$
$$5P = 2{,}075$$
$$P = 415$$
$$4P = 4(415) = 1{,}660$$

The cabinets cost $415 and $1,660.

13. $10.13 **15.** $16.16 **17.** $3.24 **19.** $661.57 **21.** 12.2 MPG **23.** 23.6 MPG

25. 41.6 MPG **27.** 19.6 MPG **29.** 49.3 MPG

31. $2\left(\begin{array}{c}\text{DISTANCE LOIS}\\\text{TRAVELED}\end{array}\right) = \left(\begin{array}{c}\text{DISTANCE CLARK}\\\text{TRAVELED}\end{array}\right)$

$2\left(\begin{array}{c}\text{LOIS'}\\\text{RATE}\end{array}\right)\left(\begin{array}{c}\text{LOIS'}\\\text{TIME}\end{array}\right) = \left(\begin{array}{c}\text{CLARK'S}\\\text{RATE}\end{array}\right)\left(\begin{array}{c}\text{CLARK'S}\\\text{TIME}\end{array}\right)$

$2(50)(2) = \left(\begin{array}{c}\text{CLARK'S}\\\text{RATE}\end{array}\right)\left(\dfrac{1}{6}\right)$

Let C = CLARK'S RATE.

$$200 = \frac{1}{6}C$$
$$1{,}200 = C$$

Superman traveled at 1,200 miles per hour.

33. 40 mph (22.5 MPG) for 500 mi; 22.2 gallons; $38.89   **a.** 50 mph; 23.3 gallons; $40.70; savings $1.81
b. 60 mph; 25.6 gallons; $44.87; savings $5.98   **c.** 70 mph; 28.9 gallons; $50.58; savings $11.69

4.5 Review Problems, Page 166

1. Answers vary; x; $x + 10$; $3x + 30$; $3x + 6$; $x + 2$; x **2.** $-20, 10, 40, 310$ **3.** 4
4. a. $a = 50$ (addition property)   **b.** $b = 24$ (multiplication property)   **c.** $c = 4$ (division property)
d. $d = -3$ (subtraction property)
5. a. $x = 2$ **b.** $x = -9$ **6. a.** $y = 18$ **b.** $y = 13$ **7. a.** $z = -12$ **b.** $z = -20$

8. $\begin{pmatrix} \text{MILES PER} \\ \text{GALLON} \end{pmatrix} \begin{pmatrix} \text{GALLONS IN} \\ \text{TANK} \end{pmatrix} = \begin{pmatrix} \text{CRUISING} \\ \text{RANGE} \end{pmatrix}$

$$(12)(23.8) = \begin{pmatrix} \text{CRUISING} \\ \text{RANGE} \end{pmatrix}$$

$$285.6 = \begin{pmatrix} \text{CRUISING} \\ \text{RANGE} \end{pmatrix}$$

The estimated cruising range is about 286 miles.
9. The 32 oz bottle is about 3.09¢ per oz; the 44 oz bottle is about 3.39¢ per oz. The 32 oz bottle is the least expensive by about 0.3¢ per oz.   **10.** 35 MPG

Problem Set 5.1, Pages 174–177

1. ——————   **3.** ————————————————
5. ————————
7. a. 4 cm   **b.** 3.6 cm   **9. a.** 2 cm   **b.** 2.4 cm   **11.** B   **13.** B   **15.** C   **17.** B   **19.** C
21. B   **23.** 948 mi   **25.** 360 mi   **27.** 7,500 mi   **29.** 640 km   **31.** 1,424 km   **33.** 8,896 km

Problem Set 5.2, Pages 180–182

1. 2 c   **3.** 11 oz   **5.** 320 mℓ   **7.** $1\frac{3}{4}$ c   **9.** 420 mℓ   **11.** 75 mℓ   **13.** A   **15.** C   **17.** A
19. C   **21.** A   **23.** A   **25.** B   **27.** 5 mℓ   **29.** 240 mℓ   **31.** 475 mℓ   **33.** 38 ℓ
35. 9.2 gal   **37.** 53 ℓ   **39.** Answers vary

Problem Set 5.3, Pages 186–189

1. A   **3.** B   **5.** C   **7.** B   **9.** C   **11.** C   **13.** A   **15.** A   **17.** D   **19.** A
21. Kilometer   **23.** Milliliter   **25.** Gram   **27.** Celsius   **29.** 30 mℓ   **31.** 151 g   **33.** 290°C
35. 454 g   **37.** 950 mℓ   **39.** $\frac{1}{4}$ c   **41.** 1 oz   **43.** 2.2 lb   **45.** 425°F
47. 6 to 8 apples, sliced (about 1 ℓ); 60 mℓ water; 180 mℓ sugar; 120 mℓ cake flour; 5 mℓ cinnamon; 90 mℓ butter; 2.5 mℓ salt; change 375° to 190°

Problem Set 5.4, Pages 196–197

1. a. $\frac{1}{12}$ **b.** $\frac{1}{36}$ **c.** $\frac{1}{63,360}$ **2. a.** 10 **b.** .01 **c.** .00001 **5. a.** 36 **b.** 3 **c.** $\frac{1}{1,760}$
6. a. 1,000 **b.** 100 **c.** .001 **9. a.** $\frac{1}{8}$ **b.** $\frac{1}{16}$ **c.** $\frac{1}{2}$ **10. a.** 1,000 **b.** 10 **c.** .001
13. a. $\frac{1}{16}$ **b.** $\frac{1}{32,000}$ **c.** $\frac{1}{2,000}$ **14. a.** 1,000 **b.** 100 **c.** .001 **17.** $\frac{1}{10,560}$ mi; $\frac{1}{6}$ yd; $\frac{1}{2}$ ft
18. .00006 km; .0006 hm; .006 dkm; .06m; .6 dm; 60 mm **21.** $\frac{5}{2,112}$ mi; $4\frac{1}{6}$ yd; $12\frac{1}{2}$ ft
22. .0015 km; .015 hm; .15 dkm; 1.5 m; 15 dm; 1,500 mm **25.** $1\frac{3}{4}$ qt; 7 c; 56 oz
26. .0035 kℓ; .035 hℓ; .35 dkℓ; 35 dℓ; 350 cℓ; 3,500 mℓ **29.** 10 pt; 20 c; 160 oz
30. .31 kℓ; 31 dkℓ; 310 ℓ; 3,100 dℓ; 31,000 cℓ; 310,000 mℓ **33.** $\frac{3}{1,000}$ T; 6 lb
34. .096 kg; .96 hg; 9.6 dkg; 960 dg; 9,600 cg; 96,000 mg **37.** 9,000 lb; 144,000 oz
38. 45 hg; 450 dkg; 4,500 g; 45,000 dg; 450,000 cg; 4,500,000 mg

5.5 Review Problems, Page 200

1. a. ———————————————— **b.** ———————— **2. a.** 3 cm **b.** 3.1 cm
3. Answers vary **a.** 100–200 cm **b.** 50–100 kg **c.** 37°C
4. a. Inch and centimeter **b.** Ounce and gram **c.** Ounce and milliliter **d.** Mile and kilometer
e. Degree Fahrenheit and degree Celsius
5. a. Capacity **b.** Length **c.** Length **d.** Weight **e.** Temperature
6. Answers vary **a.** 3,000 mi or 4,800 km **b.** 12 oz or 355 mℓ **c.** $\frac{1}{10}$ oz or 3 g **d.** 70°F or
20°C **e.** 105°F or 40°C **7.** Kilo (1,000); hecto (100); deka (10); deci $(\frac{1}{10})$; centi $(\frac{1}{100})$; milli $(\frac{1}{1,000})$
8. a. Milligram **b.** Cup **c.** Inch **d.** Centiliter **e.** Millimeter **f.** Dekagram
g. Centigram **h.** Ton **i.** Kiloliter **j.** Hectometer
9. a. Kilogram **b.** Inch **c.** Liter **d.** Mile **e.** Ounce
10. a. 4,800 **b.** 3.5 **c.** .45 **d.** .48 **e.** 52

Problem Set 6.1, Pages 204–207

1. Quadrilateral **3.** Octagon **5.** Triangle **7.** Octagon **9.** Quadrilateral **11.** Octagon
13. *Both* a young woman and an old woman
15. a. Acute **b.** Obtuse **c.** Straight **d.** Obtuse **e.** Acute **17.** $\angle AOC$ **19.** Answers vary
21. Sum should be 540° **23.**

Problem Set 6.2, Pages 211–216

1. 18 in. **3.** 150 m **5.** 40 mm **7.** 18 yd **9.** 390 m **11.** 31.4 in. **13.** 15.072 m
15. 75.36 in. **17.** 46 ft **19.** 530 cm **21.** $7\frac{1}{4}$ in. **23.** 257 cm **25.** 64.26 ft
27. 11.14 cm **29.** 65 ft **31.** Sides are 26 in., 39 in., 52 in. **33.** 66 dm **35.** 14.8 cm
37. 13.2 cm

Problem Set 6.3, Pages 223–228

1. 15 in.^{2} **3.** 1,196 m^2 **5.** 100 mm^2 **7.** 7,560 ft^2 **9.** 136.5 dm^2 **11.** 314 in.^{2}
13. 307.7 in.^{2} **15.** 353.3 cm^2 **17.** 6 cm; 37.7 cm; 113.0 cm^2 **19.** 6 m; 18.8 m; 28.3 m^2
21. 1 km; 3.1 km; .8 km^2 **23.** 13,875 ft^2 **25.** 8 ft   **27.** 286 ft^2 **29.** 132 ft^2 **31.** 136$\frac{1}{2}$ ft^2
33. \$312   **35.** \$57   **37.** 113 in.^{2} **39.** 5¢   **41.** 5.1¢   **43.** Small
45. Answers vary; \$2.90 (about 5.8¢ per square inch)   **47.** Lot C **49.** 196,020 ft^2
51. 7,680 A

Problem Set 6.4, Pages 234–240

1. 60 cm^3 **3.** 42 cm^3 **5.** 1,728 in.^{3} **7.** 3,600 in.^{3} **9.** 400 mm^3 **11.** 42 m^3
13. 4.3 gal   **15.** 500 ℓ **17.** 3,000 ℓ **19. a.** 45.375 ft^3 **b.** 26.375 ft^3 **21.** 2.5 yd^3
23. 9$\frac{1}{2}$ yd^3 **25.** 11$\frac{1}{2}$ yd^3 **27.** 6$\frac{1}{2}$ yd^3 **29.** 112 kℓ
31. a. .5 × 10^6 or 5 × 10^5 people per mi^2 **b.** .7 × 10^4 or 7,000 mi^2 (this is less than the area of New Jersey)
c. 1.5 × 10^{-2} or .015 mi^2 per person (this is about 9.6 acres per person!)
33. b. 5   **c.** 6   **d.** 6   **e.** 6; 8   **f.** 8; 12   **g.** 12; 20
h. 20; 12 Pattern: NUMBER OF SIDES + NUMBER OF VERTICES = NUMBER OF EDGES + 2

6.5 Review Problems, Pages 240–242

1. a. Pentagon   **b.** Right angle   **c.** Acute angle   **d.** Obtuse angle   **e.** 40°
2. a. 28 ft   **b.** 60 in.   **c.** 32 m   **3. a.** 40.82 dm   **b.** 43.13 in.   **c.** 16 cm   **4.** 17.5 cm
5. a. 33 ft^2 **b.** 225 in.^{2} **c.** 32 m^2 **6. a.** 132.7 dm^2 **b.** 121.8 in.^{2} **c.** 6 cm^2 **7.** 21 yd^2
8. a. 392 in.^{3} **b.** 1,848 in.^{3} **c.** 896 dm^3 **9. a.** 1.7 gal   **b.** 8 gal   **c.** 896 ℓ **10.** 6$\frac{1}{2}$ yd^3

Problem Set 7.1, Pages 248–249

1. $\frac{3}{2}$ **3.** $\frac{4}{7}$ **5.** $\frac{15}{1}$ **7.** $\frac{3}{1}$ **9.** $\frac{1}{2}$ **11.** $\frac{2}{5}$ **13.** $\frac{25}{11}$
15. 5 is to 8 as 35 is to 56; 8 and 35 are the means; 5 and 56 are the extremes
17. 5 is to 3 as 2 is to x; 2 and 3 are the means; 5 and x are the extremes   **19.** Yes   **21.** Yes   **23.** No
25. Yes   **27.** No   **29.** No   **31.** 20 to 1   **33.** 14 to 1   **35.** 18 to 1   **37.** 50 to 53
39. 39 to 100   **41.** 8 to 25

Problem Set 7.2, Pages 255–257

1. 15   **3.** 10   **5.** 8   **7.** 21   **9.** $\frac{14}{3}$ **11.** $\frac{15}{2}$ **13.** 4   **15.** 3   **17.** 9   **19.** $\frac{1}{4}$ **21.** $\frac{5}{2}$
23. 22   **27.** 286 mi   **29.** 24 gal   **31.** 38 minutes   **33.** $\frac{3}{2} = \frac{5}{x}$; 3$\frac{1}{3}$ ft
35. Since she started with 32 gallons and ended with 1 pt (of pure soft drink), the amount of soft drink served was 31 gal, 1 qt, and 1 pt.

Problem Set 7.3, Pages 263–265

1. $\frac{1}{2}$ **3.** $\frac{67}{500}$ **5.** $\frac{1,243}{1,250}$ **7.** .50; 50%   **9.** $\frac{3}{4}$; 75%   **11.** $\frac{13}{20}$; .65   **13.** $\frac{6}{5}$; 1.20   **15.** $\frac{7}{10}$; 70%
17. .$\overline{3}$; 33$\frac{1}{3}$%   **19.** .2; 20%   **21.** $\frac{1}{20}$; 5%   **23.** $\frac{1}{25}$; .04   **25.** $\frac{13}{200}$; .065   **27.** $\frac{3}{8}$; 37$\frac{1}{2}$%
29. .15; 15%   **31. a.** 70%; C   **b.** 80%; B   **c.** 75%; C   **d.** 83%; B   **e.** 95%; A
33. a. 80%; B   **b.** 75%; C   **c.** 78%; C   **d.** 60%; D   **e.** 98%; A

Problem Set 7.4, Pages 270–272

1. $\frac{P}{100} = \frac{4}{5}$; 80%   **3.** $\frac{P}{100} = \frac{9}{12}$; 75%   **5.** $\frac{14}{100} = \frac{21}{W}$; 150   **7.** $\frac{35}{100} = \frac{49}{W}$; 140   **9.** $\frac{15}{100} = \frac{A}{64}$; 9.6

11. $\frac{P}{100} = \frac{10}{5}$; 200%   **13.** $\frac{120}{100} = \frac{16}{W}$; $13\frac{1}{3}$ **15.** $\frac{33\frac{1}{3}}{100} = \frac{12}{W}$; 36   **17.** $\frac{6}{100} = \frac{A}{8,150}$; $489   **19.** 5%

21. 59 people (can't employ half a person)   **23.** 42,500 mi
25. Old wage was $1,250; new wage is $1,350   **27.** 94 points   **29.** 90%   **31.** 400%

7.5 Review Problems, Page 272

1. a. $\frac{20}{1}$ **b.** $\frac{2}{1}$ **c.** $\frac{18}{1}$ **2. a.** 5 is to 8 as A is to 2   **b.** 5 and 2   **c.** Yes
3. a. $\frac{5}{4}$ **b.** $\frac{15}{4}$ **c.** $66\frac{2}{3}$ **4.** 130 mi   **5. a.** $\frac{1}{200}$; .5%   **b.** $.1\overline{6}$; $16\frac{2}{3}$%   **6. a.** 54   **b.** 75%
7. a. 25%   **b.** 75   **8. a.** 75   **b.** 125%   **9.** $62.54   **10.** 80%

Problem Set 8.1, Pages 283–288

1. personal; business   **3.** endorsement   **5.** reconcile   **7.** Balance: $37,340.93
9. Balance: $26,840.20   **11.** Answers vary; "For deposit only, Acct 09-310"
13. (1) $510.70   (2) $3.00; $507.70*   (3) $2,211.70   (4) $0; $2,211.70   (5) $1,704.00; $507.70*
15. Balances: **a.** $2,949.57   **b.** $2,814.43   **c.** $1,413.90   **d.** $1,365.60   **e.** $1,326.15
f. $1,317.15   **g.** $1,285.15   **h.** $285.15   **i.** $85.15
17. There is an error in the subtraction for check #486; the correct amount to balance (on lines marked *) is $65.24.

Problem Set 8.2, Pages 294–296

1. $22.50   **3.** $13.75   **5.** $3.75   **7.** $5   **9.** $1.00   **11.** $218.50   **13.** $74.10   **15.** $7.50
17. $175   **19.** $9.71   **21.** 8.6%   **23.** $525   **25.** 47.5%   **27.** $13.80   **29.** $3.83
31. Answers vary; in California it is $6.60   **33.** $13.25   **35.** $4,473.70   **37.** $26.50

Problem Set 8.3, Pages 302–304

1.

Variable	
Food	$ 251.67
Car gasoline	$ 60.20
UTILITIES	
a. Electricity	$ 48.30
b. Gas	$ —
c. Water & Sewer	$ 14.30
d. Telephone	$ —
e. Cable TV	$ —
f. Other	$ —
Entertainment	$
OTHER	
a. Ron's allowance	$ 50.00
b. Lorraine's "	$ 50.00
c. Laundromat/misc.	$ 27.15
TOTAL VARIABLE EXPENSES	$ 501.62

3.

Variable	
Food	$ 293.98
Car gasoline	$ 145.38
UTILITIES	
a. Electricity	$ 61.30
b. Gas	$ —
c. Water & Sewer	$ 23.50
d. Telephone	$ 23.50
e. Cable TV	$ 9.00
f. Other	$ —
Entertainment	$ 44.70
OTHER	
a. Allowances	$ 325.00
b. Parking & tolls	$ 75.00
c.	$
TOTAL VARIABLE EXPENSES	$ 1,001.36

5. $425   **7.** $65   **9.** $95   **11.** $78   **13.** $50   **15.** $100   **17.** $208   **19.** $13
21. $25   **23.** $39
25. Answers vary—should roughly correspond to low-income spending listed in Table 8.2.
27. Answers vary—should roughly correspond to high-income spending listed in Table 8.2.

Problem Set 8.4, Pages 311–312

1.	(7)	11,380	**3.**	(7)	14,312	**5.**	(7)	17,216	**7.**	(7)	31,122

1.
(7) 11,380
(8e) 0
(9b) 0
(10) 11,380

3.
(7) 14,312
(8a) 108
(8b) 0
(8c) 108
(8d) 0
(8e) 108
(9) 0
(10) 14,420

5.
(7) 17,216
(8a) 29
(8b) 213
(8c) 242
(8d) 200
(8e) 42
(9) 0
(10) 17,258

7.
(7) 31,122
(8a) 0
(8b) 928
(8c) 928
(8d) 400
(8e) 528
(9) 0
(10) 31,650

9.
(7) 26,935
(8a) 0
(8b) 185
(8c) 185
(8d) 400
(8e) 0
(9) 0
(10) 26,935

11. $1,447   **13.** $1,835   **15.** $2,273   **17.** $4,041
19. Tom, $7,043; Sheila, $3,528; Total, $10,571   **21.** $113 refund
23. $445 refund   **25.** $477 refund   **27.** $719 refund
29. Tom, $37 refund; Sheila, $312 refund

8.5 Review Problems, Pages 313–314

1. Thirty-two and 95/100; complete the check   **2.** Balance: $110.41; Safeway balance: $53.27
3. Deposit: $927.63; Long's balance $968.61
4. (1) $968.61; (2) $3; (3) $965.61*; (3) $50.27; (4) $927.63; (5) $977.90; outstanding check of $12.29 to Long's;
total $965.61* balances   **5.** $196   **6.** $300   **7.** $17   **8.** $136   **9.** $29,478   **10.** $338 refund

Problem Set 9.1, Pages 322–323

1. $80   **3.** $288   **5.** $1,440   **7.** $168   **9.** $7,200   **11.** $42.50   **13.** $150   **15.** $4,800
17. $72,000   **19.** 140 days   **21.** 250 days   **23.** 561 days   **25.** $1,070.56   **27.** $2,577.08
29. $975

Problem Set 9.2, Pages 327–329

1. $100   **3.** $100   **5.** 9%   **7.** $1,260   **9.** $900   **11.** 13%   **13.** 18%   **15.** 10 yr
17. $13.61   **19.** 17%   **21.** $100,500   **23.** $800,000   **25.** $75,000   **27.** $122\frac{1}{2}% simple interest

Problem Set 9.3, Pages 338–339

1. $400 simple interest; $469.33, $69.33 more   **3.** $720 simple interest; $809.86, $89.86 more
5. $12,000 simple interest; $43,231.47, $31,231.47 more
7. $1,400 at simple interest; $1,469.33, $69.33 more
9. $2,720 at simple interest; $2,809.86, $89.86 more
11. $17,000 at simple interest; $48,231.47, $31,231.47 more   **13.** 9%; 5; 1.538624; $1,538.62; $538.62
15. 8%; 3; 1.259712; $629.86; $129.86   **17.** 2%; 12; 1.268242; $634.12; $134.12
19. 4.5%; 40; 5.816365; $29,081.83; $24,081.83   **21.** 5%; 40; 7.039989; $35,199.95; $30,199.95
23. 2%; 60; 3.281031; $13,124.12; $9,124.12   **25.** 4%; 5; 1.216653; $1,520.82; $270.82
27. $500 simple interest; $628.90, $128.90 more   **29.** $348 simple interest; $473.68, $125.68 more
31. $1,225.05   **33.** $10,519.63

Problem Set 9.4, Pages 342–343

1. $1.00   **3.** $1.92   **5.** $5.39   **7.** $62.12   **9.** $77.65   **11.** $388.23   **13.** $1,924.28
15. $38,485.62   **17.** $64,142.70   **19.** $112.09   **21.** $39,230.88   **23.** $84,066.17   **25.** $12.06
27. $173,633.27   **29.** $38,585.17   **31.** $8,705.51   **33.** $2,865.43   **35.** $755.15
37. Yes; about 35 years

9.5 Review Problems, Page 344

1. $3,500   **2.** $8,500   **3.** $9,835.76   **4.** $4,948.95   **5.** $8,823.53   **6.** About $8\frac{1}{3}$ yr
7. About 6 yr   **8.** $24,432.20   **9.** $18,990,105   **10.** $1,884.28

Problem Set 10.1, Pages 350–351

1. $432 interest; $45.34 payment   **3.** $144 interest; $100 payment   **5.** $330 interest; $76.25 payment
7. $240 interest; $51.67 payment   **9.** $720 interest; $100 payment   **11.** About 22%
13. About 24%   **15.** About 20%   **17.** About 22%   **19.** About 36%
21. Bank *B* offers the lowest APR (14%); both the add-on and discount rates are more than the stated interest rates.

Problem Set 10.2, Pages 355–357

1. 15%   **3.** 18%   **5.** 8%   **7.** $9.86   **9.** $15.00   **11.** Previous balance   **13.** $3.75
15. $4.50   **17.** $1.97   **19.** $49.88   **21.** $52.50   **23.** $34.52
25. Over $33 the interest would be more than 50¢

Problem Set 10.3, Pages 362–363

1. $6,054   **3.** $5,272   **5.** $7,241   **7.** $10,199   **9.** $19,918   **11.** $542   **13.** $174
15. $681   **17.** $284   **19.** $1,355   **21.** $7,176 to $7,505   **23.** $6,598 to $6,901
25. $8,718 to $9,114   **27.** $11,407 to $11,931   **29.** $22,737 to $23,800   **31.** $16\frac{1}{2}$%   **33.** $14\frac{1}{2}$%

Problem Set 10.4, Pages 371–372

1. $302 **3.** $346 **5.** $814 **7.** $2,425 **9.** $5,320 **11.** $17,000 **13.** $970 **15.** $1,596
17. $5,950 **19.** 11.86% **21.** 14.64% **23.** 15.96% **25.** $573.17 **27.** $545.99
29. $694.53 **31.** $895.56 **33.** $860.20 **35.** $1,004.30
37. a. $15,000 **b.** $60,000 **c.** $759 **d.** $213,240 **e.** $2,108
39. a. $37,500 **b.** $112,500 **c.** $1,423.13 **d.** $399,825 **e.** $3,953

10.5 Review Problems, Page 374

1. $72.50 **2.** About 26% **3. a.** $7.64 **b.** $3.21 **4. a.** $7.50 **b.** $0.75
5. a. $8.25 **b.** $8.25 **6.** $10,289.10 **7. a.** $10,803.56 **b.** $11,318.01
8. a. $8,500 **b.** $17,000 **c.** $21,250 **9. a.** $967.73 **b.** $805.80 **c.** $755.44
10. 14.67% for lender A; 14.94% for lender B; lender A is better by .02%

Problem Set 11.1, Pages 383–386

Since this section deals with experiments, the answers will, of course, vary.

Problem Set 11.2, Pages 395–398

1. Empirical probability is the result of experimentation and may vary from experiment to experiment; it is dependent on the number of times the event is repeated. The theoretical probability is the numerical value obtained by using a mathematical model. The theoretical probability should be predictive of the empirical probability for a large number of trials.
2. If an event can occur in any one of n mutually exclusive and equally likely ways, and if s of these ways are considered successful, then the probability of an event E is denoted by $P(E)$ and defined by $P(E) = \dfrac{s}{n}$.

3. $\frac{3}{8}$ **5.** $\frac{3}{8}$ **7.** $\frac{2}{9}$ **9.** $\frac{1}{52}$ **11.** $\frac{1}{4}$ **13.** About .05 **15.** .19 **17. a.** $\frac{1}{2}$ **b.** $\frac{2}{3}$ **c.** $\frac{7}{12}$
19. a. $\frac{1}{12}$ **b.** $\frac{7}{12}$ **21.** $\frac{5}{36}$ **23.** $\frac{5}{36}$ **25.** $\frac{7}{36}$ **27.** $\frac{2}{9}$ **29.** $\frac{1}{9}$

31.

	1	2	3	4
1	(1, 1)	(1, 2)	(1, 3)	(1, 4)
2	(2, 1)	(2, 2)	(2, 3)	(2, 4)
3	(3, 1)	(3, 2)	(3, 3)	(3, 4)
4	(4, 1)	(4, 2)	(4, 3)	(4, 4)

33. a. $\frac{1}{4}$ **b.** $\frac{3}{16}$ **c.** $\frac{1}{8}$

35. a.

A C	0	0	4	4	4	4	
2	(2, 0)	(2, 0)	(2, 4)	(2, 4)	(2, 4)	(2, 4)	
2	(2, 0)	(2, 0)	(2, 4)	(2, 4)	(2, 4)	(2, 4)	
2	(2, 0)	(2, 0)	(2, 4)	(2, 4)	(2, 4)	(2, 4)	
2	(2, 0)	(2, 0)	(2, 4)	(2, 4)	(2, 4)	(2, 4)	← A wins
6	(6, 0)	(6, 0)	(6, 4)	(6, 4)	(6, 4)	(6, 4)	
6	(6, 0)	(6, 0)	(6, 4)	(6, 4)	(6, 4)	(6, 4)	← C wins

Die C wins. The probability of winning is $\frac{20}{36} = \frac{5}{9}$.

b.

	A					
D	0	0	4	4	4	4
1	(1, 0)	(1, 0)	(1, 4)	(1, 4)	(1, 4)	(1, 4)
1	(1, 0)	(1, 0)	(1, 4)	(1, 4)	(1, 4)	(1, 4)
1	(1, 0)	(1, 0)	(1, 4)	(1, 4)	(1, 4)	(1, 4)	← A wins
5	(5, 0)	(5, 0)	(5, 4)	(5, 4)	(5, 4)	(5, 4)
5	(5, 0)	(5, 0)	(5, 4)	(5, 4)	(5, 4)	(5, 4)	← D wins
5	(5, 0)	(5, 0)	(5, 4)	(5, 4)	(5, 4)	(5, 4)

Die D wins. The probability of winning is $\frac{24}{36} = \frac{2}{3}$.

c. You should pick die D (your probability of winning is $\frac{2}{3}$).
37. You should pick die B. By looking at the sample spaces, we see that, by choosing die B, you would win $\frac{2}{3}$ of the time.
39. The reason this game is called WIN is that if you let your opponent choose first, you can always wind up with a probability of winning of $\frac{2}{3}$.

If your opponent picks	You pick	Probability of winning
A	D	$\frac{2}{3}$
B	A	$\frac{2}{3}$
C	B	$\frac{2}{3}$
D	C	$\frac{2}{3}$

Problem Set 11.3, Pages 405–408

1. 1 to 12   **3.** 15 to 1   **5.** $\frac{33}{34}$ **7.** Tails; $\frac{1}{2}$ **9.** Incorrect guess; $\frac{4}{5}$ **11.** White Sox lose; .43
13. $\frac{7}{8}$ **15.** $\frac{4}{5}$
17. a. BBBB; BBBG; BBGB; BBGG; BGBB; BGBG; BGGB; BGGG; GBBB; GBBG; GBGB; GBGG; GGBB; GGBG; GGGB; GGGG   **b.** $P(4 \text{ girls}) = P(4 \text{ boys}) = \frac{1}{16}$
c. $P(1 \text{ girl and } 3 \text{ boys}) = P(3 \text{ girls and } 1 \text{ boy}) = \frac{1}{4}$ **d.** $P(2 \text{ boys and } 2 \text{ girls}) = \frac{3}{8}$ **e.** 1
19. $\frac{4}{11}$ **21. a.** $P(3|2) = \frac{2}{11}$ **b.** $P(4|2) = \frac{1}{11}$ **c.** $P(2|2) = 0$ **23.** Answers vary

Problem Set 11.4, Pages 415–419

1. $0.83   **3.** The barber's expectation is about 9¢, but he is most likely to receive a nickel.
5. About $0.02   **7.** $12   **9.** $318   **11.** $42.50   **13.** $E = \$500(\frac{1}{1,000}) = \frac{1}{2} = \$0.50$
15. $E = (\$425,000 - \$25,000)(\frac{1}{40}) + (\$125,000 - \$25,000)(\frac{1}{20}) + (0 - \$25,000)(\frac{37}{40})$
$= -\$8,125$
They should not sink the test well, since the expectation is negative.
17. $E = 0$ **19.** $E = -\frac{1}{2}$; since the expectation is loss of $0.50 per play, it is not a good deal
21. $\frac{2}{13} \approx .1538$ **23.** $\frac{4}{2,197} \approx .002$ **25.** $\frac{27}{2,197} \approx .012$ **27.** $\frac{8}{2,197} \approx .004$ **29.** $E \approx .048$ or about $0.05
31. .92 coin

11.5 Review Problems, Page 421

1. a. $P(\text{Three}) = \frac{1}{6}$ **b.** $P(\text{Three } or \text{ four}) = \frac{1}{3}$ **c.** $P(\text{Three } and \text{ four}) = 0$
2. a. $P(\text{Diamond}) = \frac{1}{4}$ **b.** $P(\text{Diamond } and \text{ a two}) = \frac{1}{52}$ **c.** $P(\text{Diamond } or \text{ a two}) = \frac{4}{13}$
3. If an event E can occur in any one of n mutually exclusive and equally likely ways, and if s of those ways are considered successful, then the probability of the event is given by $P(E) = \frac{s}{n}$
4. $\frac{1}{16} = .0625$
5. Consider the sample space. Let E be the desired event. Then $P(E) = 1 - P(\bar{E}) = 1 - \frac{7}{36} = \frac{29}{36}$.
6. a. $\frac{3}{5}$ **b.** $\frac{2}{4} = \frac{1}{2}$ **7.** $P(\text{At least one a six}) = 1 - P(\text{No six}) = \frac{11}{36}$; $P(\text{Both sixes}) = \frac{1}{36}$

8. $P(\bar{A}) = 1 - P(A) = \frac{6}{11}$ **9. a.** $\frac{1}{2}$ **b.** $\frac{\frac{1}{2}}{\frac{1}{2}} = \frac{1}{1}$, or 1 to 1

10. $E = \$500 \times (\frac{3}{800}) = \$1.875$; no; should be willing to pay about 62¢ each for the tickets

Problem Set 12.1, Pages 428–435

1.

Height	Tally	Frequency
72	I	1
71	II	2
70	III	3
69	IIII	4
68	III	3
67	ＨＴ	5
66	III	3
65	III	3
64	IIII	4
63	II	2

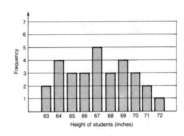

3. a. October   **b.** August

5. a. 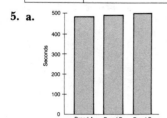   **b.** 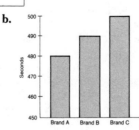   **c.** Figure b   **d.** No; answers vary

7. Answers vary; graph is worthless without scale (compare graphs a and b in Problems 5 and 6).

9.

Outcome	Tally	Frequency
Sun.	ＨＴ IIII	9
Mon.	ＨＴ ＨＴ III	13
Tue.	IIII	4
Wed.	ＨＴ	5
Thur.	ＨＴ IIII	9
Fri.	ＨＴ	5
Sat.	I	1

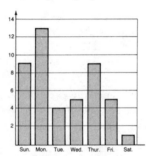

11. Answers vary   **13. a.** 1970   **b.** 1970   **c.** 1980   **15.** Answers vary

Problem Set 12.2, Pages 444–447

1. Mean = 3; median = 3; no mode; range = 4; var = 2.5; $\sigma \approx 1.58$
3. Mean = 105; median = 105; no mode; range = 4; var = 2.5; $\sigma \approx 1.58$
5. Mean = 6; median = 7; mode = 7; range = 8; var ≈ 6.67; $\sigma \approx 2.58$
7. Mean = 10; median = 8; no mode; range = 18; var = 52; $\sigma \approx 7.21$
9. Mean = 91; median = 95; mode = 95; range = 17; var = 50.5; $\sigma \approx 7.11$
11. Mean = 3; median = 3; mode = 3; range = 4; var ≈ 1.67; $\sigma \approx 1.29$
13. Same range, variance, and standard deviation
15. Mean = $7,900; median = $8,000; mode = $5,000
17. Mean ≈ 68.1; median = 70; mode = 70
19. Mean = 56; median = 65; mode = 70; range = 90   **21.** Examples vary
23. Mean = $34,000; median = $27,000; bimodal at $20,000 and $30,000; the median is most descriptive

25.

Outcome	Tally
1				
2	ЖН	
3	ЖН
4	ЖН ЖН ||||
5					
6			
7					
8	ЖН
9		
10		

mean ≈ 4.6; median = 4; mode = 4; range = 9; $\sigma \approx 2.3$

27. Answers vary   **29.** var = 96.2; $\sigma \approx 9.8$
31. var = 714; $\sigma \approx 26.7$ **33.** var = 21.6; $\sigma \approx 4.6$
35. Mean = 7; var = 6; $\sigma \approx 2.45$ **37.–41.** Answers vary

Problem Set 12.3, Pages 451–453

1. About 34   **3.** About 8   **5.** 25   **7.** 1 student   **9.** .022   **11.** .978   **13.** Exceed 40.5 inches
15. This is $\bar{x} + 2\sigma$. [*Note:* $\sigma = \sqrt{\text{var}} = \sqrt{.0004} = .02$.] The probability is about .022.
17. a. 1 shirt   **b.** 22 shirts   **c.** 136 shirts   **d.** 682 shirts   **e.** 136 shirts   **f.** 22 shirts
g. 1 shirt   **19.** 12.9 oz

Problem Set 12.4, Pages 456–457

1.–2. Answers vary

12.5 Review Problems, Pages 457–458

1.

Outcome	Tally	Frequency
6				2
7						4
8	ЖН	5
9					3
10			1

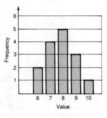

2. a.

b.

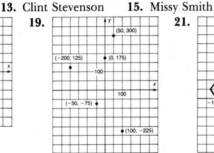

3. a. 25   **b.** 25   **c.** None   **d.** 4   **e.** 2.5
f. 1.58
4. Mean = 13.5; median = 14; mode = 16;
mode is the most useful
5. a. 79   **b.** 74   **c.** None   **d.** 19
e. 72.5   **f.** 8.5
6. Mean = \$14,667; median = \$15,000;
mode = \$18,000; median is most representative
7. a. 6,820   **b.** 220
8. Lower score is 66; upper score is 81.

9. a. The class with the smaller standard deviation would be more homogeneous and consequently easier to teach.   **b.** She did better on her math test, since she was 2 standard deviations above the mean, whereas in history she was only 1 standard deviation above the mean.   **10.** Answers vary

Problem Set 13.1, Pages 465–466

1. Telegraph Hill   **3.** Embarcadero   **5.** Ferry Building   **7.** Old Mint Building   **9.** Coit Tower
11. Niels Sovndal   **13.** Clint Stevenson   **15.** Missy Smith
17.

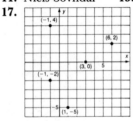

19.

21.

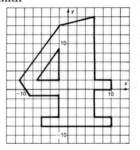

23. $(2, 9)$, $(3, 5)$, $(-2, 5,)$ $(1, 10)$, $(5, 11)$, $(5, 10)$, $(8, 10)$, $(9, 7)$, $(4, 6)$, $(5, 2)$, $(8, 0)$, $(5, -1)$, $(4, -3)$, $(4\frac{1}{2}, -5)$, $(2, -7)$, $(4, -7)$, $(5, -10)$, $(-4, -7)$, $(-3, -6)$, $(-7, -7)$, $(-9, -10)$, $(-10, -6)$, $(-8, -4)$, $(-5, -5)$, $(-5, -1)$, $(1, 5)$
25. a. Infinitely many   **b.** Two   **27.**

Problem Set 13.2, Pages 469–470

1.

3.

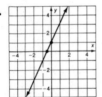

5.

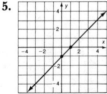

7.

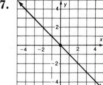

9. 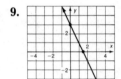 **11.** 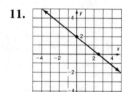 **13.** 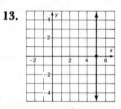 **15.**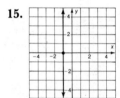

17. 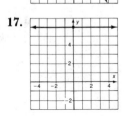 **19.** 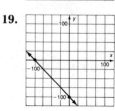 **21.** 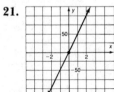 **23.**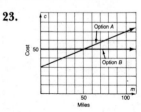

25. Option A: $c = 4h$; Option B: $c = 24$ **27.** They are the same for 6 hours **29.**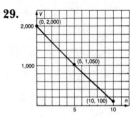

Problem Set 13.3, Pages 474–475

1. 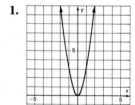 **3.** 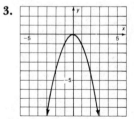 **5.** 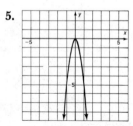 **7.**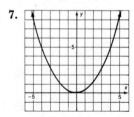

9. 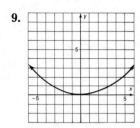 **11.** 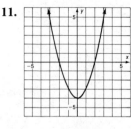 **13.** 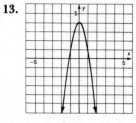 **15.**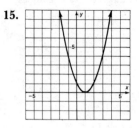

17. 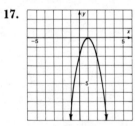 **19.** 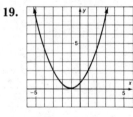 **21.**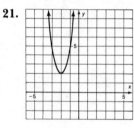

23. a. 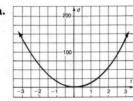 **b.** Up **c.**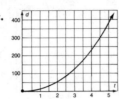

Problem Set 13.4, Pages 480–481

1. 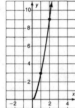 **3.** 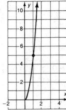 **5.** 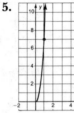 **7.**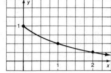

9. 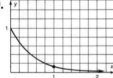 **11.** 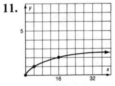 **13.** 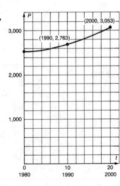 **15.**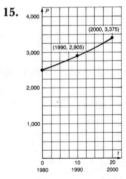

13.5 Review Problems, Pages 481–482

1. a. 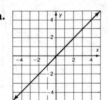 **b.** 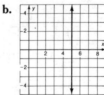 **2.** 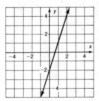 **3.**

4. a. Option A: $c = 8d$; Option B: $c = 40$ **c.** 5 days
 b.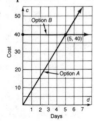

5. 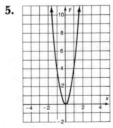 **6.**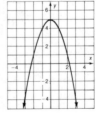

7. 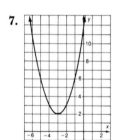 **8.** 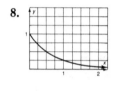 **9.** 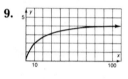 **10.**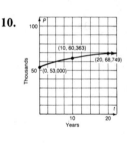

Problem Set 14.1, Pages 492–493

1. Analog: continuous measurement of data and information; Digital: discrete measurement
3. Magnetic tape; cassette tape; punched cards; CRT
5. Floating-point is a representation of a number that is written as a decimal between 1 and 10 and a power of 10.
7. a. $5.37E - 5$ **b.** 3.14159 **9. a.** $3.78596E + 5$ **b.** $5.002E + 2$ **11. a.** $4{,}560$ **b.** $.000179$
13. a. $30{,}500$ **b.** 705.6 **15. a.** $5{,}932{,}600$ **b.** $.00000\ 3217$
17. Answers vary. **a.** Analog devices: barometer, speedometer, bathroom scale, stopwatch, rain gauge
b. Digital devices: digital clock, TV channel selector, cash register, piano, calculator
19.–23. Answers vary

Problem Set 14.2, Pages 503–505

1. Step 1: Analyze the problem. **Step 2:** Prepare a plan or flowchart. **Step 3:** Put the flowchart into a language that the machine can "understand." **Step 4:** Test program. **Step 5:** Debug program.
3. a.   **b.**   **c.**   **d.**   **e.** OUTPUT

5. Answers vary **7.** GOOD MORNING.
  CAN YOU DO THIS PROBLEM CORRECTLY?
    5
9. a. $35x^2 - 13x + 2$ **b.** $6x - 7$ **11. a.** $6.29^{14} - 7$ **b.** $17x^3 - 13x^2 + \frac{15}{2}$
13. $\left(\frac{16.34}{12.5}\right)(42.1^2 - 64)$ **15.** $36x^3 + \frac{5}{2}x + \frac{135}{2}$
17. a. 5*X↑3 - 6*X↑2 + 11 **b.** 14*X↑3 + 12*X↑2 + 3
19. a. (5 - X)*(X + 3)↑2 **b.** 6*(X + 3)*(2*X - 7)↑2
21. (2*X - 3)*(3*X↑2 + 1) **23.** (2/3)*X↑2 + (1/3)*X - 17
25. 10 PRINT "I WILL DEMONSTRATE MY COMPUTATIONAL SKILL."
    20 END
27. 10 PRINT (1 + .08)↑12 **29.** 10 PRINT (23.5↑2 - 5*61.1)/2
    20 END   20 END

31. 1, 3, 5, 7, 9   **33.**

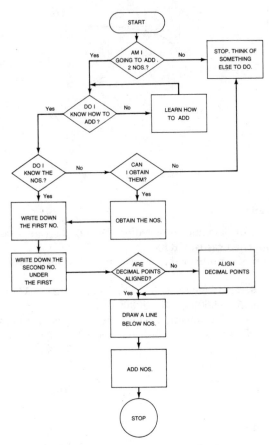

35. Answers vary

Problem Set 14.3, Pages 514–517

1. To do calculations, as in `10 PRINT 4 + 5`; to type out text, as in `10 PRINT "HELLO"`; to generate a line feed, as in `10 PRINT`.

3. INPUT, READ, or LET   **5.** 9
```
    HELLO, I LIKE YOU,
    4 + 5 =   9
```

7. 19
```
 -    5    84
```

9. No operation symbol between the factors in line 20

11.
```
I WILL CALCULATE IQ,
WHAT IS MENTAL AGE? 12
WHAT IS CHRONOLOGICAL AGE? 10

IQ IS 120
```

13.

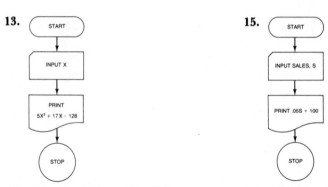

15.

```
10   PRINT "WHAT IS X";
20   INPUT X
30   PRINT 5*X↑3 + 17*X - 128
40   END
```

```
10   PRINT "WHAT ARE YOUR SALES FOR THIS WEEK";
20   INPUT S
30   PRINT .05*S + 100
40   END
```

17. Answers vary

19. 4.6188456×10^{11}, or about \$461 billion—this is more than 5 times the value of the land today; the Indians made a good deal (if they could have deposited their money at 7%)

21.
```
10   PRINT "          PPPPPP       EEEEE      AAAAAA";
15   PRINT "  CCCCCC       EEEEE"
20   PRINT "          PP    P       E          A    A";
25   PRINT "  C             E "
30   PRINT "          PPPPP       EEEE      AAAAAA";
35   PRINT "  C             EEEE"
40   PRINT "          PP            E          A    A";
45   PRINT "  C             E "
50   PRINT "          PP            E          A    A";
55   PRINT "  C             E "
60   PRINT "          PP          EEEEE      A    A";
65   PRINT "  CCCCCC       EEEEEE"
70   END
```

Problem Set 14.4, Pages 523–525

1. A loop is a repetition of one or more lines in a program.

3. `HI FRIEND.`   **5.** `10   PRINT "HI FRIEND."`   **7.** `A IS BETWEEN 1 AND 100 INCLUSIVE.`
`                        20   END`

9. `A IS LESS THAN 1.`

11. NUMBER	NUMBER CUBED	13. 5	15. 1
1	1	13	4
2	8	7	9
4	64	11	
3	27		

17.
```
10   LET N=100
20   PRINT N;
30   PRINT "BOTTLES OF BEER ON THE WALL."
40   IF N=1 THEN GOTO 70
50   LET N=N-1
60   GOTO 20
70   PRINT "1    BOTTLE OF BEER ON THE WALL."
80   PRINT "THAT'S ALL FOLKS!"
90   END
```

```
19.   10   PRINT "I WILL SOLVE ANY EQUATION OF THE FORM A*X + B = 0."
      20   PRINT "WHAT IS A";
      30   INPUT A
      40   PRINT "WHAT IS B";
      50   INPUT B
      60   IF A = 0 GOTO 90
      70   PRINT "THE ROOT IS";-B/A
      80   GOTO 130
      90   IF B = 0 GOTO 120
     100   PRINT "NO SOLUTION."
     110   GOTO 130
     120   PRINT "SOLUTION IS ALL REALS."
     130   END
```

14.5 Review Problems, Page 526

1. *Input:* means of putting information into the computer; *Storage:* the part of the computer that saves or stores information; *Accumulator:* the part of the computer in which the arithmetic operations are carried on; *Control:* the part of the computer that directs the flow of data to and from the various components of the computer; also the part in which the steps of the program are followed; *Output:* means of getting results from the computer

2. A floating-point number is a representation of a number as a decimal between 1 and 10 times an appropriate power of 10. For example, 230,000 may be written as 2.3E + 5. Floating-point numbers are needed whenever the magnitude of the number exceeds the capacity of the computer for a single number.

3. a. 5.76E + 11 **b.** 2E − 6 **4. a.** 380,000,000 **b.** .00000 00574

5. Computer programming refers to the process by which we give a computer a series of step-by-step instructions to complete a particular task or problem. The level of a program and type of instructions the computer can "understand" may vary greatly.

6. a. $5(x + 2)^2$ **b.** 3*(X + 4)*(2*X − 7)↑2

7.

```
      10   PRINT "WHAT IS X";
      20   INPUT X
      30   PRINT "18*X↑2 + 10 IS"; 18*X↑2 + 10;
      40   PRINT "WHEN X =";  X
      50   END
```

8. THE MISSISSIPPI IS WET.   **9.**
```
      10   PRINT "THE MISSISSIPPI IS WET."
      20   END
```

10.
```
      10   PRINT "WHAT IS THE RADIUS";
      20   INPUT R
      30   PRINT "THE AREA IS";
      40   PRINT 3.1416*R↑2
      50   END
```

Index

Abel, Niels, 407
Absolute value, 92
Accumulator, 487, 490
Acre, 216
Acute angle, 203
Addition
  in BASIC, 502
  of fractions, 74–83
  of integers, 94
  on a number line, 93
  property of equations, 143
Add-on interest, 347
Address, 488
Adjusted balance, 353–355
Adjusted gross income, 306
Algebra, *see* Chapter 4
  translating from English, 137
Algebraic logic, 45
ALGOL language, 499
Altered sample space, 404
Altitude, 128, 130
American League, 433
Amortized loan, 346
Amount due, 347
Amount in a percent problem, 266
Amount present, 317, 511, 512
Analog computer, 486
And, 392
Andersen, Hal, 399
Andersen, Linda, 276, 277, 280

Anderson, Bob, 463
Angles, 202–204
  classification, 203
  equal, 203
Annual interest rate, 316
Annual percentage rate, 348
  table for finding, 349
Apple computer, 485
APR, 348, 352
  table for finding, 349
Are, 216
Area, 216–223
  conversion factors, 222
  of a rectangle program, 516
Aristotle, 32
  logic according to, 32
Arithmetic logic, 45
Arithmetic operations in BASIC,
  501–502
Arithmetic sequence, 22
Arithmetic unit, 487, 490
Atari 8800 computer, 485
Atmospheric pressure, 128
Atmospheric temperatures, 184
Automobile
  price sticker, 360
  prices, 358, 359
  purchase, 357–369
  sources for loans, 362
Average, 409, 437

Average daily balance, 353–355
Axes, 426, 464
Axioms, 32

Babylonians, 168, 202
Balanced bank statement, 280
Balloon payment, 346
Banana peel problem, 154
Bank loans, 362
Bank statement, 279–283
Bar graphs, 426
Bar in a decimal, 58
Base, 102, 218
Base in a percent problem, 266
Base price, 360
Base ten logarithm, 116
Base two logarithm, 116
Baseball, 433, 446
BASIC language, 499
  mathematical symbols, 502
  programming, 495–523
Batting average, 62, 249
Bell, Alexander, 126
Bellboy problem, 265
Bell-shaped curve, 447, 448
Bense, Max, 505
Bibbo, John, 518
Bill of rights, 6
Billion, 114
Bimodal, 439

Binary coding, 489
Birthday problem, 407
Bits, 490
Body temperatures, 184
Boole, George, 33
Borrowing, 76
Boschen, Sharon, 463
Bouncing ball problem, 80
Box, volume of, 230
Boyle, Pat, 399
Brady, Jim, 340
Brahmagupta, 91
British system, 168
Brosseau, Brother, 28
Budgets, 296–302
  sample form, 297
  typical amounts, 302
Business, 447
Business checking account, 276
By the Time I Get to Phoenix, 34

Calculator
  CHS key, 98
  clear entry key, 47
  clear key, 46
  divided by key, 60
  ENTER key, 45
  equal key, 45
  fraction addition and subtraction,
    82–83
  fractions, 60
  K key, 110
  keyboard, 44
  log key, 122
  logic types, 45
  memory, 44
  negative numbers, 98
  parenthesis keys, 45
  percent key, 270
  pi key, 211
  plus/minus key, 98
  power, 110
  proportions, 254
  ratios, 247
  save key, 45
  square root key, 110
  types of, 44–45
  Y^x key, 110, 123
Canceled checks, 279
Canceling, 66
Cannonball problem, 471–472
Capacity, 177–180
  conversions, 180, 198
  volume, 231
Car rental problem, 467–468, 469
Cardano, Girolamo, 91, 407
Cards, sample space, 393
Cartesian coordinate system, 460,
  464

Cash advance, 355
Cathode ray tube, 490
Celsius, 184
Center, 209
Centi, 170, 192–193
Centimeter, 170
Central tendency, 437–440
Certificate of deposit, 321
Check, format for writing, 277
Check registers, 277
Check stub, 277
Checking accounts, 276–283
Chessboard grain problem, 115
Chisanbop, 49
Chronological age, 515
Circle, 209
  area of, 219
Circumference, 209
Closed-end credit, 346
Closing, 364, 372
Closing costs, 364
COBOL language, 499
Column check register, 277
Columns, names in a number, 55
Commands, computer, 500
Common denominators, 74
Common difference, 22
Common fractions, 64–72
Common logarithms, 118, 119, 480
Common ratio, 24
Comparison rate, 367
Complement, 291, 401
Complex decimal, 72
Component, 462
Composite, 65
Compound interest, 329–339
  comparison, 336
  table, 334–335
Compounded annually, 330
Computer applications,
  see Chapter 10
Computer components, 487
Computers, *see* Chapter 14
  career opportunities, 506
  graphics, 527
  hearing, 494
  poetry, 505
  seeing, 494
  talking, 494
Conclusion, 32
Conditional equation, 143
Conditional probability, 404–405
Conjecture, 30
Control, 487
Conventional loan, 366
Conversions, 189–196
Cost factor, 360
Cost to finance a home loan, 369
Counting, 486

Counting numbers, 40
Credit, 347
Credit card buying, 351–355
Credit union loans, 362
CRT, 487, 500
Cube, 204, 228, 239
  volume of, 229
Cubic centimeter, 228, 231, 299
Cubic decimeter, 231
Cubic inch, 228, 229
Cup, 177
Currency exchange, 163–164

Daily compounding, 336
Daniel, Jim, 6
Data, 376, 424
DATA command, 520
Data processing, 484–485
Days in the year, 319
Dealer loans, 362
Dealer's cost, 360
  American automobiles, table of,
    358
Dear Abby, 402
Debug, 496, 502
Decagon, 202
Deci, 192–193
Decibel, 120
Decimal
  change to a fraction, 71, 260
  change to a percent, 260
  division, 57
  form, 55
  fractional and percent equivalents,
    263
  fractions, 57
  numbers in BASIC, 507
  point, 55
Decimals, 52–61
  repeating, 58
Deductive reasoning, 32–33
Degree, 203
Deka, 192–193
Deposit slip, 278
Deposit ticket, 278
Depth of a well problem, 475
Descartes, René, 407
Descriptive statistics, 436–444
Destination charge for an
  automobile, 360
Diagonal, 206
Diameter, 209
Dice, 378
  craps game, 396, 397, 420
  sample space, 391
Die, 378
  sample space, 390
Difference, 137
Digital computer, 486, 487

Discount, 289–290
Discount interest, 347, 349–350
Dispersion, 440–444
Distance-rate-time formula, 158
Divided by key, 60
Dividend, 137
Division
decimal, 57
fractions, 70
integers, 98
property of equations, 147
symbol for, 137
zero, 52
Divisor, 137
Dodecagon, 202
Dodecahedron, 239
Dollar, purchasing power, 429
Domain, 135
Dot, multiplication, 137
Down payment, effect on home cost, 368
Drawee of a check, 277
Drawer of a check, 277
Dungeons and Dragons, 397
Dunker, Harvey, 463

E + exponent, 490
Earnest money, 365
Earthquakes, 124–126, 129
Edge, 229
Educational testing, 447
Egyptians, 168, 202
Elementary operations, 40
Empirical probability, 383, 389, 391
END command, 500
Endorsement, 278
Equal, 91
Equal angles, 203
Equal symbol in BASIC, 521
Equally likely, 387
Equations, 142–148, 150–152
addition property, 143
definition, 143
division property, 147
multiplication property, 146
procedure for solving, 152
Equilateral triangle, 208
Erg, 125
Escrow, 364
Euler, Leonhard, 407
Evaluating an expression, 138
Event, 387
Exact interest, 320, 366
Expectation, 408–415
Expected value, 408
Exponential curve, 475–478, 479
Exponential equation, 476
Exponential notation, 102
Exponentiation in BASIC, 502

Exponents, 102–106
Expression, 136
evaluation of, 138
Extra square centimeter problem, 228
Extremes, 245

Factor, 64, 137
Factorization, 64
prime, 65
Fahrenheit, 184
Failure, 381
Fair game, 409
Fermat, Pierre, 407
FHA loans, 360
Fibonacci, Leonardo, 26
Fibonacci Association, 28
Fibonacci numbers, 26, 30
Fibonacci sequence, 26
Finance company loans, 362
Financial ruin problem, 420
Finger multiplication, 49–51
First component, 462
Fixed-point form, 491
Floating-point form, 510
Floating-point notation, 490
Flowchart, 496
Flush, 398
Foley, Rosamond, 463
Foot, 170
Foreclose, 366
FORTRAN language, 499
Four-fours problem, 64
Four of a kind, 398
Fractions, 52–61
adding, 74–83
changing to a decimal, 57, 260
changing to a fraction, 260
common, 64–72
definition, 53
dividing, 70
multiplying, 67
percent equivalents, 263
reducing, 66
subtracting, 74–83
Frequency distributions, 424–426
Full house, 398
Fundamental property of fractions, 64

Galileo, Galilei, 474
Gallon, 177
Galois, Evariste, 407
Galton, Francis, 447
Gauss, Karl, 407
General Motors Acceptance Corporation, 361
Geometric sequence, 24
Geometry, see Chapter 6

Giles, Richard, 463
Googol, 114, 491
GOTO command, 519
Grading on a curve, 450
Graduated income tax, 310
Gram, 182, 192
Graphs, see Chapter 13
bar, 426–427
line, 427–428
Gravity, 471
Greater than, 91
Greater than or equal to symbol in BASIC, 521
Greater than symbol in BASIC, 521
Gross income, 306
Grouped data, 426

Hardware, computer, 486
Hay, Wes, 363, 364–365, 367
Hectare, 216
Hecto, 192–193
Height, 218
Heptagon, 202
Hexagon, 202
Hexahedron, 239
Hippopotamus and bird problem, 149
Hoehn, Milt, 463
Hoggatt, Verner, 28
Holidays, table of, 432
Home buying, 363–370
Home loans
comparison rate, 367
cost to finance, 369
Home-run champions, 432–433
Horizontal axis, 426, 464
Horse races, 405
Household budgeting, 296–302
HUD statement, 372–373
Hundreds, 55
Hundred thousands, 55
Hundred-thousandths, 55
Hundredths, 55

Icosahedron, 239
IF GOTO command, 520–522
IF THEN command, 520–522
Improper fraction, 53
Inch, 170, 171
Income taxes, 304–311
Independent events, 381, 418
Indexing, 310
Inductive reasoning, 30–32
Inflation, 340–342
Information retrieval, 485
Input, 487
INPUT command, 511–514
Installment buying, 346–350
Installment loan, 346

Insurance, life, 411–412
Integers, 90–91
  addition, 94
  dividing, 98
  multiplying, 95–97
  subtracting, 95
Intensity of sound, 126, 127, 129–130
Interest, 316, 350, see also Chapter 9
  add-on, 347
  comparison, 336
  compound, 329–339
  credit cards, 353–355
  discount, 347, 349–350
  formula, 316, 332, 511
  only loan, 346
  rate, 316, 366, 511
  simple, 316
  variations of, 324
Internal Revenue Service, 304, 470
Invert a fraction, 70
IQ, 448, 515
IQ test, 21
Irish Sweepstakes, 417
Irrational number, 108
IRS, 304, 470

Jastrow, Robert, 493
Juxtaposition, 81

Kemeny, John, 493
Keno, 420
Keypunch, 488
Kilo, 170, 178, 192–193
Kilogram, 183
Kilometer, 170, 173, 174, 178
Kintzi, Jim, 463
Kitchen conversions, 185
Knievel, Evel, 471
Kogelman, Stanley, 6

Lee, Josephine, 463
Lee, Phil, 312
Leibniz, Gottfried von, 33
Length, 168, 170, 218
  conversion, 198
  conversion factors, 170
Leonard, Bill, 142
Less than, 91
Less than or equal to symbol in
    BASIC, 521
Less than symbol in BASIC, 521
LET command, 511
Life insurance, 411–412
Life insurance loans, 362
Like terms, 150
Line feed, 501
Line graph, 427–428
Line number, 500

Line of credit, 347
Line printer, 488
Lines, 467–469
LIST command, 507
List price, 359, 360
Liter, 177, 178, 192, 231
Loans, 346–350
  cost to finance home loans, 369
  sources for automobile loans, 362
  sources for home loans, 366
Location, 488
Logarithm, 116, 478
  base ten, 118
  base two, 116
Logarithm table, 120
Logarithmic curve, 478–480
Logic, 32
Log-in procedure, 500
Loop, 519
Lotteries, 417
Lowest common denominator, 77
Lucas number, 28

Magnetic disk, 490
Magnetic tape, 487, 488
Manhattan Island problem, 517
Manufacturer's suggested retail
    price, 360
Map problem, 14–17
Map reading, 460
Mark-sense card, 488
Mastercard, 347, 352
Masterman, Margaret, 505
Math anxiety, 4–10
  bibliography, 8–10
  Bill of Rights, 6
Math-avoider, 4
Math myths, 6, 7
Mathematical expectation, 408–415
  definition, 410
Mathematical forms in nature, 239
McBerty, Dan, 289
Mean, 298–299, 437
Means, 245
Measure, 168
Measures of central tendency,
    437–440
Measures of dispersion, 440–444
Measuring, 486
Measuring cup, 179
Median, 437
Memory, 487
Mensa test, 21
Mental age, 515
Meter, 170, 173, 192
Metric, 168, 216, see also Chapter 5
  capacity, 177, 193
  conversions, 192–196
  countries, 169

Metric (continued)
  length, 170, 193
  measurements, 192–196
  temperature, 184
  U.S. conversions, 198–199
Microcomputer, 485, 486
Mikalson, Eva, 463
Mile, 170, 173, 174
Miles per gallon, 155–156, 164–165
Milli, 178, 192–193
Milligram, 183
Milliliter, 178, 231
Million, 114
Millions, 55
Millionths, 55
Minus symbol, 137
Mirror image, 479
Missing Dollar problem, 265
Mixed number, 54
Mixed operations, 41–43
Mode, 437
Modernizing credit plan, 357
Money management, see Chapter 8
Monthly compounding, 336
Monthly home loan payments, 364
Monthly payments, 347, 369
Monthly spending guidelines, 302
Morehouse, Otis, 463
Mortality table, 411–412
Mortgage, 365
Mpg, 155–156, 164–165
Multiple-entry check register, 277
Multiplication
  BASIC, 502
  fractions, 67
  integers, 95–97
  mixed numbers, 68
  property of equations, 146
  symbols, 137
  whole numbers and fractions, 68
Music and computers, 518
Mutually exclusive, 381, 387

Napier, John, 115
National League, 433
Natural numbers, 40
Nature, math forms in, 239
Negative sign, 90
Negotiable instrument, 278
Newton, Sir Isaac, 407
Nightingale, Florence, 447
Nonagon, 202
Nood, Robin McKinnon, 305
Normal curve, 447–451
Normal frequency curve, 448
Not equal to symbol in BASIC, 521
Number line, 91, 92
Number sequence, 22
Numbers racket, 416

Obtuse, 203
Octagon, 202
Octrahedron, 239
Odds, 399
    racetrack, 405
Ode to Billy Joe, 34–35
Olmstead, Carol, 463
Open-end loan, 347
Opens down, 473
Opens up, 473
Operations
    elementary, 40
    mixed, 41–43
Opposite operations, 106
Opposites, 90, 143
Options, 360
Or, 392
Order of operations, 43, 80
Order symbols, 91
Ordered pair, 460, 462
Ordinary interest, 320, 366
Origin, 464
Origination fee, 366
Ounce, 178, 182
Outcome, 387
Output, 487, 488
Oven temperatures, 184

Pair, 398
Parabola, 471–474
Parallelepiped
    rectangular, 229
    volume of, 230
Parallelogram, 218
    area of, 218
Parentheses, 137
    with multiplication, 97
    with negative numbers, 93
Pascal, Blaise, 407
PASCAL language, 499
Pascal's triangle, 16–18, 20–21, 28
Passbook loans, 362
Patterns, 12–13
    mathematical, 21–30
    recognition of, 484–485
Payee of a check, 277
Payments, 347, 350
Pearson, Karl, 447
Pentagon, 202
Percent, 257–263
    change to a decimal, 259, 260
    change to a fraction, 260
    definition, 258
Percent, fraction, and decimal
    equivalents, 263
Percent markdown, 289, 290
Percent of occurrence, 378
Percent problem, 266
Percentage, 266

Percents, problem solving, 266–269
Perfect square, 106
Perimeter, 207–211
    definition, 207
Peripherals, 486
Personal checking account, 276
Personal computers, 485, 486
Personal money management, *see*
    Chapter 8
pH, 130–131
Phyllotaxy, 28
Pi, 209
Pint, 177
Pizza menu, 226
Plane, 202
Plus symbol, 137
Points on a mortgage, 364, 366
Poker hands, 398
Polls, 453
Polygon, 202
Polyhedron, 207, 238–239
Population, 227–228, 238, 453
Population growth, 131, 477
Positional notation, 55
Positive sign, 90
Positive square root symbol, 107
Pound, 182
Power, 102
Powers of ten, 103
Premise, 32
Present time, 477
Previous balance, 353–355
Prime, 65
Prime factorization, 65
Principal, 316, 347, 350, 511
PRINT command, 500, 508–511
Probability, *see* Chapter 11
    conditional, 404–405
    definition, 388
    experiments, 376–383
    models, 399–405
    procedure for finding, 394
    relationship to odds, 400
Problem solving, *see* Chapter 4
    steps in, 13
    procedure for, 155
Processor, 490
Product, 137
Program, 495
Programming, 495–496
Progression, 22
Proper fraction, 53
Proportion, 245–246
    problem solving, 250
    property of, 246
Proportion machine, 249
Protractor, 203
Psychology, 447
Punched card, 487, 488

Purchase and sale agreement, 364
Purchasing power of the dollar, 429
Pyramid, 239

Quadrant, 464
Quadrilateral, 202
Quart, 177, 178
Quarterly compounding, 336
Quetelet, 447
Quotient, 137

Rabbits, birth patterns, 27
Racetrack odds, 405
Radio Shack computer, 485
Radiolaria, 239
Radius, 209
Random numbers, 383–385
    tests of randomness, 385
Range, 440
Rate, 347, 350
    in a percent problem, 266
Ratio, 244
Rational number, 107
READ command, 520
Reader's Digest Sweepstakes, 419
Real numbers, 112
Reasoning, 492
Reciprocals, 70, 143
Reconcile a bank statement, 279–283
Rectangle, 208
    area of, 218
    program for area, 516
Rectangular coordinates, 464
Reduced fractions, 64
Reduced price, 289
Reducing a fraction, 66
Relations symbols in BASIC, 521
Remainder, 52
Repeating decimals, 58
Repetitive processor, 519–522
Restrictive endorsement, 278
Return key, 501
Revolving credit, 347
Richter numbers, 124, 125
Richter scale, 124, 125
Riemann, G. F. B., 407
Right angle, 203
Root, 106–112, 143
Roulette, 420
Rounding
    procedure, 59
    with monthly payments, 347
Royal flush, 398
RPN logic, 45
RUN command, 501
Russell, Bertrand, 1, 33

SAAB advertisement, 435
Sale price, 290–292

Sales contract, 364, 365, 372
Sales tax, 292–294
  table, by state, 293
Sample, 453
Sample space, 379, 387
  altered, 404
Sampling, 453–455
San Francisco, map of, 360
Santa Rosa street problem, 215
Satisfy, 143
Saucedo, Victor, 518
Savick, Wayne, 346, 463
Scientific method, 30
Scientific notation, 105
SCR command, 507
Sears Revolving charge, 356–357
Second component, 462
Semiannual compounding, 336
Semicircle, 210
Sequence
  arithmetic, 22
  Fibonacci, 26
  geometric, 24
Settlement, 364
Shell, Terry, 463
SI system, 168
Side, 218
Signed numbers, 90–99
Similar terms, 150
Simple interest, 316–329, 336; see
    also Chapter 9
  formula, 316, 511
  variations of, 324
Simplified expression, 150
Simulation, 485
Skycycle problem, 471, 475
Slot machine, 418, 419, 420
Smart, James, 7
Smith, Linda, 189
Smith, Melissa, 328, 380, 463
Smith, Shannon, 386, 463
Snake eyes, 397
Social Security, 305
Software, 486
Solution, 143
Solving proportions, 251
Sovndal, Niels, 463
Sovndal, Rigmor, 148
Space invaders, 484, 486
Spinner, 388
Square, 208
  area of, 218
Square centimeter, 216
Square foot, 216, 217
Square inch, 216
Square meter, 216
Square mile, 216
Square root, 106
Square roots, table of, 443

Square units, 216
Square yard, 216
Squares, table of, 443
Squaring a number, 106
St. Petersburg paradox, 419
Standard deductions, 470
Standard deviation, 440, 442, 448
State retail sales taxes, 293
Statistical methods, 424
Statistics, see Chapter 12
Stevenson, Clint, 463
Sticker price, 359
Storage, 487, 488
Straight, 203, 398
Straight flush, 398
Subtraction
  BASIC, 502
  integers, 95
  property of equations, 145
Success, 381
Sum, 137
Sunflower, 28
Switzer, Steve, 463
Symmetric curve, 471
Symmetry, 473
System command, 501, 508

Tablespoon, 177
Tally marks, 376, 425
Tax table, 308–309
Taxes, income, 304–311
Teaspoon, 177
Temperature, 183–186
  conversions, 198
Tens, 55
Ten-thousands, 55
Ten-thousandths, 55
Term, 137
Tetrahedron, 239
Theorems, 32
Theoretical probability, 383, 387,
    389–391
Thinking, 492, 493
Thousands, 55
Thousandths, 55
Three card problem, 382–383, 408,
    414–415
Three of a kind, 398
Throp, Edward, 420
Time, 316
Times sign, 137
Tobias, Sheila, 4
Ton, 182
Total price, 293
Tote board, 405
Tower of Hanoi, 20
TRAC language, 505
Trailing zeros, 56
Transit number, 277

Tree diagram, 402
Triangle, 202, 208
  area of, 219
Trillion, 114
TRS-80 computer, 485
Truth-in-Lending Act, 348
Two-state device, 489
Type I error, 454, 455
Type II error, 454, 455

Unit scale, 91
Units, 55
Universal characteristic, 33
Unknown, 142
U.S. measurement system
  area, 216
  capacity, 177
  conversions, 189–192
  length, 170
  metric conversions, 198–199
  temperature, 184
  weight, 184

VA loans, 366
Vacuum tube, 490
Valid reasoning, 32
Value, present, 341
Variable, 134
Variable expression, 136
Variance, 440–441
Vertex, 202, 473
Vertical axis, 426, 464
Vertices (pl. of vertex), 202, 473
VISA, 347, 352
Volume, 228–234
  capacity, 231
  conversion factors, 232

Warren, Joseph, 6
Wassmansdorf, Mark, 250
Weight conversions, 198
Weight measurements, 182–183
White, William, 273
Whitehead, Alfred North, 33
Whole numbers, 40–47, 53
Whole quantity in a percent
    problem, 266
Width, 218
WIN, game of, 385, 397–398
Wohlert, Walt, 413
Word, 488
W-2 Form, 305

x-axis, 464

y-axis, 464
Yard, 170, 173
Year, number of days, 352

Zeros, trailing, 56